T0292900

On the Brink of Paradox

# On the Brink of Paradox

## Highlights from the Intersection of Philosophy and Mathematics

Agustín Rayo

with contributions from Damien Rochford

The MIT Press
Cambridge, Massachusetts
London, England

This book was set in Stone Serif by Westchester Publishing Services. Printed and bound in the United States of America.

Library of Congress Cataloging-in-Publication Data

Names: Rayo, Agustín, author.
Title: On the brink of paradox : highlights from the intersection of philosophy and mathematics / Agustín Rayo.
Description: Cambridge, MA : The MIT Press, [2019] | Includes bibliographical references and index.
Identifiers: LCCN 2018023088 | ISBN 9780262039413 (hardcover : alk. paper)
Subjects: LCSH: Mathematics—Problems, exercises, etc. | Paradoxes. | Logic, Symbolic and mathematical.
Classification: LCC QA11.2 .R39 2019 | DDC 510—dc23 LC record available at https://lccn.loc.gov/2018023088

10 9 8 7 6 5 4 3 2 1

To my son, Felix

# Contents

# Preface

Galileo Galilei's 1638 masterpiece, *Dialogues Concerning Two New Sciences*, has a wonderful discussion of infinity, which includes an argument for the conclusion that "the attributes 'larger,' 'smaller,' and 'equal' have no place [...] in considering infinite quantities" (pp. 77–80).

Galileo asks us to consider the question of whether there are more squares $(1^2, 2^2, 3^2, 4^2, \ldots)$ or more roots $(1, 2, 3, 4, \ldots)$. One might think that there are more roots on the grounds that, whereas every square is a root, not every root is a square. But one might also think that there are just as many roots as squares, since there is a *bijection* between the roots and the squares. (As Galileo puts it, "Every square has its own root and every root has its own square, while no square has more than one root and no root has more than one square" p. 78.)

We are left with a paradox. And it is a paradox of the most interesting sort. A boring paradox is a paradox that leads nowhere. It is due to a superficial mistake and is no more than a nuisance. An interesting paradox, on the other hand, is a paradox that reveals a genuine problem in our understanding of its subject matter. The most interesting paradoxes of all are those that reveal a problem interesting enough to lead to the development of an improved theory.

Such is the case of Galileo's infinitary paradox. In 1874, almost two and a half centuries after *Two New Sciences*, Georg Cantor published an article that describes a rigorous methodology for comparing the sizes of infinite sets. Cantor's methodology yields an answer to Galileo's paradox: it entails that the roots are just as many as the squares. It also delivers the arresting conclusion that *there are different sizes of infinity*. Cantor's work was the turning point in our understanding of infinity. By treading on the brink of paradox, he replaced a muddled pre-theoretic notion of infinite size with a rigorous and fruitful notion, which has become one of the essential tools of contemporary mathematics.

In part I of this book, I'll tell you about Cantor's revolution (chapters 1 and 2) and about certain aspects of infinity that are yet to be tamed (chapter 3). In part II I'll turn to decision theory and consider two different paradoxes that have helped deepen our understanding of the subject. The first is the so-called Grandfather Paradox, in which someone attempts to travel back in time to kill his or her grandfather before the grandfather has any children (chapter 4); the second is Newcomb's Problem, in which probabilistic dependence and causal dependence come apart (chapter 5). We'll then focus on probability theory and measure theory more generally (chapters 6 and 7). We'll consider ways in which these theories lead to bizarre results, and explore the surrounding philosophical terrain. Our discussion will culminate in a proof of the Banach-Tarski Theorem, which states that it is possible to decompose a ball into a finite number of pieces and then reassemble the pieces (without changing their size or shape) so as to get two balls, each the same size as the original (chapter 8).

Part III is a journey through computability theory (chapter 9), leading to Gödel's Incompleteness Theorem, the amazing result that arithmetic is so complex that no computer, no matter how powerful, could possibly be programmed to output every arithmetical truth and no falsehood (chapter 10). We'll prove the theorem and discuss some of its remarkable philosophical consequences.

Although many of the topics we'll be exploring are closely related to paradox, this is not a book about paradoxes. It is a book about awe-inspiring topics at the intersection between philosophy and mathematics. For many years now, I have been lucky enough to teach Paradox and Infinity at MIT. One of the chief aims of the class is to get MIT's brilliant science- and engineering-oriented students to see the wonders of philosophy. Over the years, I have settled on a list of topics that succeed in feeding my students' sense of wonder. Those are the topics we will discuss in this book.

This book does not aim to be comprehensive. The goal is to introduce you to some exceptionally beautiful ideas. I'll give you enough detail to put you in a position to understand the ideas, rather than watered-down approximations, but not so much detail that it would dampen your enthusiasm. For readers who would like to go further, I offer lists of recommended readings.

Figure 0.1 depicts the philosophical and mathematical demandingness of each chapter in the book. On the philosophical side, a demandingness level of 100% means the ideas discussed are rather subtle; you won't need philosophical training to understand them, but you'll have to think about them very carefully. On the mathematical side, a demandingness level of 100% means that the chapter is written for someone who is familiar with college-level mathematics or is otherwise experienced in mathematical proof.

The various chapters needn't be read sequentially. Figure 0.2 suggests different possible paths through the book by explaining how the chapters presuppose one another.

This book originated as a series of columns I wrote for *Investigación y Ciencia*, which is the Spanish edition of *Scientific American*. Writing for a general audience turned out to be an incredibly rewarding experience and reminded me of the importance of bringing

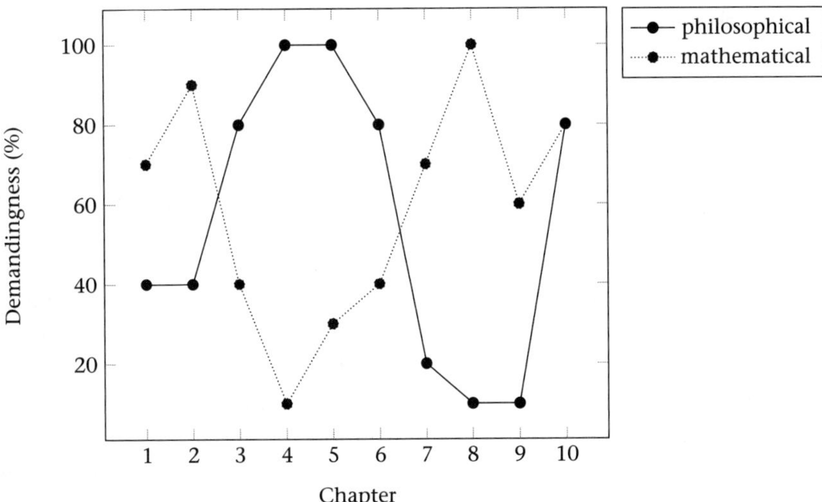


**Figure 0.1**
Levels of philosophical and mathematical demandingness corresponding to each chapter of the book.

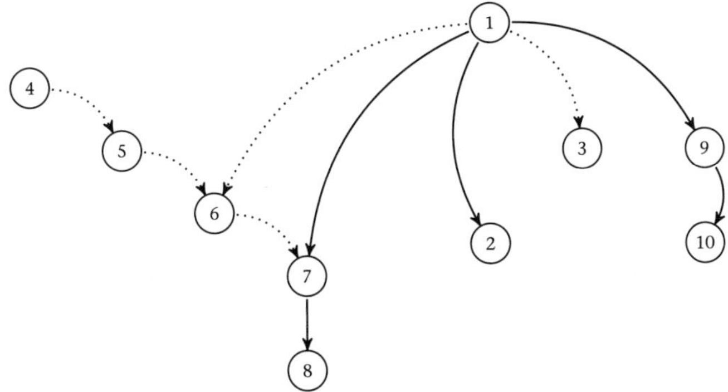


**Figure 0.2**
Different paths through the book. A solid arrow from $X$ to $Y$ means that chapter $X$ is presupposed by chapter $Y$; a dotted arrow means that reading chapter $X$ before chapter $Y$ could be helpful. Chapters closer to the top of the page are more desirable starting points, assuming you have no prior exposure to the material.

academic work outside academia. I am grateful to Gabriel Uzquiano, for getting me involved in the project and improving the drafts I would send him every couple of months; to my editors, J. P. Campos and Ernesto Lozano, for years of advice and encouragement; to Puri Mayoral, for her unfailing efficiency and kindness; and to Laia Torres, for arranging for *Prensa Científica S. A.* to grant me the rights to use this material, which in some cases appears almost in its entirety. (*Prensa Científica S. A.* owns the intellectual and industrial property of all contents published in *Investigación y Ciencia.*)

Over the years, the columns were transformed into lecture notes for my class Paradox and Infinity. The lecture notes then became the basis for an MITx massive open online course (MOOC) of the same name, which first ran in the summer of 2015, with generous funding from MIT's Office of Digital Learning. A number of people were involved in that project. Damien Rochford worked as a postdoctoral fellow for MIT's Office of Digital Learning and was the heart of our operation. He enhanced existing materials in countless ways and added materials of his own. Descendants of many of Damien's materials are included here, with his permission. (The relevant materials are credited in footnotes, with the exception of answers to exercises, which are not individually credited but often build on Damien's work.) I am grateful to Rose Lenehan and Milo Philips-Brown, who helped improve many of the course materials and were instrumental in setting up assignments for the MOOC. I also benefited from many hours of support from the MITx team. Dana Doyle managed the overall project, Shira Fruchtman was in charge of the technical aspects of MOOC creation, and Jessica Kloss led our audiovisual efforts in collaboration with Damien. The MOOC team's dedication to making higher education accessible to the world was on full display. I owe them all an enormous debt of gratitude.

The residential version of Paradox and Infinity has created many opportunities to work on these materials. My teaching assistants are responsible for countless improvements. Many thanks to David Balcarras, Nathaniel Baron-Schmitt, David Boylan, David Builes, Nilanjan Das, Kelly Gaus, Cosmo Grant, Samia Hesni, Matthias Jenny, Rose Lenehan, Milo Philips-Brown, and Kevin Richardson. I am also grateful to my students, who were not only filled with helpful suggestions but also identified a number of embarrassing mistakes. I am especially grateful to Evan Chang, Peter Griggs, Rogério Guimaraes Júnior, Anthony Liu, Pedro Mantica, Patrick McClure, Xianglong Ni, Cattalyya Nuengsigkapian, Marcos Pertierra, Aleksejs Popovs, Matt Ryback, Basil Saeed, Crystal Tsui, Jerry Wu, and Josephine Yu. Thanks are also due to Alejandro Pérez Carballo and Steve Yablo. Alejandro created some beautiful problem sets when he taught Paradox and Infinity and allowed me to use them in my version of the class. I have drawn from some of those materials in preparing the book. Steve introduced me to the joys of Paradox and Infinity when I was a graduate student and has been a constant source of inspiration since.

I am grateful to Philip Laughlin, from MIT Press, for his seemingly endless reservoir of patience, encouragement, and understanding, and for arranging for four excellent readers to review the material and send in their suggestions. Special thanks are due to Salvatore Florio and Øystein Linnebo. They each did a careful reading of the entire manuscript and sent me pages and pages of excellent comments. It's hard to quantify the extent to which those comments have enriched the manuscript.

Finally, my greatest debt of gratitude is to my wife, Carmen, for her love and support.

# I Infinity

# 1 Infinite Cardinalities

In this chapter we'll discuss some important tools for thinking about infinity and use them to show that some infinities are bigger than others.

## 1.1 Hilbert's Hotel[1]

We begin our journey with Hilbert's Hotel, an imaginary hotel named after the great German mathematician David Hilbert. Hilbert's Hotel has infinitely many rooms, one for each natural number (0, 1, 2, 3, ...), and every room is occupied: guest 0 is staying in room 0, guest 1 is staying in room 1, and so forth:

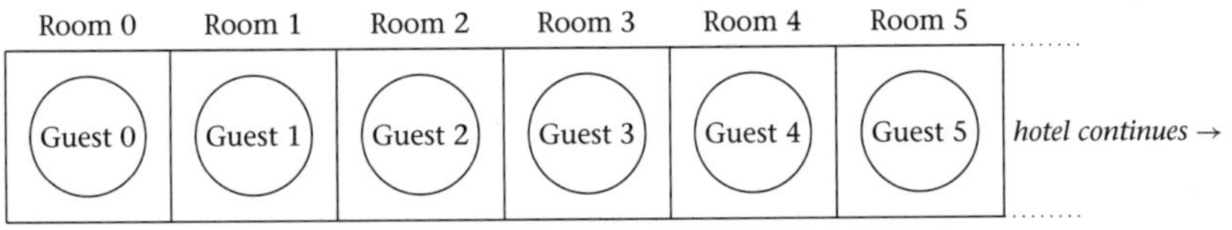

| Room 0 | Room 1 | Room 2 | Room 3 | Room 4 | Room 5 | |
|---|---|---|---|---|---|---|
| Guest 0 | Guest 1 | Guest 2 | Guest 3 | Guest 4 | Guest 5 | *hotel continues →* |

When a *finite* hotel is completely full, there is no way of accommodating additional guests. In an *infinite* hotel, however, the fact that the hotel is full is not an impediment to accommodating extra guests. Suppose, for example, that Oscar shows up without a reservation. If the hotel manager wants to accommodate him, all she needs to do is make the following announcement: "Would guest $n$ kindly relocate to room $n+1$?" Assuming everyone abides by this request, guest 0 will end up in room 1, guest 1 will end up in room 2, and so forth. Under the new arrangement each of the original guests has her own room, and room 0 is available for Oscar!

1. An earlier version of this material appeared in Rayo, "El infinito," *Investigación y Ciencia*, December 2008.

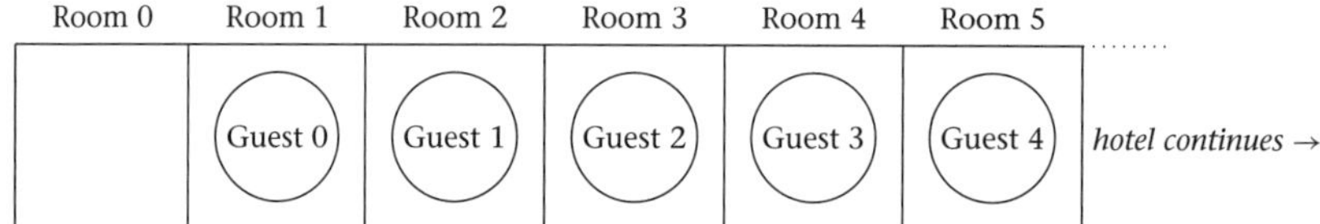


"But what about the guest in the *last* room?" you might wonder, "What about the guest in the room with the *biggest* number?" The answer is that there is no such thing as the last room, because there is no such thing as the biggest natural number. Every room in Hilbert's Hotel is followed by another.

Now suppose 5 billion new guests arrive at our completely full but infinite hotel. We can accommodate them too! This time, the manager's announcement is this: "Would guest $n$ kindly relocate to room $n+5$ billion?" In a similar way, we can accommodate *any* finite number of new guests! We can even accommodate *infinitely* many new guests. For suppose the manager announced, "If you are in room $n$, please relocate to room $2n$." Assuming everyone complies, this will free infinitely many rooms while accommodating all of the hotel's original guests, as shown:

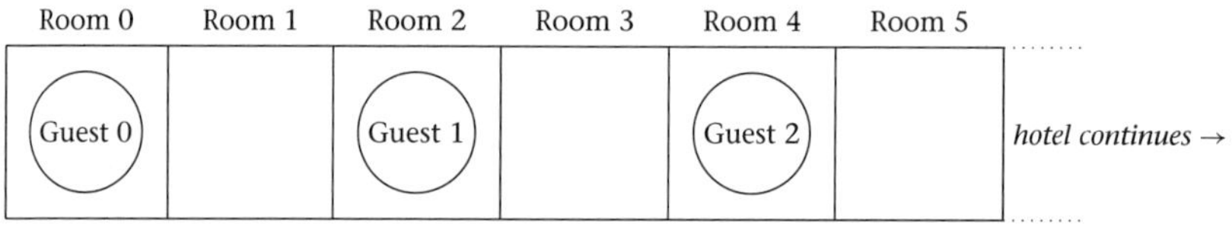

Room 0 | Room 1 | Room 2 | Room 3 | Room 4 | Room 5
Guest 0 | Guest 1 | Guest 2
*hotel continues →*

### 1.1.1 Paradox?

Hilbert's Hotel might seem to give rise to paradox. Say that the "old guests" are the guests who were occupying the hotel prior to any new arrivals and that the "new guests" are the old guests plus Oscar. We seem to be trapped between conflicting considerations. On the one hand, we want to say that there are *more* new guests than old guests. (After all, the new guests include every old guest plus Oscar.) On the other hand, we want to say that there are *just as many* new guests as there are old guests. (After all, the new guests and the old guests can be accommodated using the exact same rooms, without any multiple occupancies or empty rooms.)

You'll notice that this is the same problem I introduced in the preface, where I talked about Galileo's treatment of infinity. Consider the squares $1^2, 2^2, 3^2, 4^2, \ldots$ and their roots $1, 2, 3, 4, \ldots$. On the one hand, we want to say that there are *more* roots than squares. (After all, the roots include every square and more.) On the other hand, we want to say that there are *just as many* roots as there are squares. (As Galileo put it, "Every square has its own root and every root has its own square, while no square has more than one root and no root has more than one square" [p. 78].)

Here is a more abstract way of thinking about the matter. There are two seemingly sensible principles that turn out to be incompatible with one another in the presence of infinite sets. The first principle is this:

**The Proper Subset Principle** Suppose $A$ is a **proper subset** of $B$ (in other words: everything in $A$ is also in $B$, but not vice-versa). Then $A$ and $B$ are *not* of the same size: $B$ has more members than $A$.

For example, suppose that $A$ is the set of kangaroos and $B$ is the set of animals. Since every kangaroo is an animal but not vice-versa, the Proper Subset Principle tells us that the set of kangaroos and the set of animals are not of the same size: there are more animals than kangaroos.

The second principles is this:

**The Bijection Principle** Set $A$ has the same size as set $B$ if and only if there is a *bijection* from $A$ to $B$.

What is a bijection? Suppose that you have some beetles and some boxes and that you pair the beetles and the boxes by placing each beetle in a different box and leaving no empty boxes. A bijection is a pairing of this kind. (More generally, a **bijection** from set $A$ to set $B$ is a function from $A$ to $B$ such that (*i*) each member of $A$ is assigned to a different member of $B$ and (*ii*) no member of $B$ is left without an assignment from $A$. A **function** from $A$ to $B$ is simply an assignment of one member of $B$ to each member of $A$.)

If both of our principles seem true, it is presumably because they *are* true when attention is restricted to finite sets, and because our intuitions about size are almost exclusively developed in the context of finite sets. But the lesson of Hilbert's Hotel and of Galileo's argument is that the principles cannot both be true when it comes to *infinite sets*. For whereas the Bijection Principle entails that the set of old guests and the set of new guests have the same size, the Proper Subset Principle entails that they do not.

As it turns out, the most fruitful way of developing a theory of infinite size is to give up on the Proper Subset Principle and to keep the Bijection Principle. This crucial insight came in 1873, when the great German mathematician Georg Cantor discovered that there are infinite sets between which there can be no bijections. Cantor went on to develop a rigorous theory of infinite size based on the Bijection Principle. This work made it possible, for the first time in more than two centuries, to find our way out of the paradox that Galileo had articulated so powerfully in 1638. (David Hilbert described Cantor's work on the infinite as "the finest product of mathematical genius and one of the supreme achievements of purely intellectual human activity" [1925].)

Throughout this book, I will follow Cantor's strategy: I will keep the Bijection Principle and discard the Proper Subset Principle. Accordingly, when I say that set $A$ and set $B$ "have the same size," or when I say that $A$ has "just as many" members as $B$, I mean that there is a bijection from $A$ to $B$.

### 1.1.2 Bijections Between Infinite Sets

When the Bijection Principle is in place, many different infinite sets can be shown to have the same size. Three prominent examples are the natural numbers, the integers, and the rational numbers:

| Set | Symbol | Members |
|---|---|---|
| Natural numbers | $\mathbb{N}$ | $0, 1, 2, 3, \ldots$ |
| Integers | $\mathbb{Z}$ | $\ldots -2, -1, 0, 1, 2, \ldots$ |
| Rational numbers | $\mathbb{Q}$ | $a/b$ (for $a, b \in \mathbb{Z}$ and $b \neq 0$) |

It is easy to verify that the set $\mathbb{N}$ of natural numbers has the same size as the set $\mathbb{Z}$ of integers. Here is one example of a bijection from the natural numbers to the integers: assign the even natural numbers to the non-negative integers and the odd natural numbers to the negative integers. More precisely:

$$f(n) = \begin{cases} \frac{n}{2}, & \text{if } n \text{ is even} \\ -\frac{(n+1)}{2}, & \text{if } n \text{ is odd} \end{cases}$$

In the next section we will verify that the set $\mathbb{N}$ of natural numbers has the same size as the set $\mathbb{Q}$ of rational numbers.

#### Exercises

1. Define a bijection $f$ from the natural numbers to the even natural numbers.
2. Define a bijection $f$ from the natural numbers to the integers that are multiples of 7.
3. Verify each of the following facts about bijections:
   a) **Reflexivity** For any set $A$, there is a bijection from $A$ to $A$.
   b) **Symmetry** For any sets $A$ and $B$, if there is a bijection from $A$ to $B$, then there is a bijection from $B$ to $A$.
   c) **Transitivity** For any sets $A$, $B$, and $C$, if there is a bijection from $A$ to $B$, and if there is a bijection from $B$ to $C$, then there is a bijection from $A$ to $C$.
   Because of the symmetry property, one often says that there is a bijection *between* $A$ and $B$ rather than saying that there is a bijection *from* $A$ to $B$.

### 1.1.3 The Rational Numbers

Recall that the set of **rational numbers**, $\mathbb{Q}$, is the set of numbers equal to some fraction $a/b$, where $a$ and $b$ are integers and $b$ is distinct from 0. (For instance, $17/3$ is a rational number and so is $-4 = 4/-1$.) The **non-negative** rational numbers, $\mathbb{Q}^{\geq 0}$, are simply the

| $\frac{0}{1}$ | $\frac{0}{2}$ | $\frac{0}{3}$ | $\frac{0}{4}$ | $\frac{0}{5}$ | … |
|---|---|---|---|---|---|
| $\frac{1}{1}$ | $\frac{1}{2}$ | $\frac{1}{3}$ | $\frac{1}{4}$ | $\frac{1}{5}$ | … |
| $\frac{2}{1}$ | $\frac{2}{2}$ | $\frac{2}{3}$ | $\frac{2}{4}$ | $\frac{2}{5}$ | … |
| $\frac{3}{1}$ | $\frac{3}{2}$ | $\frac{3}{3}$ | $\frac{3}{4}$ | $\frac{3}{5}$ | … |
| $\frac{4}{1}$ | $\frac{4}{2}$ | $\frac{4}{3}$ | $\frac{4}{4}$ | $\frac{4}{5}$ | … |
| ⋮ | ⋮ | ⋮ | ⋮ | ⋮ | |

**Figure 1.1**
The non-negative rational numbers arranged on an infinite matrix. (Note that each number occurs infinitely many times, under different labels. For instance, $\frac{1}{2}$ occurs under the labels $\frac{1}{2}$, $\frac{2}{4}$, $\frac{3}{6}$, etc.)

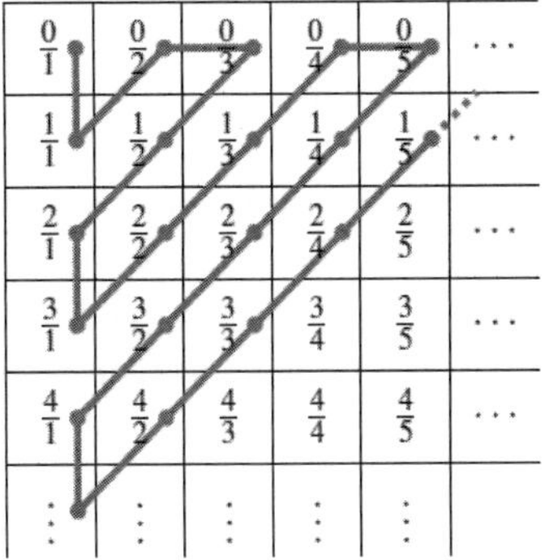

**Figure 1.2**
A path that goes through each cell of the matrix exactly once.

rational numbers greater than or equal to 0. In this section we'll prove an astonishing result: there is a bijection between the natural numbers and the non-negative rational numbers. In other words, *there are just as many natural numbers as there are non-negative rational numbers*. (In one of the exercises below, I'll ask you to show that there is also a bijection between the natural numbers and the full set of rational numbers, $\mathbb{Q}$.)

The proof is arrestingly simple. Consider the matrix in figure 1.1, and note that every (non-negative) rational number appears somewhere on the matrix. (In fact, they all appear multiple times, under different labels. For instance, 1/2 occurs under the labels 1/2, 2/4, 3/6, etc.) Now notice that the gray path depicted in figure 1.2 traverses each cell of the matrix exactly once. We can use this observation to define a bijection from natural numbers to matrix cells: to each natural number $n$, assign the cell at the $n$th step of the path.

If each rational number appeared only once on the matrix, this would immediately yield a bijection from natural numbers to rational numbers. But we have seen that each rational number appears multiple times, under different labels. Fortunately, it is easy to get around the problem. We can follow the same route as before, except that this time we skip any cells corresponding to rational numbers that have already been counted. The resulting assignment is a bijection from the natural numbers to the (non-negative) rational numbers. (I was so excited when I first learned this result I decided I wanted to spend the rest of my life thinking about this sort of thing!)

### Exercises

1. The construction above shows that there is a bijection between the natural numbers and the *non-negative* rational numbers. Show that there is also a bijection between the natural numbers and the full set of (negative and non-negative) rational numbers.
2. Show that there is a bijection between the natural numbers, $\mathbb{N}$, and as many copies of the natural numbers as there are natural numbers.

   I can be more precise by introducing you to **set notation**. We will use $x \in y$ to say that $x$ is a member of set $y$. In addition, we will say that $\{a_1, a_2, \ldots, a_n\}$ is the set whose members are $a_1, a_2, \ldots, a_n$, and that $\{x : \phi(x)\}$ is the set whose members are all and only the objects that satisfy condition $\phi$. (For example, $\{x : x \in \mathbb{N}\}$ is the set of natural numbers.)

   Using this notation, we can restate the problem as follows: For each natural number $k$, let $\mathbb{N}_k$ be the result of "labeling" each member of $\mathbb{N}$ with $k$. In other words: $\mathbb{N}_k = \{n_k : n \in \mathbb{N}\}$. Let $\bigcup_{k\in\mathbb{N}} \mathbb{N}_k$ be the result of combining the $\mathbb{N}_k$ for $k \in \mathbb{N}$. In other words: $\bigcup_{k\in\mathbb{N}} \mathbb{N}_k = \{x : x \in \mathbb{N}_k \text{ for some } k \in \mathbb{N}\}$. Show that there is a bijection between $\mathbb{N}$ and $\bigcup_{k\in\mathbb{N}} \mathbb{N}_k$.
3. $A$ is said to be a **subset** of $B$ (in symbols: $A \subset B$) if and only if every member of $A$ is also a member of $B$. (As noted earlier, $A$ is a *proper* subset of $B$ if an additional condition is also met: not every member of $B$ is a member of $A$.) Show that there is a bijection from the natural numbers to a subset of the rational numbers between 0 and 1. (Equivalently, show that there is a function that assigns to each natural number a different rational number between 0 and 1.)

## 1.2 Size Comparisons

Let me introduce you to a nice piece of notation. If $A$ is a set, we will use $|A|$ to talk about the size, or **cardinality**, of $A$.

In exercise 3 of section 1.1.2 we verified that bijections are reflexive, symmetrical, and transitive. This means that the relation $|A| = |B|$ is itself reflexive, symmetrical, and transitive. In other words, for any sets $A$, $B$, and $C$ we have:

**Reflexivity** $|A| = |A|$

**Symmetry** If $|A| = |B|$, then $|B| = |A|$.

**Transitivity** If $|A| = |B|$ and $|B| = |C|$, then $|A| = |C|$.

(A reflexive, symmetric, and transitive relation is said to be an **equivalence relation**.) By relying on the fact that $|A| = |B|$ is an equivalence relation, we can provide this beautifully succinct summary of the results so far:

$$|\mathbb{N}| = |\mathbb{Z}| = |\mathbb{Q}^{\geq 0}| = |\mathbb{Q}|$$

These results make it tempting to suppose that any infinite set has the same cardinality as any other. But we'll soon verify that this is not so: in fact, we will find that *there are infinite sets with different cardinalities.*

To prepare for these results, we'll need to get clear about what it means to say that one cardinality is greater than another. That will be the task of the present section.

Start by recalling the Bijection Principle, which I restate here using our new notation:

**The Bijection Principle** $|A| = |B|$ if and only if there is a bijection from $A$ to $B$.

How might we extend this idea to explain what it takes for the cardinality of $A$ to be *smaller than or equal to* the cardinality of $B$ (in symbols: $|A| \leq |B|$)? The answer we will be working with here is based on the following variant of the Bijection Principle:

**The Injection Principle** $|A| \leq |B|$ if and only if there is a bijection from $A$ to a subset of $B$.

(Why call it the Injection Principle? Because an **injection** from $A$ to $B$ is a bijection from $A$ to a subset of $B$.)

Together, the Bijection and Injection Principles give us everything we need to make cardinality comparisons between infinite sets. Notice, in particular, that $|A| < |B|$ (read: "the cardinality of $A$ is *strictly smaller* than the cardinality of $B$") can be defined in terms of $|A| \leq |B|$ and $|A| = |B|$:

$$|A| < |B| \leftrightarrow_{df} |A| \leq |B| \text{ and } |A| \neq |B|$$

(I use "$\leftrightarrow_{df}$" to indicate that the sentence to the left of the biconditional is to be defined in terms of the sentence to its right. The table in figure 1.3 summarizes our notational conventions and introduces a few more.)

One of the reasons the Injection Principle is so attractive is that $\leq$ is a **partial order**. In other words, $\leq$ satisfies the following principles for any sets $A$, $B$, and $C$:

| Notation | How it's defined | Informal notion |
|---|---|---|
| $\|A\| = \|B\|$ | Bijection from $A$ to $B$ | Just as many members in $A$ as in $B$ |
| $\|A\| \leq \|B\|$ | Injection from $A$ to $B$ | At most, as many members in $A$ as in $B$ |
| $\|A\| < \|B\|$ | $\|A\| \leq \|B\|$ and $\|A\| \neq \|B\|$ | Fewer members in $A$ than in $B$ |
| $\|A\| \geq \|B\|$ | $\|B\| \leq \|A\|$ | At least as many members in $A$ as in $B$ |
| $\|A\| > \|B\|$ | $\|A\| \geq \|B\|$ and $\|A\| \neq \|B\|$ | More members in $A$ than in $B$ |

**Figure 1.3**
Notation for comparing the cardinality of infinite sets.

**Reflexivity** $|A| \leq |A|$

**Anti-symmetry** If $|A| \leq |B|$ and $|B| \leq |A|$, then $|A| = |B|$.

**Transitivity** If $|A| \leq |B|$ and $|B| \leq |C|$, then $|A| \leq |C|$.

You'll be asked to prove the first and third of these principles in the exercises below. The second principle requires a lengthy proof, which we won't reconstruct here. It is sometimes referred to as the **Cantor-Schroeder-Bernstein Theorem**, and it is easy to learn more about it online.

There is a fourth principle that is intuitively correct, which we'd also very much like to prove:

**Totality** For any sets $A$ and $B$, either $|A| \leq |B|$ or $|B| \leq |A|$.

Unfortunately, one can prove totality only if one assumes a controversial set-theoretic axiom: the Axiom of Choice. We'll come across this axiom again in chapter 3, and I'll offer a more thorough discussion in chapter 7.

### Exercises

1. Show that a function $f$ from $A$ to $B$ is an injection (as defined above) if and only if $f$ assigns a different member of $B$ to each member of $A$.
2. For $f$ to be a **surjection** from $A$ to $B$ is for there to be no member of $B$ to which $f$ fails to assign some member of $A$. Show that $f$ is a bijection from $A$ to $B$ if and only if $f$ is both an injection from $A$ to $B$ and a surjection from $A$ to $B$.
3. Show that $\leq$ is reflexive and transitive.
4. Show that the following is true: if $|A| < |B|$ and $|B| \leq |C|$, then $|A| < |C|$. *Hint:* Use the Cantor-Schroeder-Bernstein Theorem.

## 1.3 Cantor's Theorem

In this section we will verify an amazing result I mentioned earlier: that there are infinite sets with different cardinalities.

If $A$ is a set, we define $A$'s **powerset**, $\wp(A)$, as the set of $A$'s subsets. (For example, if $A = \{0, 1\}$, then $\wp(A) = \{\{\}, \{0\}, \{1\}, \{0, 1\}\}$.) The following result was first proved by Georg Cantor:

**Cantor's Theorem** For any set $A$, $|A| < |\wp(A)|$.

In other words, a set always has more subsets than it has members.

Cantor's Theorem is the fundamental theorem of infinite cardinalities. It entails that *there are infinitely many sizes of infinity*. In particular, it entails the following

$$|\mathbb{N}| < |\wp(\mathbb{N})| < |\wp(\wp(\mathbb{N}))| < |\wp(\wp(\wp(\mathbb{N})))| < \ldots$$

This is the sort of result that should take one's breath away.

### 1.3.1 A Proof of Cantor's Theorem[2]

To prove that $|A| < |\wp(A)|$, we'll need to verify each of the following two statements:

1. $|A| \leq |\wp(A)|$
2. $|A| \neq |\wp(A)|$

The first of these statements is straightforward since the function $f(x) = \{x\}$ is an injection from $A$ to $\wp(A)$. So what we really need to verify is the second statement: $|A| \neq |\wp(A)|$.

We'll prove $|A| \neq |\wp(A)|$ by using a technique that is sometimes called *reductio ad absurdum* (from the Latin for "reduction to absurdity.") The basic idea is very simple. Suppose you want to verify a proposition $P$. A proof by *reductio* assumes not-$P$ and uses that assumption to prove a contradiction. If not-$P$ entails a contradiction, it must be false. So we may conclude that $P$ is true.

To prove $|A| \neq |\wp(A)|$ by *reductio* we start by assuming $|A| = |\wp(A)|$. We assume, in other words, that there is a bijection $f$ from $A$ to $\wp(A)$.

Note that for each $a$ in $A$, $f(a)$ must be a member of $\wp(A)$ and therefore a subset of $A$. So we can consider the question of whether $a$ is a member of $f(a)$. In symbols:

$$a \in f(a)?$$

This question might be answered positively or negatively depending on which $a$ we select. It will be answered negatively if we select an $a$ such that $f(a)$ is the empty set.

2. Based on text by Damien Rochford.

And it will be answered positively if we select an $a$ such that $f(a)$ is $A$ itself, assuming $A$ is non-empty.

Let $D$ be the set of members of $A$ for which the question is answered negatively. In other words:

$$D = \{x \in A : x \notin f(x)\}$$

Since $D$ is a subset of $A$, $f$ must map some member of $A$ to $D$. We'll call that member $d$, so we have $f(d) = D$. Now consider the question of whether $d$ is a member of $D$. The definition of $D$ tells us that $d$ is a member of $D$ if and only if $d$ is not a member of $f(d)$. In symbols:

$$d \in D \text{ if and only if } d \notin f(d)$$

But we know from the definition of $d$ that $f(d) = D$. So we may substitute $D$ for $f(d)$ above:

$$d \in D \text{ if and only if } d \notin D$$

This is a contradiction, because it says that something is the case if and only if it is not the case. So we have concluded our proof. We have shown that $|A| = |\wp(A)|$ entails a contradiction and is therefore false. So $|A| \neq |\wp(A)|$ must be true.

## 1.4 Further Cardinality Questions

We have seen that there are infinitely many sizes of infinity. In particular:

$$|\mathbb{N}| < |\wp(\mathbb{N})| < |\wp(\wp(\mathbb{N}))| < |\wp(\wp(\wp(\mathbb{N})))| < \ldots$$

But there are all sorts of interesting cardinality questions that remain unsettled. Consider, for example, the set $\mathfrak{F}$ of finite sequences of natural numbers or the set $\mathbb{R}$ of real numbers. What are the cardinalities of these sets? The answers turn out to be as follows:

$$|\mathfrak{F}| = |\mathbb{N}| \qquad\qquad |\mathbb{R}| = |\wp(\mathbb{N})|$$

We'll use the remainder of this chapter to verify these results and a few others.

## 1.5 Finite Sequences of Natural Numbers[3]

Let us start by verifying that there are just as many finite sequences of natural numbers as there are natural numbers: in other words, let us verify that $|\mathfrak{F}| = |\mathbb{N}|$.

We could, if we wanted, try to prove this result directly, by defining a bijection from $\mathfrak{F}$ to $\mathbb{N}$. But I'd like to tell you about a trick that greatly simplifies the proof. We'll start by verifying each of the following:

3. I am grateful to Anthony Liu for suggesting that I include this material.

1. $|\mathfrak{F}| \leq |\mathbb{N}|$
2. $|\mathbb{N}| \leq |\mathfrak{F}|$

We'll then use the Cantor-Schroeder-Bernstein Theorem of section 1.2 to conclude that $|\mathfrak{F}| = |\mathbb{N}|$.

To verify $|\mathbb{N}| \leq |\mathfrak{F}|$, we need an injective function from $\mathbb{N}$ to $\mathfrak{F}$. But this is totally straightforward, since one such function is the function that maps each natural number $n$ to the one-membered sequence $\langle n \rangle$. To verify $|\mathfrak{F}| \leq |\mathbb{N}|$, we need an injective function $f$ from $\mathfrak{F}$ to $\mathbb{N}$. Here is one way of defining such a function:

$$f(\langle n_1, n_2, \ldots, n_k \rangle) = p_1^{n_1+1} \cdot p_2^{n_2+1} \ldots p_k^{n_k+1}$$

where $p_i$ is the $i$th prime number. For example:

$$\begin{aligned} f(\langle 4, 0, 1 \rangle) &= 2^{4+1} \cdot 3^{0+1} \cdot 5^{1+1} \\ &= 32 \cdot 3 \cdot 25 = 2{,}400 \end{aligned}$$

Our function $f$ function certainly succeeds in assigning a natural number to each finite sequence of natural numbers. And it is guaranteed to be injective, because of the following:

**Fundamental Theorem of Arithmetic** Every positive integer greater than 1 has a unique decomposition into primes.

We have now verified $|\mathfrak{F}| \leq |\mathbb{N}|$ and $|\mathbb{N}| \leq |\mathfrak{F}|$. So the Cantor-Schroeder-Bernstein Theorem gives us $|\mathfrak{F}| = |\mathbb{N}|$. This concludes our proof!

## 1.6 The Real Numbers

We will now turn our attention to the set of real numbers, $\mathbb{R}$. We'll eventually show that $|\mathbb{R}| = |\wp(\mathbb{N})|$. But I'd like to start by proving a weaker result: $|\mathbb{N}| < |\mathbb{R}|$. *In other words, let us prove that the set of natural numbers is strictly smaller than the set of real numbers.*

### 1.6.1 Preliminaries

The set of **real numbers**, $\mathbb{R}$, is the set of numbers that you can write down in decimal notation by starting with a numeral (e.g., 17), adding a decimal point (eg., 17.), and then adding an infinite sequence of digits (e.g., 17.8423…). Any symbol between 0 and 9 is an admissible digit.

Every natural number is a real number, since every natural number can be written as a numeral followed by an infinite sequence of zeroes (e.g., 17=17.0000…). The rational numbers are also real numbers. In fact, they are real numbers with the special feature of having *periodic* decimal expansions: after a certain point, the expansions are just repetitions of some finite sequence of digits. For instance, $1{,}318/185 = 7.12(432)$, where the parentheses surrounding "432" indicate that these three digits are to be repeated indefinitely, so as to get 7.12432432432432.... Not every real number is a rational number, however. There are irrational real numbers, such as $\pi$ and $\sqrt{2}$. Unlike the decimal expansions of rational numbers, the decimal expansions of irrational numbers are never periodic: they never end with an infinitely repeating sequence of digits. (I'll ask you to verify the periodicity of rational numbers and the non-periodicity of irrational numbers in the exercises below.)

Some real numbers have multiple names in decimal notation. For example, the number $1/2$ is named both by the numeral $0.5(0)$ and by the numeral $0.4(9)$. To see this, note that each of the following inequalities must hold:

$$0.5 - 0.4(9) < 0.01 \text{ (since } 0.49 + 0.01 = 0.5 \text{ and } 0.49 < 0.4(9))$$

$$0.5 - 0.4(9) < 0.001 \text{ (since } 0.499 + 0.001 = 0.5 \text{ and } 0.499 < 0.4(9))$$

$$0.5 - 0.4(9) < 0.0001 \text{ (since } 0.4999 + 0.0001 = 0.5 \text{ and } 0.4999 < 0.4(9))$$

$$\vdots$$

This means that the difference between 0.5 and 0.4(9) must be smaller than each of $0.01, 0.001, 0.0001$, and so forth. Since the absolute difference between real numbers must be a non-negative real number, and since the only non-negative real number smaller than each of $0.01, 0.001, 0.0001, \ldots$ is 0, it follows that the absolute difference between 0.5 and 0.4(9) must be 0. So $0.5 = 0.4(9)$.

Having real numbers with multiple names in decimal notation turns out to be a nuisance in the present context. Fortunately, the problem is easily avoided. The crucial observation is that the only real numbers with multiple names are those named by numerals that end in an infinite sequence of 9s. Such numbers always have exactly two names: one ending in an infinite sequence of 9s and one ending in an infinite sequence of 0s. So we can avoid having numbers with multiple names by treating numerals that end in an infinite sequence of 9s as invalid and naming the relevant numbers using numerals that end in an infinite sequence of 0s.

### Exercises

1. I suggested avoiding the problem of multiple numerals by ignoring numerals that end in an infinite sequence of 9s. Could we have also avoided the problem by ignoring numerals that end in an infinite sequence of 0s?
2. Show that distinct decimal expansions $0.\delta_1^1\delta_2^1\delta_3^1\ldots$ and $0.\delta_1^2\delta_2^2\delta_3^2\ldots$ can only refer to the same number if one of them is of the form $0.s(9)$ and the other is of the form $0.s'(0)$, where $s$ is a sequence of digits and $s'$ is the result of raising the last digit in $s$ by one.
3. Show that every rational number has a periodic decimal expansion.

4. If $p$ is a period of length $n$, show that $0.(p) = \frac{p}{10^n - 1}$, and therefore that $0.(p)$ is a rational number.
5. Show that every real number with a periodic decimal notation is a rational number.

### 1.6.2 The Proof

Now for the main event: we will prove $|\mathbb{N}| < |\mathbb{R}|$.

We'll focus our attention on the subset $[0, 1)$ of $\mathbb{R}$, which consists of real numbers greater than or equal to 0 but smaller than 1. Proving $|\mathbb{N}| < |[0, 1)|$ is enough for present purposes because we know that $|(0, 1]| \leq |\mathbb{R}|$, and it follows from exercise 4 of section 1.2 that $|\mathbb{N}| < |[0, 1)|$ and $|(0, 1]| \leq |\mathbb{R}|$ together entail $|\mathbb{N}| < |\mathbb{R}|$.

In order to prove $|\mathbb{N}| < |[0, 1)|$, we'll need to verify each of the following two claims:

1. $|\mathbb{N}| \leq |[0, 1)|$
2. $|\mathbb{N}| \neq |[0, 1)|$

The first of these claims is totally straightforward. (In fact, you've already proved it if you completed exercise 3 from section 1.1.3.) So all we need to prove is $|\mathbb{N}| \neq |[0, 1)|$.

As in our proof of Cantor's Theorem, we'll proceed by *reductio*. We want to verify $|\mathbb{N}| \neq |[0, 1)|$. So we'll assume $|\mathbb{N}| = |[0, 1)|$ and use it to prove a contradiction. This will show that $|\mathbb{N}| = |[0, 1)|$ is false, and therefore that $|\mathbb{N}| \neq |[0, 1)|$ is true.

$|\mathbb{N}| = |[0, 1)|$ is the claim that there is a bijection $f$ from $\mathbb{N}$ to $[0, 1)$. So let us assume that such an $f$ exists. The first thing to note is that $f$ can be used to make a complete *list* of real numbers between 0 and 1. The zeroth member of our list is $f(0)$, the first member of our list is $f(1)$, and so forth. Since $f$ is a bijection, the list must be complete: every member of $[0, 1)$ must occur somewhere on our list.

The resulting list can be represented as in figure 1.4. This is because each real number in $[0, 1)$ is named by a numeral of the form $0.\delta_0\delta_1\delta_2\ldots$ where each $\delta_i$ is a digit between 0 and 9. (Note that because we are excluding duplicate names, each number in $[0, 1)$ corresponds to a unique numeral.)

I have highlighted certain digits in figure 1.4 using boldface: the zeroth digit of $f(0)$, the first digit of $f(1)$, and, in general, the $n$th digit of $f(n)$. These digits can be used to define a "diagonal" number:

Diagonal number: $0 \,.\; a_0 \; b_1 \; c_2 \; d_3 \; e_4 \ldots$

We will now transform the diagonal number into its "evil twin" by replacing each zero digit in its decimal expansion by a one and each non-zero digit in its decimal expansion by a zero. Suppose, for example, that function $f$ is as shown in figure 1.5. Then we have the following:

Diagonal number: $0 \,.\; 3\;0\;1\;0\;4 \ldots$

Evil twin: $0 \,.\; 0\;1\;0\;1\;0 \ldots$

| | | | | | | |
|---|---|---|---|---|---|---|
| $f(0) = 0.$ | $\boldsymbol{a_0}$ | $a_1$ | $a_2$ | $a_3$ | $a_4$ | $\ldots$ |
| $f(1) = 0.$ | $b_0$ | $\boldsymbol{b_1}$ | $b_2$ | $b_3$ | $b_4$ | $\ldots$ |
| $f(2) = 0.$ | $c_0$ | $c_1$ | $\boldsymbol{c_2}$ | $c_3$ | $c_4$ | $\ldots$ |
| $f(3) = 0.$ | $d_0$ | $d_1$ | $d_2$ | $\boldsymbol{d_3}$ | $d_4$ | $\ldots$ |
| $f(4) = 0.$ | $e_0$ | $e_1$ | $e_2$ | $e_3$ | $\boldsymbol{e_4}$ | $\ldots$ |
| $\vdots$ | $\vdots$ | $\vdots$ | $\vdots$ | $\vdots$ | $\vdots$ | |

**Figure 1.4**
We are assuming the existence of a function $f$, which induces a complete list of real numbers in $[0, 1)$.

| | | | | | | |
|---|---|---|---|---|---|---|
| $f(0) = 0.$ | **3** | 5 | 7 | 0 | 1 | $\ldots$ |
| $f(1) = 0.$ | 4 | **0** | 7 | 3 | 4 | $\ldots$ |
| $f(2) = 0.$ | 1 | 0 | **1** | 1 | 1 | $\ldots$ |
| $f(3) = 0.$ | 6 | 2 | 8 | **0** | 9 | $\ldots$ |
| $f(4) = 0.$ | 2 | 7 | 7 | 5 | **4** | $\ldots$ |
| $\vdots$ | $\vdots$ | $\vdots$ | $\vdots$ | $\vdots$ | $\vdots$ | |

**Figure 1.5**
An example of the list of real numbers in $[0, 1)$ that might be induced by $f$.

Note that regardless of what the diagonal sequence turns out to be, its evil twin will always be greater than or equal to 0, and smaller than 1. So it will always be a number in $[0, 1)$.

The climax of our proof is the observation that even though the evil twin is in $[0, 1)$, it cannot appear anywhere on our list. To see this, note that the evil twin's zeroth digit differs from the zeroth digit of $f(0)$'s decimal expansion, the evil twin's first digit differs from the first digit of $f(1)$'s decimal expansion, and so forth. (In general, the evil twin's

$k$th digit differs from the $k$th digit of $f(k)$'s decimal expansion.) So, *the evil twin is distinct from every number on our list.*

We have reached our contradiction. We began by assuming that $|\mathbb{N}| = |[0,1)|$. It follows from this assumption that we can generate a complete list of real numbers in $[0,1)$. But we have seen that the evil twin is a member of $[0,1)$ that cannot be on that list, contradicting our earlier claim that the list is complete. Since $|\mathbb{N}| = |[0,1)|$ entails a contradiction, it is false. So $|\mathbb{N}| \neq |[0,1)|$ is true, which is what we wanted to prove. The infinite size of the real numbers is *bigger* than the infinite size of the natural numbers. Amazing!

### 1.6.3 Additional Results

We have shown that $|\mathbb{N}| < |[0,1)|$ and therefore that $|\mathbb{N}| < |\mathbb{R}|$. Notice, however, that this does not immediately settle the question of whether $[0,1)$ and $\mathbb{R}$ have the same cardinality, since nothing we've proved so far rules out $|\mathbb{N}| < |[0,1)| < |\mathbb{R}|$. As you'll be asked to prove below, they do have the same cardinality:

$$|[0,1)| = |\mathbb{R}|$$

This is a nontrivial result. It means that just like there are different infinite sets that have the same size as the natural numbers, so there are different infinite sets that have the same size as the real numbers.

The exercises below will ask you to show that the following sets have the same size as the real numbers:

| Set | Also known as | Members |
|---|---|---|
| $(0,1)$ | Unit interval (open) | Real numbers larger than 0 but smaller than 1 |
| $[0,1]$ | Unit interval (closed) | Real numbers larger or equal to 0 but smaller or equal to 1 |
| $[0,a]$ | Arbitrarily sized interval | Real numbers larger or equal to 0 but smaller or equal to $a$ ($a > 0$) |
| $[0,1] \times [0,1]$ | Unit square | Pairs of real numbers larger or equal to 0 but smaller or equal to 1 |
| $[0,1] \times [0,1] \times [0,1]$ | Unit cube | Triples of real numbers larger or equal to 0 but smaller or equal to 1 |
| $\underbrace{[0,1] \times \ldots \times [0,1]}_{n \text{ times}}$ | $n$-dimensional hypercube | $n$-tuples of real numbers larger or equal to 0 but smaller or equal to 1 |

As you work through the exercises, you'll find it useful to avail yourself of the following result: *adding up-to-natural-number-many members to an infinite set doesn't change the set's cardinality*. More precisely:

**No Countable Difference Principle** $|S| = |S \cup A|$, whenever $S$ is infinite and $A$ is countable.

(In general, $A \cup B = \{x : x \in A \text{ or } x \in B\}$. For a set $S$ to be **infinite** is for it to be the case that $|\mathbb{N}| \leq |S|$; for a set $A$ to be **countable** is for it to be the case that $|A| \leq |\mathbb{N}|$.)

## Exercises

1. Prove the No Countable Difference Principle. *Hint:* Use the following two steps:

   **Step 1** Since $S$ is infinite (and therefore $|\mathbb{N}| \leq |S|$), we know that there is an injective function from the natural numbers to $S$. Let the range of that function be the set $S^{\mathbb{N}} = s_0, s_1, s_2, \ldots$ and show that $|S^{\mathbb{N}}| = |S^{\mathbb{N}} \cup A|$.

   (In general, if $f$ is a function from $A$ to $B$, we say that the **domain** of $f$ is $A$ and that the **range** of $f$ is the subset $\{f(x) : x \in A\}$ of $B$.)

   **Step 2** Use the fact that $|S^{\mathbb{N}}| = |S^{\mathbb{N}} \cup A|$ to show that $|S| = |S \cup A|$.

2. Recall that the open interval $(0, 1)$ is the set of real numbers $x$ such that $0 < x < 1$. Show that $|(0, 1)| = |\mathbb{R}|$.

   *Hint:* There is a nice geometric proof of this result. Start by representing $(0, 1)$ as an "open" line segment of length 1 (i.e., a line segment from which the endpoints have been removed). Bend the line segment into the bottom half of a semicircle with center $c$. Next, represent $\mathbb{R}$ as a real line placed below the semicircle:

   Use this diagram to find a bijection from the set of points in the "open" semicircle to the set of points on the real line.

3. Recall that the closed interval $[0, 1]$ is the set of real numbers $x$ such that $0 \leq x \leq 1$, and that $[0, 1)$ is the set of real numbers $x$ such that $0 \leq x < 1$. Show that $|[0, 1]| = |\mathbb{R}|$ and $|[0, 1)| = |\mathbb{R}|$. *Hint:* Use the No Countable Difference Principle.

4. Show that $|[0, a]| = |\mathbb{R}|$ for $a \in \mathbb{R}$ and $a > 0$.

5. Just as one can think of the interval [0, 1] as representing a line of length 1, can one think of the points in the Cartesian product $[0,1]\times[0,1]$ as representing a square of side-length 1. (In general, the **Cartesian product** $A\times B$ is the set of pairs $\langle a,b\rangle$ such that $a\in A$ and $b\in B$.) Show that there are just as many points on the unit line as there are points on the unit square:

   $$|[0,1]| = |[0,1]\times[0,1]|$$

   *Hint:* Use the following four steps:

   **Step 1** Let $D$ be the set of infinite sequences $\langle d_0, d_1, d_2, \ldots\rangle$ where each $d_i$ is a digit between 0 and 9, and show that $|D| = |D\times D|$.

   **Step 2** Show that $|[0,1]| = |D|$.

   **Step 3** Show that $|[0,1]\times[0,1]| = |D\times D|$.

   **Step 4** Use steps 1–3 to show that $|[0,1]| = |[0,1]\times[0,1]|$.

6. Just as one can think of the points in the Cartesian product $[0,1]\times[0,1]$ as representing a square of side-length 1, so can one think of the points in $\underbrace{[0,1]\times\ldots\times[0,1]}_{n \text{ times}}$ as an $n$-dimensional *hypercube*. Show that $|[0,1]| = |\underbrace{[0,1]\times\ldots\times[0,1]}_{n \text{ times}}|$.

## 1.7 The Power Set of the Natural Numbers

We have seen that there are more real numbers than natural numbers: $|\mathbb{N}| < |\mathbb{R}|$. And we know from Cantor's Theorem that there are more subsets of natural numbers than natural numbers: $|\mathbb{N}| < |\wp(\mathbb{N})|$. These results entail that $\wp(\mathbb{N})$ and $\mathbb{R}$ both have cardinalities greater than $\mathbb{N}$. But they do not settle the relative sizes of $\wp(\mathbb{N})$ and $\mathbb{R}$. In this section we will show that the two sets have exactly the same size: $|\wp(\mathbb{N})| = |\mathbb{R}|$.

### 1.7.1 Binary Notation

Our proof will rely on representing real numbers in binary notation. In particular, we will be using **binary expansions** of the form $0.b_1b_2b_3\ldots$, where each digit $b_i$ is either a 0 or a 1. (For instance, 0.00100100001….)

It will be useful to start by saying a few words about binary expansions. The way to tell which binary expansion corresponds to which number is to apply the following recipe:

The binary name $0.b_1b_2b_3\ldots$ represents the number $\frac{b_1}{2^1}+\frac{b_2}{2^2}+\frac{b_3}{2^3}+\ldots$.

Consider, for example, the binary expansion 0.001(0) (i.e., 0.00100000000…). This is a name for the number 1/8, since according to our recipe:

The binary name 0.001(0) represents the number $\frac{0}{2^1}+\frac{0}{2^2}+\frac{1}{2^3}+\sum_{n=4}^{\infty}\left(\frac{0}{2^n}\right)=\frac{1}{8}$.

As you'll be asked to prove in the exercises below, every real number in [0, 1] is named by some binary expansion of the form $0.b_1b_2b_3\ldots$. It is also worth noting that binary notation shares some important features with decimal notation. In particular, the binary expansions of rational numbers are always periodic: they end with an infinitely repeating string of digits. And the binary expansions of irrational numbers are never periodic. Also, some real numbers are named by more than one binary expansion. For instance, 1/8 is represented not just by 0.001(0), but also by 0.000(1), because of the following:

The binary name 0.000(1) represents the number $\frac{0}{2^1}+\frac{0}{2^2}+\frac{0}{2^3}+\sum_{n=4}^{\infty}\left(\frac{1}{2^n}\right)=\frac{1}{8}$.

Fortunately, the only real numbers with multiple names are those named by binary expansions that end in an infinite sequence of 1s. Such numbers always have exactly two binary names: one ending in an infinite sequence of 1s and one ending in an infinite sequence of 0s. Since only rational numbers have periodic binary expansions, this means that there are only countably many real numbers with more than one binary name. (The proofs of these results are analogous to the corresponding proofs in the answers to exercises at the end of section 1.6.1.)

#### Exercises

1. Identify the two binary expansions of 11/16.
2. Show that every real number in [0, 1] has a name in binary notation.

### 1.7.2 The Proof

Now for our proof of $|\wp(\mathbb{N})|=|\mathbb{R}|$. What we will actually prove is $|\wp(\mathbb{N})|=|[0,1]|$, but this suffices to give us what we want because we know that $|\mathbb{R}|=|[0,1]|$.

We will proceed in two steps. The first step is to show that $\wp(\mathbb{N})$ has the same cardinality as an "intermediary" set $B$. The second step is to show that $B$ has the same cardinality as [0, 1]. (Combining the two steps yields the result that $\wp(\mathbb{N})$ has the same cardinality as [0,1], which is what we need.)

Our intermediary set $B$ will be the set of binary expansions $0.b_0b_1\ldots$. In other words:

$$B=\{x : x \text{ is an expression "}0.b_0b_1\ldots\text{," where each } b_i \text{ is a binary digit}\}$$

**Step 1** It is easy to define a bijection $f$ from $\wp(\mathbb{N})$ to $B$. For each $A\in\wp(\mathbb{N})$, let $f(A)$ be the expression "$0.b_0b_1\ldots$," where:

$$b_i = \begin{cases} 1, & \text{if } i \in A \\ 0, & \text{if } i \notin A \end{cases}$$

For example, if $A$ is the set of odd numbers, then $f(A) = 0.01(01)$ (which is a binary name for the number 1/3).

**Step 2** We now need to show that $B$ has the same cardinality as [0, 1]. Recall that that $B$ contains multiple notations for certain numbers in [0, 1]. (For instance, the expressions "0.1(0)" and "0.0(1)" are both in $B$, even though they are both names for 1/2.) Let $B^-$ be the result of eliminating repeated notations from $B$. So we immediately get the result that $B^-$ has the same cardinality as [0,1].

Notice, however, that one can get $B$ by adding countably many members to $B^-$ (because only rational numbers have multiple notations and they never have more than two). So it follows from the No Countable Difference Principle (section 1.6.3) that adding countably many members to an infinite set doesn't change the set's cardinality. So $B$ must have the same cardinality as $B^-$. But we noted above that $B^-$ has the same cardinality as [0,1]. So $B$ must have the same cardinality as [0,1].

### Exercises

1. I gave a relatively informal proof for the claim that there is a bijection from [0,1] to $B_0$. Give a more rigorous version of the proof.
2. Where $F$ is the set of functions from natural numbers to natural numbers, show that $|\mathbb{N}| < |F|$.

## 1.8 Conclusion

Our discussion of infinity began with Hilbert's Hotel, which illustrates the fact that not all of our intuitions about finite cardinalities carry over to the infinite. We saw, however, that there is a way of taming infinity. One can theorize rigorously about infinite size by measuring the relative sizes of infinite sets on the basis of whether there are bijections and injections between them. Using this approach, we concluded that there are infinitely many different sizes of infinity. In particular:

$$|\mathbb{N}| < |\wp(\mathbb{N})| < |\wp(\wp(\mathbb{N}))| < |\wp(\wp(\wp(\mathbb{N})))| < \ldots$$

We then compared the cardinalities of the first two sets in this sequence, $\mathbb{N}$ and $\wp(\mathbb{N})$, with the cardinalities of a number of other sets, and established the following:

$$|\mathbb{N}| = |\mathbb{Z}| = |\mathbb{Q}| = |\mathfrak{F}|$$

$$|\wp(\mathbb{N})| = |\mathbb{R}| = |(0,1)| = |[0,1]| = |\underbrace{[0,1] \times \ldots \times [0,1]}_{n \text{ times}}|$$

These are exciting results. And they are only the beginning. In the next chapter we will see that there are infinite sets of much greater cardinality than any member of the above sequence. We will also see that there is a lot that is unknown. For example, the question of whether there is any set $A$ such that $|\mathbb{N}| < |A| < |\wp(\mathbb{N})|$ turns out to be independent of the standard axioms of set theory.

## 1.9 Further Resources

- If you'd like to delve further into the material in this section (or if you'd like to do some additional exercises), you should have a look at a set theory textbook. A good option is Karel Hrbáek and Thomas Jech's *Introduction to Set Theory*. For a more advanced treatment of the issues, I recommend Drake and Singh's *Intermediate Set Theory*.
- If you'd like to know more about the history of Cantor's discovery, I recommend Joan Bagaria's "Set Theory" in the *Stanford Encyclopedia of Philosophy*, which is available online.

## Appendix: Answers to Exercises

### Section 1.1.2

1. One way to do so is to let $f(n) = 2n$.
2. One way to do so is to combine the technique used to match natural numbers with integers and the technique used to match natural numbers with even natural numbers. If $n$ is an even natural number, let $f(n) = \frac{7n}{2}$, and if $n$ is an odd natural number, let $f(n) = -\frac{7(n+1)}{2}$.
3. (a) *Reflexivity*: We want to show that for any set $A$, there is a bijection from $A$ to $A$. Let $f$ be the identity function; that is, $f(x) = x$ for any $x$. In this case, $f$ is obviously a bijection from $A$ to itself.

   (b) *Symmetry*: Suppose that $f$ is a bijection from $A$ to $B$ and consider the *inverse* function $f^{-1}$ (i.e., the function $g$ such that for each $b \in B$, $g(b)$ is the $a \in A$ such that $f(a) = b$). We verify that $f^{-1}$ is a bijection from $B$ to $A$ with the following:
   - $f^{-1}$ is a *function* because $f$ is injective.
   - $f^{-1}$ is an *injective* function because $f$ is a function.
   - $f^{-1}$ is a function *from B* because $f$ is a surjective function *to B*.
   - $f^{-1}$ is a *surjective* function to $A$ because $f$ is function *from A*.

   (c) *Transitivity*: Suppose that $f$ is a bijection from $A$ to $B$ and that $g$ is a bijection from $B$ to $C$. Now consider the *composite* function $g \circ f$ (i.e., the function $g(f(x))$). It is straightforward to verify that $g \circ f$ is a function and that it is defined for every member of $A$. Let us now verify that it is a *bijection* from $A$ to $C$:
   - $g \circ f$ is *injective*: For $a_1, a_2 \in A$, let $a_1 \neq a_2$. Since $f$ is injective, $f(a_1) \neq f(a_2)$, and since $g$ is injective, $g(f(a_1)) \neq g(f(a_2))$. So $g \circ f(a_1) \neq g \circ f(a_2)$. So $g \circ f$ is injective.
   - $g \circ f$ is *surjective*: Let $c \in C$. Since $g$ is surjective, there is some $b \in B$ such that $g(b) = c$, and since $f$ is surjective, there is some $a \in A$ such that $f(a) = b$. So $g \circ f(a) = c$, for arbitrary $c \in C$. So $g \circ f$ is surjective.

### Section 1.1.3

1. We know that there is a bijection $f$ from the natural numbers to the non-negative rational numbers. A similar construction can be used to show that there is a bijection $g$ from the natural numbers to the negative rational numbers.

   We can then show that there is a bijection $h$ from the natural numbers to the rational numbers by combining these two results. For we can let:

$$h(n) = \begin{cases} f\left(\frac{n}{2}\right), & \text{if } n \text{ is even} \\ g\left(\frac{(n-1)}{2}\right), & \text{if } n \text{ is odd} \end{cases}$$

2. One way to do so is to follow the route in figure 1.2, but interpret the matrix differently. Instead of thinking of cells as corresponding to rational numbers, think of each column as corresponding to one copy of the natural numbers. (More precisely, treat the cell labeled $n/k$ as corresponding to $n_k \in \mathbb{N}_k$.)
3. $f(n) = \frac{1}{n+1}$ is a bijection from $\mathbb{N}$ to a subset of the set of rational numbers between 0 and 1.

## Section 1.2

1. Suppose, first, that $f$ is an injection from $A$ to $B$. In other words, $f$ is a bijection from $A$ to some subset $B'$ of $B$. By the definition of bijection, this means that $A$ assigns a different element of $B'$ (and therefore $B$) to each element of $A$.

   Now suppose that $f$ assigns a different element of $B$ to each element of $A$. If $B'$ is the set of elements of $B$ to which $f$ assigns some element of $A$, $f$ is a bijection from $A$ to $B'$.
2. A bijection from $A$ to $B$ is a function from $A$ to $B$ such that (*i*) each element of $A$ is assigned to a different element of $B$, and (*ii*) no element of $B$ is left without an assignment from $A$. Part (*i*) of this definition is satisfied if and only if $f$ is an injection, and part (*ii*) is satisfied if and only if $f$ is a surjection.
3. The identity function $f(x) = x$ is an injection from $A$ to $A$, so $A$ is reflexive for any $A$.

   To show that transitivity holds for arbitrary sets $A$, $B$, and $C$, assume that $f$ is an injection from $A$ to $B$ and that $g$ is an injection from $B$ to $C$. Since the composition of injections is injective, $g \circ f$ is an injection from $A$ to $C$.
4. We will assume that $|A| < |B|$ and $|B| \leq |C|$, and prove $|A| < |C|$. This requires proving two things: $|A| \leq |C|$ and $|A| \neq |C|$.

   $|A| \leq |C|$ follows from transitivity, which was verified in the previous exercise. To prove that $|A| \neq |C|$, we shall assume that $|A| = |C|$ and use this assumption to derive a contradiction. Since we are assuming that $|A| = |C|$, we are assuming that there is a bijection $f$ from $A$ to $C$ and therefore a bijection $f^{-1}$ from $C$ to $A$. Since $|B| \leq |C|$, we know that there is an injection $g$ from $B$ to $C$. So $h(x) = f^{-1}(g(x))$ is an injection from $B$ to $A$. But our initial assumptions give us $|A| < |B|$ (and therefore $|A| \leq |B|$). So, by the Cantor-Schroeder-Bernstein Theorem, $|A| = |B|$, which contradicts $|A| < |B|$.

## Section 1.6.1

1. No. The number 0 is named by a numeral that ends in an infinite sequence of 0s but not by a numeral that ends in an infinite sequence of 9s.
2. Since $0.\delta_1^1\delta_2^1\delta_3^1\ldots$ and $0.\delta_1^2\delta_2^2\delta_3^2\ldots$ are distinct, they must differ in at least one digit. Let $k$ be the smallest index such that $\delta_k^1 \neq \delta_k^2$. This means that the truncated expansions $0.\delta_1^1\ldots\delta_k^1$ and $0.\delta_1^2 l\ldots\delta_k^2$ must differ in value by at least $1/10^k$.

   Let us now focus on the remaining digits, $\delta_{k+1}^i\delta_{k+2}^i\delta_{k+3}^i\ldots$ (where $i$ is either 1 or 2). This sequence of digits contributes the following value to $0.\delta_1^i\delta_2^i\delta_3^i\ldots$:

   $$\frac{\delta_{k+1}^i}{10^{k+1}} + \frac{\delta_{k+2}^i}{10^{k+2}} + \frac{\delta_{k+3}^i}{10^{k+3}} + \ldots$$

   This sum must yield a non-negative number smaller or equal to $1/10k$. The sum will be 0 if every digit in $\delta_{k+1}^i\delta_{k+2}^i\delta_{k+3}^i\ldots$ is 0, since

   $$\frac{0}{10^{k+1}} + \frac{0}{10^{k+2}} + \frac{0}{10^{k+3}} + \ldots = 0$$

   And the number will be $1/10k$, if every digit in $\delta_{k+1}^i\delta_{k+2}^i\delta_{k+3}^i\ldots$ is 9, since

   $$\frac{9}{10^{k+1}} + \frac{9}{10^{k+2}} + \frac{9}{10^{k+3}} + \ldots = \frac{1}{10^k}$$

   Notice, moreover, that the *only* way for the sum to be 0 is for every digit in $\delta_{k+1}^i\delta_{k+2}^i\delta_{k+3}^i\ldots$ to be 0, and that the *only* way for the sum to be $1/10k$ is for every digit in $\delta_{k+1}^i\delta_{k+2}^i\delta_{k+3}^i\ldots$ to be 9.

   We are now in a position to verify that $0.\delta_1^1\delta_2^1\delta_3^1\ldots$ and $0.\delta_1^2\delta_2^2\delta_3^2\ldots$ can refer to the same number only if the following two conditions are met:

   (a) The difference between $0.\delta_1^1\ldots\delta_k^1$ and $0.\delta_1^2\ldots\delta_k^2$ must be smaller or equal to $1/10^k$. (The reason is that this difference must match the difference between the value contributed by the sequence of digits $\delta_{k+1}^1\delta_{k+2}^1\delta_{k+3}^1\ldots$ and the value contributed by the sequence of digits $\delta_{k+1}^2\delta_{k+2}^2\delta_{k+3}^2\ldots$, which we know to be at most $1/10^k$.) We have also seen, however, that the difference between $0.\delta_1^1\ldots\delta_k^1$ and $0.\delta_1^2\ldots\delta_k^2$ must be at least $1/10^k$. Putting the two together gives us the result that the difference must be *exactly* $1/10^k$. Assuming, with no loss of generality, that $0.\delta_1^1\ldots\delta_k^1 < 0.\delta_1^2\ldots\delta_k^2$, this means that the digit $\delta_k^2$ is the result of raising digit $\delta_k^1$ by one.

   (b) The value contributed by $\delta_{k+1}^1\delta_{k+2}^1\delta_{k+3}^1\ldots$ must be the value contributed by $\delta_{k+1}^2\delta_{k+2}^2\delta_{k+3}^2\ldots$ plus $1/10^k$, in order to compensate for the fact that $0.\delta_1^2\ldots\delta_k^2 = 0.\delta_1^1\ldots\delta_k^1 + 1/10^k$. But we know that the value contributed by $\delta_{k+1}^2\delta_{k+2}^2\delta_{k+3}^2\ldots$ and the value contributed by $\delta_{k+1}^1\delta_{k+2}^1\delta_{k+3}^1\ldots$ must be both smaller than or equal to

$1/10^k$. It follows that the value contributed by $\delta^1_{k+1}\delta^1_{k+2}\delta^1_{k+3}\ldots$ must be $1/10^k$ and the value contributed by $\delta^2_{k+1}\delta^2_{k+2}\delta^2_{k+3}\ldots$ must be 0. And we know that the only way for this to happen is for $\delta^1_{k+1}, \delta^1_{k+2}, \delta^1_{k+3}, \ldots$ to all be 9s and $\delta^2_{k+1}, \delta^2_{k+2}, \delta^2_{k+3}, \ldots$ to all be 0s.

3. Note that $d_0.d_1d_2\ldots$ is a decimal expansion for $a/b$, where
   - $d_0$ is the integral part of $\frac{a}{b}$, with the remainder $r_0$;
   - $d_1$ is the integral part of $\frac{10r_0}{b}$, with the remainder $r_1$;
   - $d_2$ is the integral part of $\frac{10r_1}{b}$, with the remainder $r_2$;
   - and so forth.

   Notice, moreover, that there are only finitely many distinct values that the $r_k$ can take, since they must always be smaller than $b$. Because the values of $d_{i+1}$ and $r_{i+1}$ depend only on the value of $r_i$, this means that the $d_k$ values will eventually start repeating themselves. So our decimal expansion is periodic.

4. Note that the first line of the transformation below is a simple algebraic truth and that successive lines are obtained by substituting the first line's right-hand term for bold-faced occurrences of $p$:

$$p=\frac{(10^n-1)p}{10^n}+\frac{\mathbf{p}}{10^n}$$

$$p=\frac{(10^n-1)p}{10^n}+\frac{\frac{(10^n-1)p}{10^n}+\frac{p}{10^n}}{10^n}$$

$$p=\frac{(10^n-1)p}{10^n}+\frac{(10^n-1)p}{10^{2n}}+\frac{\mathbf{p}}{10^{2n}}$$

$$p=\frac{(10^n-1)p}{10^n}+\frac{(10^n-1)p}{10^{2n}}+\frac{\frac{(10^n-1)p}{10^n}+\frac{p}{10^n}}{10^{2n}}$$

$$p=\frac{(10^n-1)p}{10^n}+\frac{(10^n-1)p}{10^{2n}}+\frac{(10^n-1)p}{10^{3n}}+\frac{\mathbf{p}}{10^{3n}}$$

$$\vdots$$

$$p=\frac{(10^n-1)p}{10^n}+\frac{(10^n-1)p}{10^{2n}}+\frac{(10^n-1)p}{10^{3n}}+\frac{(10^n-1)p}{10^{4n}}+\ldots$$

$$p=(10^n-1)\left(\frac{p}{10^n}+\frac{p}{10^{2n}}+\frac{p}{10^{3n}}+\frac{p}{10^{4n}}+\ldots\right)$$

From this we may conclude,

$$\frac{p}{10^n-1}=\frac{p}{10^n}+\frac{p}{10^{2n}}+\frac{p}{10^{3n}}+\ldots=0.(p)$$

5. Every real number with a periodic decimal notation is a number of the form

$$\frac{a+0.(p)}{10^n}$$

where $a$ is an integer, $n$ is a natural number, and $p$ is a period. But we know that this must be a rational number, because we verified in the previous exercise that $0.(p)$ is a rational number.

## Section 1.6.3

1. **Step 1:** We wish to show that $|S^{\mathbb{N}}| = |S^{\mathbb{N}} \cup A|$. So we need to show that there is a bijection $f$ from $S^{\mathbb{N}}$ to $S^{\mathbb{N}} \cup A$.

   Since $A$ is countable, we know that the following possibilities are exhaustive: (i) every element of $A$ is also in $S^{\mathbb{N}}$; (ii) there are finitely many things in $A$ that are not also in $S^{\mathbb{N}}$; and (iii) there are a countable infinity of things in $A$ that are not also in $S^{\mathbb{N}}$.

   First suppose we are in case (i). Then we can let $f$ be the identity function. Now suppose we are in case (ii). Let $B = b_0, b_1, b_2, \ldots, b_n$ be the set of things in $A$ that are not also in $\mathbb{N}$, and recall that $S^{\mathbb{N}} = s_0, s_1, s_2, \ldots$. We can define $f$ as follows:

$$f(s_k) = \begin{cases} f(s_k) = b_k, & \text{for } k \leq n \\ f(s_k) = s_{k-(n+1)}, & \text{for } k > n \end{cases}$$

   Finally, suppose we are in case (iii). Let $B = b_0, b_1, b_2, \ldots$ be the set of things in $A$ that are not also in $S^{\mathbb{N}}$, and recall that $S^{\mathbb{N}} = s_0, s_1, s_2, \ldots$. We can let $f(s_{2n}) = s_n$ and $f(s_{2n+1}) = b_n$.

   **Step 2:** We wish to show that $|S| = |S \cup A|$. We do so by using the bijection $f$ from $S^{\mathbb{N}}$ to $S^{\mathbb{N}} \cup A$ to define a bijection $h$ from $S$ to $S \cup A$:

$$h(x) = \begin{cases} h(x) = f(x), & \text{if } x \in S^{\mathbb{N}} \\ h(x) = x, & \text{if otherwise} \end{cases}$$

2. Our task is to show that there is a bijection from $(0, 1)$ to $\mathbb{R}$. We do so twice, once geometrically and once analytically.

   Here is the geometric proof: As in the main text, start by representing $(0, 1)$ as an "open" line segment of length 1 (i.e., a line segment from which the endpoints have been removed). Bend the line segment into the bottom half of an "open" semicircle with center $c$. Next, represent $\mathbb{R}$ as a real line placed below the semicircle. Now consider the family of "rays" emanating from $c$ that cross the open semicircle and reach the real line. The dotted line in figure A.1.1 is an example of one such ray: it crosses the semicircle at point $p$ and reaches the real line at point $p'$. Note that each

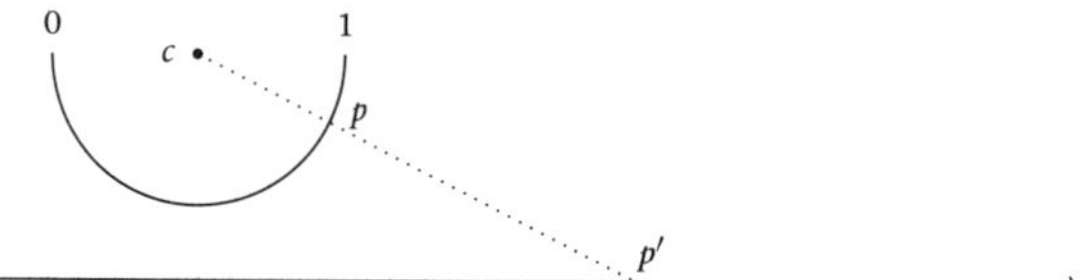


**Figure A.1.1**
A bijection from (0, 1) to ℝ.

ray in our family crosses exactly one point on the open semicircle and one point on the real line. Notice, moreover, that each point on the open semicircle and each point on the real line is crossed by exactly one ray. This means that one can define a bijection from (0, 1) to ℝ by assigning each point on the open semicircle to the point on the real line with whom it shares ray. (Note that it is essential to this proof that we work with the open interval (0, 1) rather than the closed interval [0, 1], since a ray crossing the semicircle at point 0 or 1 would fail to reach the real line.)

A version of the same idea can be used to define a bijection from (0, 1) to ℝ analytically rather than geometrically. First note that the tangent function *tan*($x$) is a bijection from $(-\pi/2, \pi/2)$ to the real numbers. Next, note that the function $f(x) = \pi(x - \frac{1}{2})$ is a bijection from (0, 1) to $(-\pi/2, \pi/2)$. Putting the two together, we get the result that $tan \circ f$ is a bijection from (0, 1) to the real numbers.

3. It is easy to verify that (0, 1) is infinite. (For instance, $f(n) = \frac{1}{n+2}$ is an injective function from ℕ to (0, 1).) Since $|(0,1)| = |(0,1) \cup \{0,1\}|$, it follows from the No Countable Difference Principle that $|(0,1)| = |[0,1]|$. Since $|(0,1)| = |(0,1) \cup \{0\}|$, it follows from the No Countable Difference Principle that $|(0,1)| = |[0,1)|$.

   But in an earlier exercise, we showed that $|(0,1)| = |\mathbb{R}|$. So it follows from the symmetry and transitivity of bijections that $|[0,1]| = |\mathbb{R}|$ and $|[0,1)| = |\mathbb{R}|$.

4. $f(x) = ax$ is a bijection from [0, 1] to [0, $a$]. Since we know that $|[0,1]| = |\mathbb{R}|$, it follows from the symmetry and transitivity of bijections that $|[0,a]| = |\mathbb{R}|$

5. **Step 1** The following $f$ is a bijection from $D$ to $D \times D$:

   $$f(\langle d_0, d_1, d_2, \ldots \rangle) = \langle \langle d_0, d_2, d_4, \ldots \rangle, \langle d_1, d_3, d_5, \ldots \rangle \rangle$$

   **Step 2** Think of each sequence $\langle d_0, d_1, d_2, \ldots \rangle$ in $D$ as the decimal expansion

   $$0.d_0, d_1, d_2, \ldots$$

   Then every real number in [0, 1] is named by some sequence in $D$, and every sequence in $D$ names some real number in [0, 1]. Unfortunately, this does not yet give us a bijection between $D$ and [0, 1] because $D$ includes decimal expansions that end in an infinite sequence of 9s. So there are numbers in [0, 1] that are named by two different sequences in $D$.

Let $D^-$ be the result of removing from $D$ every sequence ending in an infinite sequence of 9s. Since $D^-$ consists of exactly one name for each number in $[0,1]$, we have $|[0,1]| = |D^-|$. By the symmetry and transitivity of bijections, this means that in order to prove $|[0,1]| = |D|$, it suffices to show that $|D| = |D^-|$.

Notice, however, that there are only countably many sequences in $D$ that are not in $D^-$. (There are only countably many rational numbers, and every decimal expansion ending in an infinite sequence of 9s is a rational number.) So it follows from the No Countable Difference Principle that $|D| = |D^-|$.

**Step 3** The upshot of step 2 is that there is a bijection $f$ from $[0,1]$ to $D$. One can use $f$ to define a bijection $g$ from $[0,1] \times [0,1]$ to $D \times D$, as follows:

For each $\langle x, x' \rangle$ in $[0,1] \times [0,1]$, let $g(\langle x, x' \rangle) = \langle f(x), f(x') \rangle$.

**Step 4** The result follows immediately from the symmetry and transitivity of bijections.

6. We know from the previous result that there is a bijection $f$ from $[0,1]$ to $[0,1] \times [0,1]$. Notice that $f$ can be used to define a bijection $f'$ from $[0,1]$ to $[0,1] \times [0,1] \times [0,1]$, as follows:

   If $f(x) = \langle x_1, x_2 \rangle$, then $f'(x) = \langle x_1, f_2(x_2) \rangle$.

   This technique can be iterated to generate a bijection from $[0,1]$ to $\underbrace{[0,1] \times \ldots \times [0,1]}_{n \text{ times}}$ for any finite $n$.

### Section 1.7.1

1. The two binary expansions of $11/16$ are $0.1011(0)$ and $0.1010(1)$. This can be verified as follows:

$$\begin{aligned}
0.1011(0) &= \frac{1}{2^1} + \frac{0}{2^2} + \frac{1}{2^3} + \frac{1}{2^4} \\
&= \frac{8}{16} + \frac{2}{16} + \frac{1}{16} \\
&= \frac{11}{16} \\
0.1010(1) &= \frac{1}{2^1} + \frac{0}{2^2} + \frac{1}{2^3} + \frac{0}{2^4} + \frac{1}{2^5} + \frac{1}{2^6} + \frac{1}{2^7} + \ldots \\
&= \frac{1}{2^1} + \frac{0}{2^2} + \frac{1}{2^3} + \frac{0}{2^4} + \frac{1}{2^4} \\
&= \frac{8}{16} + \frac{2}{16} + \frac{1}{16} \\
&= \frac{11}{16}
\end{aligned}$$

2. Let $r$ be a real number in [0, 1]. We will verify that $r$ is named by the binary expression "$0.b_1b_2b_3\ldots$," whose digits are defined as follows:

$$b_1 = \begin{cases} 0, & \text{if } r < \frac{1}{2} \\ 1, & \text{if otherwise} \end{cases}$$

$$b_{k+1} = \begin{cases} 0, & \text{if } r - 0.b_1b_2\ldots b_k < \frac{1}{2^{n+1}} \\ 1, & \text{if otherwise} \end{cases}$$

These definitions guarantee that $r - 0.b_0\ldots b_n < \frac{1}{2^n}$, for each $n$. From this it follows that the difference between $r$ and $0.b_1b_2b_3\ldots$ must be smaller than any positive real number, and therefore equal to 0. So $r = 0.b_1b_2b_3\ldots$.

### Section 1.7.2

1. As in the main text, $B_0$ is the set of binary names starting in 0. Let $B_0^-$ be a subset of $B$ that includes exactly one name for each number in [0,1]. (For example, $B_0^-$ might be the result of eliminating from $B_0$ every name that ends with an infinite sequence of 1s, except 0.(1) which is the unique name for 1 in $B_0$.)

   Since $|[0,1]| = |B_0^-|$, all we need to do to verify $|[0,1]| = |B_0|$ is to show that $|B_0^-| = |B_0|$. But recall the No Countable Difference Principle (section 1.6.3): adding countably many members to an infinite set can never change the set's cardinality. Since $B_0^-$ is infinite, this means that all we need to do is identify a countable set $R$ such that $|B_0| = |B_0^- \cup R|$. But we can simply let $R = B_0 - B_0^-$. Since there are countably many rational numbers, and since only rational numbers have periodic binary notations, $R$ must be countable.

2. It suffices to show that there are more functions from the natural numbers to $\{0, 1\}$ than there are natural numbers. But every function from natural numbers to $\{0, 1\}$ can be used to determine a different subset of the natural numbers: the subset corresponding to numbers that the function assigns to 1. And we have seen that there are more subsets of natural numbers than natural numbers.

# 2 The Higher Infinite

Chapter 1 was mostly focused on infinite sets of two different sizes: the size of the natural numbers, $|\mathbb{N}|$, and the size of the real numbers, $|\mathbb{R}|$. In this chapter we'll talk about infinite sets that are much bigger than that—much, *much* bigger. Some of the infinities we'll be talking about are so vast that we'll need to develop a new tool to keep track of their size: the notion of an ordinal.

## 2.1 Living Large

There is an important strategy for building infinite sets of bigger and bigger cardinalities. It is based on the deployment of two basic resources: the power set operation and the union operation. We'll begin by reviewing these resources.

### 2.1.1 The Power Set Operation

In Chapter 1 we proved Cantor's Theorem, which shows that a set's power set always has more members than the set itself. This result entails that there is an infinite hierarchy of bigger and bigger infinities:

$$|\mathbb{N}| < |\wp^1(\mathbb{N})| < |\wp^2(\mathbb{N})| < |\wp^3(\mathbb{N})| < \dots$$

where

$$\wp^n(S) = \underbrace{\wp(\wp(\dots\wp(S)\dots))}_{n \text{ times}}$$

The sequence above includes some very big sets. For example, the set $\wp^{10^{10^{10}}}(\mathbb{N})$ is fantastically bigger than $\mathbb{R}$. But could we go bigger still? Could we characterize a set that is bigger than $\wp^k(\mathbb{N})$ for *every* natural number $k$? One might be tempted to introduce a set by way of the following definition:

$$\wp^\infty(\mathbb{N}) = \underbrace{\wp(\wp(\dots\wp(\mathbb{N})\dots))}_{\infty \text{ times}}$$

Unfortunately, it is not clear that such a definition would make sense. To see the problem, note that $\wp^{\infty}(\mathbb{N})$ is supposed to be defined by iterating the ordinary power set operation, $\wp$. But each application of $\wp$ requires a definite input. When $k$ is a positive integer, the input of $\wp^{k}(\mathbb{N})$ is $\wp^{k-1}(\mathbb{N})$. But what input should one use at the supposed "infinite-th" stage of the process? There is not a clear answer to this question. (Notice, in particular, that $\wp^{\infty-1}(\mathbb{N})$ won't do as an answer, since it is not clear what $\wp^{\infty-1}(\mathbb{N})$ is supposed to be.)

The union operation will help us get around this problem.

### 2.1.2 The Union Operation

In chapter 1 we encountered a version of the union operation that takes finitely many sets as input and delivers a single set as output. More specifically, we took $A_1 \cup A_2 \cup \ldots \cup A_n$ to be the set of individuals $x$ such that $x$ is in at least one of $A_1, A_2, \ldots, A_n$. We will now consider a variant of the union operation that takes as input a set $A$ of arbitrarily many sets. Let $A$ be a set of sets, as it might be:

$$A = \{S_1, S_2, S_3, \ldots\}$$

Then the union of $A$ (in symbols: $\bigcup A$) is the result of pooling together the members of each of the sets in $A$. In other words, $\bigcup A = S_1 \cup S_2 \cup S_3 \cup \ldots$. (Formally, we define $\bigcup A$ as the set of individuals $x$ such that $x$ is a member of some member of $A$.)

The key advantage of this new version of the union operation is that $\bigcup A$ is well-defined even if $A$ has infinitely many sets as members. This makes the union operation incredibly powerful. To illustrate this point, consider the set $\{\mathbb{N}, \wp^1(\mathbb{N}), \wp^2(\mathbb{N}), \ldots\}$. Even though it only has as many members as there are natural numbers, its union—

$$\bigcup\{\mathbb{N}, \wp^1(\mathbb{N}), \wp^2(\mathbb{N}), \ldots\}$$

—is far bigger than the set of natural numbers. In fact, it is bigger than $\wp^{k}(\mathbb{N})$ for each natural number $k$, since it includes everything in $\wp^{k+1}(\mathbb{N})$.

#### Exercises

1. Give an example of a set $A$ such that $|\bigcup A| < |A|$.

### 2.1.3 Bringing the Two Operations Together

This is what we've learned so far:

$$|\mathbb{N}| < |\wp^1(\mathbb{N})| < |\wp^2(\mathbb{N})| < \ldots < |\bigcup\{\mathbb{N}, \wp^1(\mathbb{N}), \ldots\}|$$

Can we construct sets that are bigger still? Of course we can! Let us use $\mathcal{U}$ to abbreviate $\bigcup\{\mathbb{N}, \wp^1(\mathbb{N}), \ldots\}$. Cantor's Theorem tells us that if we apply the power set operation to $\mathcal{U}$, we will get something even bigger. And, of course, each successive application of the power set operation gives us something bigger still:

$$|\mathcal{U}| < \ldots |\wp^1(\mathcal{U})| < |\wp^2(\mathcal{U})| < \ldots$$

And we can keep going. We can apply the union operation to the set of everything we've built so far. And then we can apply further iterations of the power set operation. And then apply the union operation to everything we've built so far. And then apply further iterations of the power sets. And so forth.

We have now sketched a procedure for constructing sets of greater and greater cardinality: after applying the power set operation countably many times, apply the union operation. And repeat. The main objective of this chapter is to show you how to develop the idea properly, by introducing you to one of the most beautiful tools in the whole of mathematics: the notion of an ordinal.

## 2.2 Ordinals

We'll build up to the notion of an ordinal in stages, by transitioning through each of the following notions:

Ordering → Total Ordering → Well-Ordering → Well-Order Type → Ordinal

Taking these notions in reverse order: an *ordinal* is a representative of a *well-order type*, which corresponds to a class of *well-orderings*, which are a special kind of *total ordering*, which is a special kind of *ordering*. This may not make much sense to you now, but it will soon!

### 2.2.1 Orderings

To order a set is to establish a relation of *precedence* between members of the set. When I order my family members by birth date, for example, I take family member $a$ to *precede* family member $b$ if and only if $a$ was born before $b$.

In general, we will say that a set $A$ is **(strictly) ordered** by a precedence relation $<$, or that $<$ is a **(strict) ordering** on $A$ if and only if $<$ satisfies the following two conditions for any $a, b$, and $c$ in $A$:

**Asymmetry** If $a < b$ , then not-$(b < a)$.

**Transitivity** If $a < b$ and $b < c$, then $a < c$.

These conditions are, to some extent, implicit in our informal notion of precedence: I don't precede those who precede me (Asymmetry), and I am preceded by those who precede my predecessors (Transitivity).

Taken together, our two conditions rule out precedence loops. For example, if a group of people is sitting around a table, Asymmetry and Transitivity rule out an ordering in which each person precedes the person to her left.

### Exercises

1. Show that Asymmetry entails the following:

**Irreflexivity** not-($a < a$).

2. A **relation** $R$ on a set $A$ is a subset of $A \times A$. If $R$ is a relation on $A$ and $a, b \in A$, we say that $aRb$ holds true (or that $Rab$ holds true) if and only if $\langle a, b \rangle \in R$. Now consider the following definitions:

- $R$ is **reflexive** on $A$ $\leftrightarrow_{\text{df}}$ for any $a \in A$, $aRa$ .
- $R$ is **irreflexive** on $A$ $\leftrightarrow_{\text{df}}$ for any $a \in A$, not-($aRa$).
- $R$ is **anti-symmetric** on $A$ $\leftrightarrow_{\text{df}}$ for any $a, b \in A$ if $aRb$ and $bRa$, then $a = b$.
- $R$ is **transitive** on $A$ $\leftrightarrow_{\text{df}}$ for any $a, b, c \in A$, if $aRb$ and $bRc$, then $aRc$.

$R$ is said to be a **non-strict partial order** on $A$ if it is reflexive, anti-symmetric, and transitive on $A$, and it is said to be a **strict partial order** on $A$ if it is irreflexive, anti-symmetric, and transitive on $A$.

**a)** Show that a set $A$ is (strictly) ordered by $<$ if and only if $<$ is a strict partial order on $A$.

**b)** Show that if $R$ is a non-strict partial order on $A$, then the relation $R'$ such that $aR'b$ if and only if $aRb$ and $a \neq b$ is a strict partial order on $A$.

**c)** Show that if $R$ is a strict partial order on $A$, then the relation $R'$ such that $aR'b$ if and only if $aRb$ or $a = b$ is a non-strict partial order on $A$.

### 2.2.2 Total Orderings

Even though Asymmetry and Transitivity are nontrivial, they are relatively weak ordering conditions. For example, they are both satisfied by an "empty" precedence relation $<$, according to which $a < b$ is never the case. They are also compatible with orderings that have tree-like structures. Consider, for example, the set of Queen Victoria's descendants, ordered as follows: $a$ precedes $b$ if and only if $a$ is $b$'s (direct-line) ancestor. This ordering can be represented as a tree, with Victoria at the base, each of her children branching off from the base, her grandchildren branching off from their parents' nodes, and so forth.

A feature of tree-like orderings is that they allow for individuals who don't bear the precedence relation to one another. This is true, for example, of any two of Victoria's children. Since neither is an ancestor of the other, neither of them precedes the other in our ancestry-based ordering. In what follows, we will restrict our attention to orderings in which any two objects are such that one precedes the other. Formally speaking, we will restrict our attention to *total* orderings. A **(strict) total ordering** $<$ on

$A$ is a (strict) ordering that satisfies the following additional condition for any $a$ and $b$ in $A$:

**Totality** $a < b$ or $b < a$, whenever $a$ and $b$ are distinct objects.

Here is an example of a total ordering on the set of Victoria's descendants: *a precedes b if and only if a was born before b*. Since no two of Victoria's descendants were born at exactly the same time, any two of them are such that one of them precedes the other. Another example of a total ordering is the standard ordering of the integers, $<_{\mathbb{Z}}$:

$$\ldots <_{\mathbb{Z}} -2 <_{\mathbb{Z}} -1 <_{\mathbb{Z}} 0 <_{\mathbb{Z}} 1 <_{\mathbb{Z}} 2 <_{\mathbb{Z}} 3 <_{\mathbb{Z}} \ldots$$

Integer $a$ precedes integer $b$ on this ordering if and only if $b = a + n$ for $n$, a positive integer. Since any two integers are such that there is some positive difference between them, any two of them are such that one of them precedes the other according to $<_{\mathbb{Z}}$.

### Exercises

Determine whether each of the following orderings is a total ordering.

1. The integers $\ldots, -2, -1, 0, 1, 2, \ldots$ under an "inverse" ordering $<^{-1}$ such that $a <^{-1} b$ if and only if $b <_{\mathbb{Z}} a$, where $<_{\mathbb{Z}}$ is the standard ordering of the integers, as characterized in the main text.
2. The power set of $\{0, 1\}$ under an ordering $\subsetneq$ such that $a \subsetneq b$ if and only if $a$ is a proper subset of $b$. (Note that $a$ is a proper subset of $b$ if and only if $a \neq b$ and every member of $a$ is an member of $b$.)
3. The real numbers in the interval $[0, 1]$ under the standard ordering $<_{\mathbb{R}}$, which is such that $a <_{\mathbb{R}} b$ if and only if $b = r + a$ for $r$, a positive real number.
4. The set of countably infinite sequences $\langle d_0, d_1, \ldots \rangle$ (where each $d_i$ is a digit $0, 1, \ldots, 9$) under an ordering $<$ such that $\langle d_0^1, d_1^1, \ldots \rangle < \langle d_0^2, d_1^2 \ldots \rangle$ if and only if $0.d_0^1 d_1^1 \ldots <_{\mathbb{R}} 0.d_0^2 d_1^2 \ldots$ (where $<_{\mathbb{R}}$ is defined as in the previous question).
5. The set of soldiers, under an ordering $<$ such that $x < y$ if and only if $x$ outranks $y$.

### 2.2.3 Well-Orderings

The standard ordering of the natural numbers, $<_{\mathbb{N}}$, is a total ordering:

$$0 <_{\mathbb{N}} 1 <_{\mathbb{N}} 2 <_{\mathbb{N}} \ldots$$

But it is a total ordering with an important special feature: it is a *well-ordering*. What this means is that any non-empty set of natural numbers has a $<_{\mathbb{N}}$-smallest member: a member that precedes all others according to $<_{\mathbb{N}}$. (The set of prime numbers, for example, has 2 as its $<_{\mathbb{N}}$-smallest member, and the set of perfect numbers has 6.)

Not every total ordering is a well-ordering. For example, the standard ordering of the integers, $<_{\mathbb{Z}}$, is not a well-ordering, since there are non-empty subsets of $\mathbb{Z}$ with no $<_{\mathbb{Z}}$-smallest integer. (One example of such a subset is $\mathbb{Z}$ itself.)

The set $[0, 1]$ under its standard ordering, $<_{\mathbb{R}}$, also fails to be well-ordered. For even though the entire set $[0, 1]$ has 0 as its $<_{\mathbb{R}}$-smallest member, $[0, 1]$ has subsets with no $<_{\mathbb{R}}$-smallest member. One example of such a subset is the set $(0, 1]$ (which is defined as $[0, 1] - \{0\}$).

Formally, we shall say that a set $A$ is **well-ordered** by $<$ if $<$ is a total ordering of $A$ that satisfies the following additional condition:

**Well-Ordering** Every non-empty subset $S$ of $A$ has a $<$-smallest member (that is, a member $x$ such that $x < y$ for every $y$ in $S$ other than $x$).

#### Exercises

Of each of the following orderings, determine whether it is a well-ordering:

1. The positive rational numbers under the standard ordering $<_{\mathbb{Q}}$, which is such that $a <_{\mathbb{Q}} b$ if and only if $b = q + a$ for $q$ a positive real number.
2. The natural numbers under an unusual ordering, $<_{\star}$, which is just like the standard ordering, except that the order of 0 and 1 is reversed:

   $$1 <_{\star} 0 <_{\star} 2 <_{\star} 3 <_{\star} 4 <_{\star} 5 <_{\star} \ldots$$

3. The natural numbers under an unusual ordering, $<_0$, in which 0 is counted as bigger than every positive number but the remaining numbers are ordered in the standard way:

   $$1 <_0 2 <_0 3 <_0 4 <_0 \ldots <_0 0$$

4. A finite set of real numbers under the standard ordering $<_{\mathbb{R}}$.

### 2.2.4 The Shapes of Well-Orderings

There are many different ways of well-ordering the natural numbers. There is, of course, the standard ordering, $<_{\mathbb{N}}$. But one can also well-order the natural numbers using an ordering, $<_{\frown}$, which is like the standard ordering except that it reverses the position of each even number and its successor:

$$1 <_{\frown} 0 <_{\frown} 3 <_{\frown} 2 <_{\frown} 5 <_{\frown} 4 <_{\frown} \ldots$$

Although $<_{\mathbb{N}}$ and $<_{\frown}$ correspond to different ways of ordering the natural numbers, they have exactly the same structure. The way to see this is to think of each of the two orderings as consisting of a sequence of "positions," each occupied by a particular number:

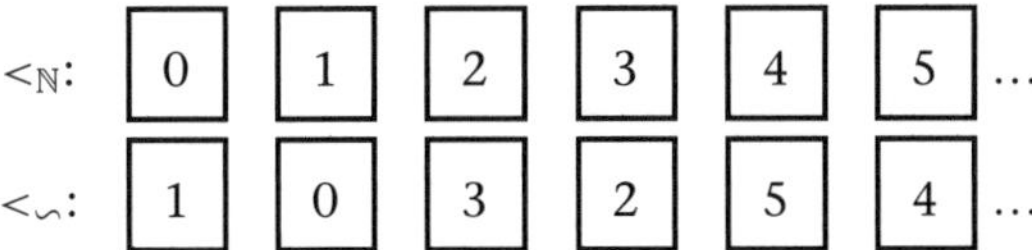


When we abstract away from the numbers that occupy the positions and consider only the positions themselves, we get the exact same structure in both cases:

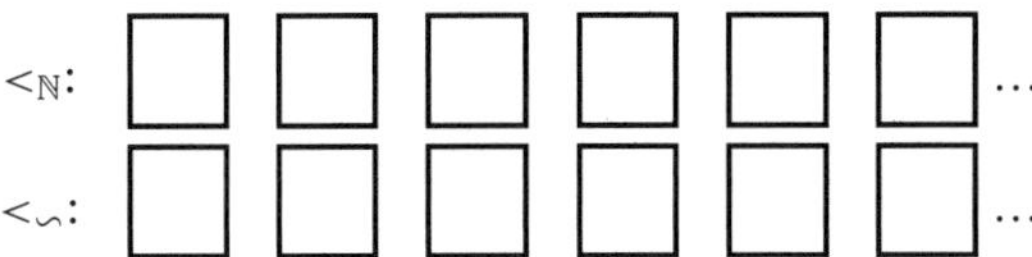


Accordingly, we can say that $<_{\mathbb{N}}$ and $<_{\backsim}$ are well-orderings with the same "shape." I will sometimes represent this shape schematically by shrinking the width of squares until they look like lines:

||||||...

Formally, we say that the well-orderings $<_1$ and $<_2$ have the same structure—or the same "shape"—if they are isomorphic to one another in the following sense:

> Let $<_1$ be an ordering on $A$ and $<_2$ be an ordering on $B$. Then $<_1$ is **isomorphic** to $<_2$ if and only if there is a bijection $f$ from $A$ to $B$ such that, for every $x$ and $y$ in $A$, $x <_1 y$ if and only if $f(x) <_2 f(y)$.
>
> (Consider, for example, the two orderings of the natural numbers we considered above: $<_{\mathbb{N}}$ and $<_{\backsim}$. They are isomorphic because the function $f$ that takes each even number $n$ to $n+1$ and each odd number $n$ to $n-1$ is a bijection from $\mathbb{N}$ to $\mathbb{N}$ such that $x <_{\mathbb{N}} y$ if and only if $f(x) <_{\backsim} f(y)$.)

Not all well-orderings of $\mathbb{N}$ are isomorphic to one another. To see this, consider the well-ordering $<_0$, which is included in an exercise at the end of section 2.2.3. It is just like the standard ordering, except that it takes every positive number to precede 0:

$1 <_0 2 <_0 3 <_0 4 <_0 \ldots <_0 0$

Note that $<_0$ is not isomorphic to $<_{\mathbb{N}}$, since its shape is |||||...|, which is different from the shape of $<_{\mathbb{N}}$, which is |||||....

In this chapter we won't be interested in individual well-orderings, like $<_{\mathbb{N}}$ or $<_0$. Instead, we will be focusing on *types* of well-orderings, or **well-order types**. Intuitively, each well-order type corresponds to a particular shape of well-ordering. Formally, a well-order type is an isomorphism-class of well-orderings: a class that consists of a well-ordering $<$, and every well-ordering that $<$ is isomorphic to. For instance, $<_{\mathbb{N}}$ and

$<_{\backsim}$ fall under the same well-order type, because they are isomorphic to each other. But $<_{\mathbb{N}}$ and $<_0$ do not fall under the same well-order type, because they are not isomorphic to each other.

#### Exercises

Let $<_{\mathbb{N}}$ be the standard ordering of the natural numbers. Determine whether each of the following pairs of sets corresponds to the same well-order type when ordered by $<_{\mathbb{N}}$:

1.

**a)** $\{7, 2, 13, 12\}$ **b)** $\{412, 708, 20081\}$

2.

**a)** the set of natural numbers **b)** the set of natural numbers greater than 17

### 2.2.5 Well-Ordering Finite and Infinite Sets

As long as a set has more than one member, it can be well-ordered in multiple ways. Notice, however, that the *well-orderings* of a *finite* set are always isomorphic to one another. For example, the finite set $\{a, b, c\}$ has six different well-orderings:

$$a < b < c \quad b < a < c \quad c < a < b$$
$$a < c < b \quad b < c < a \quad c < b < a$$

But they all fall under a single well-order type—the type corresponding to the shape |||. This is true in general: every well-ordering of an $n$-member finite set is isomorphic to any other.

In contrast, there are always non-isomorphic ways of well-ordering an infinite set. We saw an example of this above, when we noted that there are well-orderings of the set of natural numbers with the following two shapes:

|||||...
|||||...|

There are also well-orderings of the set of natural numbers with each of the following shapes:

|||||...||
|||||...|||
|||||...||||
⋮
|||||...|||||...
|||||...|||||...|
|||||...|||||...||

|||||...|||||...|||

⋮

|||||...|||||...|||||...

|||||...|||||...|||||...|

⋮

Note that although the shapes above all differ from one another, any two are such that one of them is an initial segment of the other. This is no accident: *any two well-orderings are such that one of them is isomorphic to an initial segment of the other.*

Notice also that even though each of the shapes above corresponds to a different well-order type, the sets that are ordered by well-orderings of these types *all have the same cardinality*: they all have the cardinality of the natural numbers. (We'll talk about well-orderings of uncountable sets in section 2.4.2.)

### Exercises

**1.** Define well-orderings of the natural numbers with each of the following shapes:

**a)** |||||...||| **b)** |||||...|||||...

### 2.2.6 Ordinals: Introduction

We've been doing our best to keep track of well-order types using diagrams of lines and dots. But this notation soon becomes unmanageable. A better way of keeping track of big well-order types is to use **ordinals**, which will be used as *representatives* for well-order types.

The best way of getting a sense of how the ordinals work is to consider the first few members of the ordinal hierarchy. I have depicted them in figure 2.1.

| Ordinal | Name of ordinal | Well-order type represented |
|---|---|---|
| $\{\}$ | $0$ | |
| $\{0\}$ | $0'$ | \| |
| $\{0, 0'\}$ | $0''$ | \|\| |
| $\{0, 0', 0''\}$ | $0'''$ | \|\|\| |
| ⋮ | ⋮ | |
| $\{0, 0', 0'', 0''', \ldots\}$ | $\omega$ | \|\|\|\|\|... |
| $\{0, 0', 0'', 0''', \ldots, \omega\}$ | $\omega'$ | \|\|\|\|\|...\| |
| $\{0, 0', 0'', 0''', \ldots, \omega, \omega'\}$ | $\omega''$ | \|\|\|\|\|...\|\| |
| ⋮ | ⋮ | ⋮ |

**Figure 2.1**
The first few members of the ordinal hierarchy.

I'd like to highlight two important features of the ordinal hierarchy. First, note that each ordinal is the set of its predecessors. The zeroth ordinal, which we call 0 (read: "zero"), is the empty set, because it has no predecessors. The next ordinal, which we call $0'$ (read: "successor of zero"), is the set containing only 0, because 0 is its only predecessor. The next ordinal after that, which we call $0''$ (read: "successor of successor of zero"), is the set containing 0 and $0'$, because those are its predecessors. The next ordinal after each of $0, 0', 0'', \ldots$, which we call $\omega$ (read: "omega"), is the set containing $0, 0', 0'', \ldots$, because those are its predecessors. And so forth.

The second feature of the ordinal hierarchy I'd like to highlight is that each ordinal represents the well-ordering type that it itself instantiates when its members are ordered as they appear in figure 2.1. Consider, for example, the ordinal $\omega' = \{0, 0', 0'', 0''', \ldots, \omega\}$. When one orders its members as they appear in figure 2.1, one gets the well-ordering $0 < 0' < 0'' < 0''' < \ldots < \omega$, which is of type $||||| \ldots |$. So $\omega'$ instantiates the same well-order type it represents.

### 2.2.7 Ordinals: The Intuitive Picture

In this section, I'll be a little more explicit about the intuitive picture underlying the ordinal hierarchy. The key idea is that the ordinals are introduced in "stages," in accordance with the following principle:

**Construction Principle** At each stage, we introduce a new ordinal; namely, the set of all ordinals that have been introduced at previous stages.

Accordingly, one can think of each row of the table in figure 2.1 as corresponding to a different stage of the process.

For the Construction Principle to do its job, however, we need to make sure that the hierarchy of stages has the right structure, since otherwise it will yield a hierarchy of ordinals that fails to mirror the hierarchy of the well-order types it was meant to represent. Informally speaking, what we need is for the stages to be well-ordered and for them to go on forever. The following principle delivers the desired results:

**Open-Endedness Principle** However many stages have occurred, there is always a "next" stage; that is, a first stage after every stage considered so far.

(It is important to interpret the Open-Endedness Principle as entailing that there is no such thing as "all" stages—and therefore deliver the result that there is no such thing as "all" ordinals. We will return to this idea in section 2.5.)

Together, the Construction Principle and the Open-Endedness Principle can be used to generate all the ordinals in figure 2.1 and many more. Let us consider the first few steps of the process in detail. (They are also summarized in figure 2.2.)

**Stage 0** The Open-Endedness Principle entails that there must be a zeroth stage to the process (i.e., the "next" stage after nothing has happened). At this zeroth stage,

| Stage | Ordinal introduced | Name of ordinal |
|---|---|---|
| 0 | $\{\}$ | 0 |
| 1 | $\{0\}$ | $0'$ |
| 2 | $\{0, 0'\}$ | $0''$ |
| 3 | $\{0, 0', 0''\}$ | $0'''$ |
| $\vdots$ | $\vdots$ | $\vdots$ |
| $\omega$ | $\{0, 0', 0'', 0''', \ldots\}$ | $\omega$ |
| $\omega+1$ | $\{0, 0', 0'', 0''', \ldots, \omega\}$ | $\omega'$ |
| $\omega+2$ | $\{0, 0', 0'', \ldots, \omega, \omega'\}$ | $\omega''$ |
| $\vdots$ | $\vdots$ | $\vdots$ |

**Figure 2.2**
Ordinals are built in stages, with each stage bringing together everything previously introduced.

no ordinals have been introduced. So the Construction Principle tells us to introduce the set with no members, as an ordinal: $\{\}$. As before, we refer to this set as the ordinal 0.

**Stage 1** The Open-Endedness Principle tells us that there must be a next stage to the process. At this next stage, 0 has been introduced. So the Construction Principle tells us to introduce the set $\{0\}$. As before, we refer to this set as the ordinal $0'$.

**Stage 2** The Open-Endedness Principle tells us that there must be a next stage to the process. At this next stage, 0 and $0'$ have been introduced. So the Construction Principle tells us to introduce the set $\{0, 0'\}$. As before, we refer to this set as the ordinal $0''$.

**Stage $\omega$** We carry out the process once for each natural number, yielding the sets $0, 0', 0'', \ldots$. But the Open-Endedness Principle tells us that there is a next stage to the process. And the Construction Principle tells us to introduce the set of everything that's been introduced so far at that next stage. So we introduce the set $\{0, 0', 0'', \ldots\}$. As before, we refer to this set as the ordinal $\omega$.

**Stage $\omega + 1$** The Open-Endedness Principle tells us that there must be a next stage to the process. At this next stage, we introduce the set $\{0, 0', 0'', \ldots, \omega\}$. We refer to this set as the ordinal $\omega'$.

Before bringing this section to a close, I'd like to highlight an important piece of notation, which was introduced in passing: the **successor operation**, $'$. This operation can be defined rigorously, as follows: $\alpha' = \alpha \cup \{\alpha\}$. Notice, moreover, that the Construction Principle delivers ordinals of two different kinds, *successor ordinals* and *limit ordinals*.

A **successor ordinal** is an ordinal $\alpha$ such that $\alpha = \beta'$ for some $\beta$.

A **limit ordinal** is an ordinal that is not a successor ordinal.

0 and $\omega$ are both limit ordinals, since neither of them can be reached by applying the successor operation to an ordinal; $0''$ and $\omega'$ are both successor ordinals. The distinction between successor ordinals and limit ordinals will be important in section 2.4, when we return to the project of using ordinals to build sets of greater and greater cardinality.

### 2.2.8 Ordinals: Representing Well-Order Types

In the preceding section we considered an intuitive picture in which the ordinals are introduced in stages. As you'll be asked to verify below, the Open-Endedness Principle entails that the stages are well-ordered. More precisely:

**Stage Well-Ordering** Every set of stages is well-ordered by the relation of occurring earlier-than.

And, as you'll also be asked to verify below, an immediate consequence of this result is that every set of ordinals is well-ordered. More precisely:

**Ordinal Well-Ordering** Every set of ordinals is well-ordered by the membership relation, $\in$.

This makes it natural to use the membership relation as the canonical ordering relation for ordinals. We will make this explicit by introducing an ordinal-precedece relation, $<_o$, which is defined as follows:

$$\alpha <_o \beta \leftrightarrow_{df} \alpha \in \beta$$

We can therefore say that *each ordinal represents the well-order type that it itself instantiates under* $<_o$. (See figure 2.3.)

It is worth noting that Ordinal Well-Ordering entails that every *ordinal* is well-ordered by $<_o$, because every ordinal is a set of ordinals. This is important because

| Ordinal | Ordering under $<_o$ | Well-order type represented |
|---|---|---|
| $0$ | | |
| $0'$ | $0$ | \| |
| $0''$ | $0 <_o 0'$ | \|\| |
| $0'''$ | $0 <_o 0' <_o 0''$ | \|\|\| |
| $\vdots$ | $\vdots$ | |
| $\omega$ | $0 <_o 0' <_o 0'' \ldots$ | \|\|\|\|\|... |
| $\omega'$ | $0 <_o 0' <_o 0'' \ldots <_o \omega$ | \|\|\|\|\|...\| |
| $\omega''$ | $0 <_o 0' <_o 0'' \ldots <_o \omega <_o \omega'$ | \|\|\|\|\|...\|\| |
| $\vdots$ | $\vdots$ | $\vdots$ |

**Figure 2.3**
Each ordinal represents the well-order type that it itself instantiates under $<_o$.

it ensures that the following simple and elegant principle can be used to explain how ordinals represent well-order types:

**Representation Principle** Each ordinal represents the well-order type that it itself instantiates under $<_o$.

Ordinals are such natural representatives for well-order types that I will sometimes blur the distinction between an ordinal and the well-order type it represents by speaking of ordinals as if they were themselves well-order types.

#### Exercises

1. Use the intuitive picture of ordinals we have developed so far to justify Stage Well-Ordering.
2. Show that Stage Well-Ordering entails Ordinal Well-Ordering.

### 2.2.9 Ordinals: The Official Definition

Every mathematically significant feature of the ordinals follows from Ordinal Well-Ordering, together with the following principle:

**Set-Transitivity** Every member of a member of an ordinal $\alpha$ is a member of $\alpha$.

As a result, set theorists sometimes use the following as their official definition of *ordinal*:

**Ordinal** A set that is set-transitive and well-ordered by $<_o$.

(This assumes that we restrict our attention to pure sets: sets that are such that all their members are sets, all the members of their members are sets, all the members of their members of their members are sets, and so forth.)

Unless you want to go deep into the material, you won't need to worry about the official definition of *ordinal*. Just think of the ordinals as generated by the intuitive picture of section 2.2.7.

#### Exercises

1. Justify Set-Transitivity on the basis of the intuitive picture of ordinals we developed in section 2.2.7.

## 2.3 Ordinal Arithmetic

A nice feature of the hierarchy of ordinals is that it allows for arithmetical operations. In this section I'll define ordinal addition and ordinal multiplication.

### 2.3.1 Ordinal Addition

We begin with ordinal addition. The intuitive idea is that a well-ordering of type $(\alpha+\beta)$ is the result of starting with a well-ordering of type $\alpha$ and appending a well-ordering of type $\beta$ at the end.

Consider, for example, the ordinal $(\omega+0')$. What well-order type does it represent? It is the well-order type that one gets to by starting with a well-ordering of type $\omega$ (i.e., $||||\ldots$) and appending an ordering of type $0'$ (i.e., $|$) at the end:

$||||\ldots|$

So $(\omega+0')$ represents the well-order type $|||||\ldots|$. And since $\omega'=\{0,0',0'',\ldots,\omega\}$, which is of type $|||||\ldots|$, this means that $\omega+0'$ is just $\omega'$.

Now consider the ordinal $(\omega+\omega)$. It is the result of appending a well-ordering of type $\omega$ (i.e., $|||||\ldots$) at the end of itself, as follows:

$|||||\ldots|||||\ldots$

So $(\omega+\omega)$ represents the well-order type $|||||\ldots|||||\ldots$.

An important difference between ordinal addition and regular addition is that ordinal addition is not **commutative**. In other words, it is not generally the case that $\alpha+\beta=\beta+\alpha$. Notice, for example, that $(0'+\omega)\neq(\omega+0')$. For whereas $(\omega+0')$ represents well-orderings of type $|||||\ldots|$, $(0'+\omega)$ represents well-orderings of type $|||||\ldots$, with the result that $(0'+\omega)=\omega$.

#### Exercises

1. Even though ordinal addition is not generally commutative, it does have the following property:

**Associativity** For any ordinals $\alpha$, $\beta$, and $\gamma$, $(\alpha+\beta)+\gamma=\alpha+(\beta+\gamma)$.

Verify that Associativity holds in the following special case: $(\omega+\omega)+\omega=\omega+(\omega+\omega)$.

2. Although ordinal addition is not commutative in general, it is always commutative for finite ordinals. Verify that ordinal addition is commutative in the following special case: $0'+0'''=0'''+0'$.

3. Determine whether the following is true: $(\omega+0')+(\omega+\omega)=(\omega+\omega)+(\omega+0')$.

### 2.3.2 Ordinal Multiplication

Let us now turn to ordinal multiplication. Here the intuitive idea is that a well-ordering of type $(\alpha\times\beta)$ is the result of starting with a well-ordering of type $\beta$ and replacing each position in the ordering with a well-ordering of type $\alpha$.

Consider, for example, the ordinal $(\omega\times 0'')$. What well-order type does it represent? It is the well-order type that one gets by starting with a well-ordering of type $0''$ (i.e., $||$)

and replacing each position with a well-ordering of type $\omega$ (i.e., |||||...). Schematically,

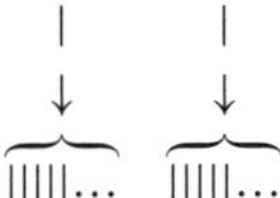

This means that $(\omega + \omega)$ is of type |||||...|||||..., and therefore that $(\omega \times 0'') = (\omega + \omega)$.

As in the case of ordinal addition, ordinal multiplication is not **commutative**: it is not generally the case that $\alpha \times \beta = \beta \times \alpha$. Notice, for example, that $(\omega \times 0'') \neq (0'' \times \omega)$. For whereas we have seen that $(\omega \times 0'') = (\omega + \omega)$, it is not the case that $(0'' \times \omega) = (\omega + \omega)$. To see this, recall that $(0'' \times \omega)$ is the result of starting with a well-ordering of type $\omega$ (i.e., |||||...) and replacing each position in that ordering with a well-ordering of type $0''$ (i.e., ||). Schematically:

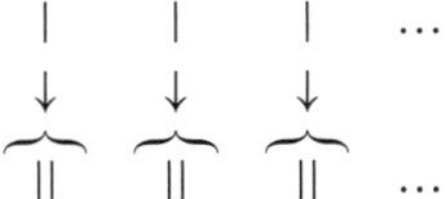

The result is just a well-ordering of type |||||.... So $(0'' \times \omega)$ is $\omega$, rather than $\omega + \omega$.

Let me mention a final example of ordinal multiplication. Consider the ordinal $\omega \times \omega$. What order type does it represent? It is the well-order type that one gets to by starting with a well-ordering of type $\omega$ (i.e., |||||...) and replacing each position | with an ordering of type $\omega$:

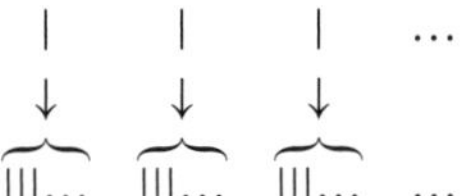

So $\omega \times \omega$ is an ordering of this type:

|||||... |||||... |||||... ...

$\omega$-many times

## Exercises

1. As in the case of ordinal addition, ordinal multiplication has the following property:

**Associativity** For any ordinals $\alpha$, $\beta$, and $\gamma$, $(\alpha \times \beta) \times \gamma = \alpha \times (\beta \times \gamma)$.

Verify that Associativity holds in the following special case: $(\omega \times \omega) \times \omega = \omega \times (\omega \times \omega)$.

2. Although ordinal multiplication is not commutative in general, it is always commutative for finite ordinals. Verify that ordinal multiplication is commutative in the following special case: $(0'' \times 0''') = (0''' \times 0'')$.

3. Unlike regular addition and multiplication, ordinal addition and multiplication fail to be **distributive**; that is, it is not generally the case for ordinals $\alpha$, $\beta$, and $\gamma$ that $(\alpha + \beta) \times \gamma = \alpha \times \gamma + \beta \times \gamma$. Prove this by showing that the following is false: $(\omega + 0''') \times 0'' = (\omega \times 0'') + (0''' \times 0'')$.

4. Ordinals are, however, **distributive on the left**. In other words, for all ordinals $\alpha$, $\beta$, and $\gamma$, $\alpha \times (\beta + \gamma) = \alpha \times \beta + \alpha \times \gamma$. Verify that this is true in the following special case: $0'' \times (\omega + 0''') = (0'' \times \omega) + (0'' \times 0''')$.

### 2.3.3 Optional: A More Rigorous Characterization of Ordinal Addition and Multiplication

In the preceding subsections I gave you an informal characterization of ordinal addition and multiplication. Fortunately, it is easy to characterize these operations more rigorously, as follows:

$$\begin{aligned} \alpha + 0 &= \alpha \\ \alpha + \beta' &= (\alpha + \beta)' \\ \alpha + \lambda &= \bigcup\{\alpha + \beta : \beta < \lambda\} \text{ ($\lambda$ a limit ordinal)} \end{aligned}$$

$$\begin{aligned} \alpha \times 0 &= 0 \\ \alpha \times \beta' &= (\alpha \times \beta) + \alpha \\ \alpha \times \lambda &= \bigcup\{\alpha \times \beta : \beta < \lambda\} \text{ ($\lambda$ a limit ordinal)} \end{aligned}$$

Notice, incidentally, that there is no reason to stop at multiplication. We could use a similar strategy to define exponentiation:

$$\begin{aligned} \alpha^0 &= 0' \\ \alpha^{\beta'} &= (\alpha^\beta) \times \alpha \\ \alpha^\lambda &= \bigcup\{\alpha^\beta : \beta < \lambda\} \text{ ($\lambda$ a limit ordinal)} \end{aligned}$$

And we could go further still, defining ordinal tetration and beyond.

You may have noticed that the definitions above are all circular. The definition of ordinal addition, for example, uses ordinal addition to define ordinal addition. But it is important to observe that the definitions are not circular in a problematic way: they all succeed in fixing the meanings of the operations they are meant to define.

Let me explain how this works by focusing on the case of addition. The basic insight is that, although our definition of $\alpha + \beta$ might presuppose that certain other instances of ordinal addition have been defined, the relevant instances will always be of the form $\alpha + \gamma$ for $\gamma <_o \beta$. To grasp this idea more fully, it will be useful to work through some examples. Notice, first, that $\omega + 0$ is well-defined, since the first clause of our definition tells us that

(0) $\omega + 0 = \omega$.

Result (0) can then be used to define $\omega+0'$, because the second clause of our definition tells us that $\omega+0'=(\omega+0)'$. So result (0) yields

(1) $\omega+0'=\omega'$.

Result (1) can then be used to define $\omega+0''$, because the second clause of our definition tells us that $\omega+0''=(\omega+0')'$. So result (1) yields

(2) $\omega+0''=\omega''$.

This technique can be iterated to prove the following for each natural number $n$:

$$(n) \quad \omega+0^{\overbrace{\prime\cdots\prime}^{n\text{ times}}}=\omega^{\overbrace{\prime\cdots\prime}^{n\text{ times}}}$$

Results (0), (1), ...can then be used together to define $\omega+\omega$. Since $\omega$ is a limit ordinal, the third clause of our definition tells us that

$$\omega+\omega=\bigcup\{\omega+\beta:\beta<\omega\}.$$

But $\bigcup\{\omega+\beta:\beta<\omega\}$ is just $\bigcup\{\omega+0,\omega+0',\omega+0'',\ldots\}$. By the definition of $\bigcup$, this means that $\omega+\omega$ consists of the members of $\omega+0^{\overbrace{\prime\cdots\prime}^{n\text{ times}}}$, for each $n$. Since an ordinal is just the set of its predecessors, this gives us

$$(\omega) \quad \omega+\omega=\{0,0',\ldots,\omega,\omega',\omega'',\ldots\}.$$

### 2.3.4 Cardinal Arithmetic

It is important not to confuse ordinal arithmetic with cardinal arithmetic. Ordinal addition and multiplication are operations on *ordinals*: they take ordinals as input and yield ordinals as output. Cardinal addition and multiplication, in contrast, are operations on the *cardinalities* (or sizes) of sets: they take cardinalities as input and yield cardinalities as output.

Cardinal arithmetic is much simpler than ordinal arithmetic. Here are the basic definitions. (As usual, we take $|A|$ to be the cardinality of set $A$.)

**Cardinal Addition** $|A|\oplus|B|=|A\cup B|$

(This assumes that $A$ and $B$ have no members in common. If they do have members in common, simply find sets $A'$ and $B'$ with no members in common such that $|A|=|A'|$ and $|B|=|B'|$, and let $|A|\oplus|B|=|A'\cup B'|$.)

**Cardinal Multiplication** $|A|\otimes|B|=|A\times B|$

(Here $A\times B$ is the Cartesian product of $A$ and $B$. In other words, $A\times B=\{\langle a,b\rangle: a\in A, b\in B\}$.)

Cardinal addition and multiplication are both fairly boring: whenever $A$ is infinite and $|B|\leq|A|$, we have $|A|\oplus|B|=|A|=|A|\otimes|B|$ (unless $B$ is empty, in which case

$|A| \otimes |B| = |B|$). Notice, in contrast, that ordinal addition and multiplication are not boring in this way. We have seen, in particular, that $\omega + \omega$ and $\omega \times \omega$ are both distinct from $\omega$.

(The claim that $|A| \otimes |B| = |A|$ whenever $A$ is infinite, $B$ is nonempty, and $|B| \leq |A|$ assumes the **Axiom of Choice**. You won't need to worry about the Axiom of Choice for now, but it states that every set $S$ of non-empty, non-overlapping sets has a choice set (i.e., a set that contains exactly one member from each member of $S$); for further details, see section 7.2.1. The Axiom of Choice has some very surprising consequences and is somewhat controversial; for further discussion, see section 7.2.3.2.)

### Exercises

Show that cardinal addition and multiplication are commutative, associative, and distributive. In other words, verify each of the following:

1. **Cardinal commutativity (addition)** $|A| \oplus |B| = |B| \oplus |A|$
2. **Cardinal commutativity (multiplication)** $|A| \otimes |B| = |B| \oplus |A|$
3. **Cardinal associativity (addition)** $|A| \oplus (|B| \oplus |C|) = (|A| \oplus |B|) \oplus |C|$
4. **Cardinal associativity (multiplication)** $|A| \otimes (|B| \otimes |C|) = (|A| \otimes |B|) \otimes |C|$
5. **Cardinal distributivity** $(|A| \oplus |B|) \otimes |C| = (|A| \otimes |C|) \oplus (|B| \otimes |C|)$

## 2.4 Ordinals as Blueprints

In section 2.1 we talked about generating sets of bigger and bigger cardinality by combining applications of the power set operation with applications of the union operation. Now that we know about ordinals, we'll be able to develop this idea properly by thinking of each ordinal as a "blueprint" for a sequence of applications of the power set and union operations. The farther up an ordinal is in the hierarchy of ordinals, the longer the sequence and the greater the cardinality of the end result.

### 2.4.1 The Construction

In this section I'll give you a more detailed explanation of how ordinals might be used as blueprints. For a given ordinal $\alpha$, we start by considering the sequence of ordinals smaller than or equal to $\alpha$. Suppose, for example, that $\alpha = \omega'$. Then the sequence of ordinals to consider is as follows:

$$\langle 0, 0', 0'', \ldots, \omega, \omega' \rangle$$

Now imagine substituting an empty box for each member of the sequence:

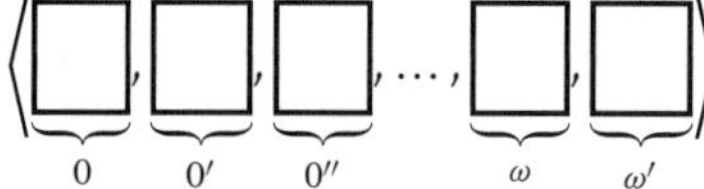

This is our blueprint. We can use it to create a sequence of sets of bigger and bigger cardinality by filling out each of the boxes in accordance with the following recipe:

- Fill box 0 with the set of natural numbers, $\mathbb{N}$.
- For $\beta'$ a successor ordinal, fill box $\beta'$ with the result of applying the power set operation to the contents of box $\beta$.
- For $\lambda$ a limit ordinal greater than 0, fill box $\lambda$ with the union of the set of every set that's been placed in a box so far.

In our example, this yields the following result:

$$\left\langle \underbrace{\mathbb{N}}_{0}, \underbrace{\wp^1(\mathbb{N})}_{0'}, \underbrace{\wp^2(\mathbb{N})}_{0''}, \ldots \underbrace{\bigcup\{\mathbb{N}, \wp^1(\mathbb{N}), \ldots\}}_{\omega}, \underbrace{\wp^1\left(\bigcup\{\mathbb{N}, \wp^1(\mathbb{N}), \ldots\}\right)}_{\omega'} \right\rangle$$

The crucial feature of this construction is that each box in the sequence is filled with a set of greater cardinality than the sets in any preceding box. And, of course, one can generate longer blueprints—longer sequences of boxes—by using ordinals that occur later in the ordinal hierarchy. So one can generate sets of larger and larger cardinality by choosing ordinals farther and farther up the hierarchy of ordinals and looking at the sets that get placed at the end-points of the resulting blueprints.

This point can be made more precisely by introducing some notation. For each ordinal $\alpha$, let $\mathfrak{B}_\alpha$ be the set that gets placed at the endpoint of the blueprint generated by $\alpha$. Formally:

$$\mathfrak{B}_\alpha = \begin{cases} \mathbb{N}, \text{ if } \alpha = 0 \\ \wp(\mathfrak{B}_\beta), \text{ if } \alpha = \beta' \\ \bigcup\{\mathfrak{B}_\gamma : \gamma <_o \alpha\} \text{ if } \alpha \text{ is a limit ordinal greater than } 0 \end{cases}$$

Cantor's Theorem then guarantees that for any ordinals $\alpha$ and $\beta$, we have the following:

If $\alpha <_o \beta$, then $|\mathfrak{B}_\alpha| < |\mathfrak{B}_\beta|$.

Here it is important to keep in mind that $<_o$ and $<$ mean two very different things:

- $<_o$ is the precedence relation for ordinals. So $\alpha <_o \beta$ means that $\alpha$ precedes $\beta$ in the hierarchy of ordinals.
- $<$ is an ordering of set cardinality. So $|A| < |B|$ means that there is an injection from $A$ to $B$ but no bijection.

Consider, for example, the ordinals $\omega$ and $\omega + \omega$. We know that $\omega <_o \omega + \omega$ because $\omega$ precedes $\omega + \omega$ in the hierarchy of ordinals (and therefore $\omega \in \omega + \omega$). But this does not mean that $\omega + \omega$ has more members than $\omega$. In fact, they are both countable sets, as shown here:

$$|\omega| = |\omega + \omega| = |\mathbb{N}|$$

In contrast, $\mathfrak{B}_\omega$ and $\mathfrak{B}_{\omega+\omega}$ are both uncountable sets, and $\mathfrak{B}_{\omega+\omega}$ has many more members than $\mathfrak{B}_\omega$:

$$|\mathbb{N}| < |\mathfrak{B}_\omega| < |\mathfrak{B}_{\omega+\omega}|$$

So we can use countable ordinals like $\omega+\omega$ and $\omega\times\omega$ to describe sets like $\mathfrak{B}_{\omega+\omega}$ and $\mathfrak{B}_{\omega\times\omega}$, which are sensationally bigger than the set of natural numbers. This makes the notion of an ordinal a very powerful device for describing sizes of infinity.

### 2.4.2 How Far Can You Go?

Suppose you wanted to characterize a really big size of infinity. How would you go about it? We have reviewed one strategy. You can identify an ordinal $\alpha$ that is really far up on the hierarchy of ordinals and use it as a blueprint to characterize $|\mathfrak{B}_\alpha|$, a correspondingly large size of infinity.

And how should one go about characterizing the ordinal that is to be used as a blueprint? A natural first thought is to use some of the arithmetical operations we've talked about in earlier subsections. For instance, you could try the following operation:

$$\underbrace{\omega\times\omega\times\ldots\times\omega}_{10^{100}\text{ times}}$$

Or, if you looked at the optional subsection 2.3.3, you could try this:

$$\omega^{\omega^{\omega^{\omega}}}$$

Since $\omega^{\omega^{\omega^{\omega}}}$ occurs significantly farther up the hierarchy of ordinals than any ordinal we've considered so far, $|\mathfrak{B}_{\omega^{\omega^{\omega^{\omega}}}}|$ is a significantly bigger size of infinity than anything we've considered so far. If you wanted to go farther still, you could try defining new arithmetical operations for the ordinals, such as tetration or other hyper-operations. Additional arithmetical operations will only take us so far, however. If we're to generate seriously big sets, the ordinals we use as blueprints will need to have uncountably many members (and therefore be farther up the ordinal hierarchy than any ordinal with countably many members). But we'll never get to an uncountable ordinal if we stay on our current trajectory. The problem is that ordinal arithmetic can't take us from countable ordinals to an uncountable one, and every ordinal we've considered so far is countable. Even the mighty $\omega^{\omega^{\omega^{\omega}}}$ is a countable set.

In order to identify an uncountable ordinal, we turn to the notion of an **initial ordinal**: an ordinal that precedes all other ordinals of the same cardinality. For example, $\omega$ is an initial ordinal because it is countably infinite and precedes every other countably infinite ordinal. Cantor introduced some nice notation to keep track of initial ordinals: the **ב-hierarchy**. (The character ב is the second letter of the Hebrew alphabet and is pronounced "beth.") The basic idea is straightforward. For $\alpha$, an ordinal, we define an ordinal $\beth_\alpha$ (read: "beth-alpha") as follows:

$\beth_\alpha$ is the first ordinal of cardinality $|\mathfrak{B}_\alpha|$

The $\beth$-hierarchy is simply the collection of initial ordinals $\beth_\alpha$, for $\alpha$ an ordinal. (The Axiom of Choice is needed to show that there is an ordinal of cardinality $|\mathfrak{B}_\alpha|$ for $0 <_o \alpha$.)

Cantor's notation can be used to characterize some very big ordinals. Consider, for example, $\beth_\omega$, which is the smallest ordinal of size $|\mathfrak{B}_\omega|$. You might recall that $\mathfrak{B}_\omega$ is $\bigcup\{\mathbb{N}, \wp^1(\mathbb{N}), \wp^2(\mathbb{N}), \ldots\}$, which is an infinite set of rather large cardinality. Accordingly, $\beth_\omega$ is the first infinite ordinal of that rather large cardinality. Since every infinite ordinal we've considered until now is of countable cardinality, it follows that $\beth_\omega$ is much farther up the ordinal hierarchy than anything we've considered so far. And, of course, $\beth_{\beth_\omega}$ is much farther up the hierarchy than that—much, much farther.

It is convenient to use each initial ordinal $\kappa$ as a representative of its own cardinality, $|\kappa|$. (When initial ordinals play this role, they are called **cardinals**.) We can therefore think of the $\beth$-hierarchy as a hierarchy of infinite cardinals. For instance, $\beth_{\beth_{\beth_\omega}}$ represents $|\beth_{\beth_{\beth_\omega}}|$, which is a cardinality of truly voracious dimensions—to say nothing of the cardinality represented by

$$\beth_{\beth_{\beth_{\underbrace{\cdots}_{10^{100}\text{ times}}\beth_\omega}}}$$

### 2.4.3 Optional: The $\aleph$-Hierarchy and the Continuum Hypothesis

I can't resist a detour. Cantor actually introduced an additional piece of notation to keep track of initial ordinals: the $\aleph$-hierarchy. (The character $\aleph$ is the first letter of the Hebrew alphabet and is pronounced "aleph.") For $\alpha$ an ordinal, we define $\aleph_\alpha$ (read: "aleph-alpha") as follows:

> $\aleph_\alpha$ is the first infinite ordinal of cardinality greater than every $\aleph_\beta$, for $\beta <_o \alpha$.

The $\aleph$-hierarchy is the collection of initial ordinals $\aleph_\alpha$, for $\alpha$ an ordinal. (Like $\beth_\alpha$, $\aleph_\alpha$ is taken to represent its own cardinality and is therefore thought of as a cardinal.)

The $\aleph$-hierarchy is exceptionally useful because it can be shown to contain the cardinality of every infinite ordinal. In fact, it can be shown to contain the cardinality of *every infinite set* (assuming the Axiom of Choice).

What about the $\beth$-hierarchy? Every member of the $\beth$-hierarchy is the cardinality of some set $\mathfrak{B}_\alpha$ (for $\alpha$, an ordinal). So the only way for every cardinality to be in the $\beth$-hierarchy is for a nontrivial condition to be satisfied: the condition that every set have the same cardinality as some $\mathfrak{B}_\alpha$.

Unfortunately, it is very hard to tell whether this condition holds. To get a sense of the problem, start with a special case: the question of whether there is a set $A$ such that $\beth_0 < |A| < \beth_1$. One hypothesis is that the answer is no:

**Continuum Hypothesis** There is no set $A$ such that $\beth_0 < |A| < \beth_1$.

Note that $\beth_0 = |\mathbb{N}|$ and $\beth_1 = |\mathbb{R}|$ (since $\mathfrak{B}_1 = \wp(\mathbb{N})$ and $|\wp(\mathbb{N})| = |\mathbb{R}|$). So the Continuum Hypothesis could also be stated as the claim that there is no set $A$ such that $|\mathbb{N}| < |A| < |\mathbb{R}|$. (The real numbers are sometimes referred to as "the continuum," hence the name of the hypothesis.)

As it turns out, the Continuum Hypothesis is *independent* of the standard axioms of set theory. In other words, neither it nor its negation is entailed by the axioms (assuming the axioms are consistent). In fact, for *any* ordinal $\alpha$, the question of whether there is a set $A$ such that $\beth_\alpha < |A| < \beth_{\alpha+1}$ is independent of the standard axioms of set theory.

So how are the $\aleph$- and $\beth$-hierarchies related? On the one hand, we know that every member of the $\beth$-hierarchy is a member of the $\aleph$-hierarchy and that $\aleph_\alpha \leq \beth_\alpha$ for each $\alpha$. On the other hand, it is compatible with the standard axioms that the two hierarchies are one and the same, and it is compatible with the standard axioms that there are members of the $\aleph$-hierarchy between any two members of the $\beth$-hierarchy.

It is remarkable that set-theoretic questions as central as these should turn out to be left open by the standard axioms.

### 2.4.4 Large Cardinals

We have been considering some very big sets ($\beth_{\beth_{\beth_\omega}}$, for example). But, as it turns out, one can go much bigger—much, much bigger. Vertiginously bigger. In fact, we've barely begun to scratch the surface. We're not even close to the sizes of infinity that set theorists refer to as "small" large cardinals—to say nothing of the sizes they refer to as "large" large cardinals.

The standard way of introducing cardinalities that a set theorist would count as large is to specify a potential property of initial ordinals and set forth the hypothesis that some initial ordinal with that property exists.

Here is an example. A **strongly inaccessible cardinal** is defined as an (uncountable) initial ordinal $\kappa$, whose cardinality cannot be reached by applying the power set operation to a set of cardinality smaller than $\kappa$, or by applying the union operation to fewer than $\kappa$ sets of cardinality smaller than $\kappa$. The claim that an initial ordinal with this property exists is independent of the standard axioms of set theory: neither it nor its negation is entailed by the axioms (assuming the axioms are consistent). But if strongly inaccessible cardinals do exist, they are breathtakingly larger than anything we have discussed so far. They are the first of the "small" large cardinals.

The existence of cardinalities that a set theorist would count as large is always independent of the standard axioms of set theory. Because of this, the search for larger and larger cardinalities cannot avoid venturing into uncharted territory. To set forth a large cardinal hypothesis is, in effect, to make a novel claim about the nature of the structure of the set-theoretic universe.

## 2.5 Paradox in Paradise

We have been considering the wonderful world of bigger and bigger ordinals—"the paradise that Cantor has created for us," as mathematician David Hilbert famously put it [1925]. But Cantor's paradise has a terrible instability looming at its core: *one can prove that there can be no such thing as the set of all ordinals.* More specifically, one can show that if the set of all ordinals existed, it would have inconsistent properties: it would have to both be an ordinal and not be an ordinal. This result is known as the **Burali-Forti Paradox**, because the Italian mathematician Cesare Burali-Forti was one of the first people to realize that the order-type of all ordinals is problematic.

When I introduced the ordinals informally in section 2.2.7, I cautioned that the Open-Endedness Principle was to be understood as entailing that there is no such thing as "all" stages of the construction process. I did so in order to avoid commitment to the existence of a definite totality of "all" ordinals, and thereby avoid the Burali-Forti Paradox.

Here is an informal version of the basic result. Suppose, for *reductio*, that $\Omega$ is the set of all ordinals. That supposition would lead us to the following:

- Since $\Omega$ consists of every ordinal, it consists of every ordinal that's been introduced (and therefore every ordinal that's been introduced so far). But by the Construction Principle of section 2.2.7, one introduces a new ordinal by forming the set consisting of every ordinal that's been introduced so far. So, $\Omega$ **is an ordinal.**
- If $\Omega$ was itself an ordinal, it would be a member of itself (and therefore have itself as a predecessor). But no ordinal can be its own predecessor. So, $\Omega$ **is not an ordinal.**

(One can carry out a rigorous version of this argument using the official definition of an ordinal, which I introduced in section 2.2.9: one shows that $\Omega$ is an ordinal by verifying that it is well-ordered and set-transitive, and one shows that $\Omega$ is not an ordinal by appealing to a set-theoretic axiom that entails that no set is a member of itself.)

One reason the Burali-Forti Paradox is important is that it constrains our understanding of sets. It is a natural thought that whenever it makes sense to speak of the *F*s, one can introduce the set of *F*s. But the Burali-Forti Paradox shows that there is at least one instance in which this cannot be the case: when the *F*s are the ordinals, there is no such thing as the set of *F*.

The Burali-Forti Paradox is part of a broader family of set-theoretic paradoxes, leading to similar conclusions. The most famous of these paradoxes is **Russell's Paradox**, which shows that there is no such thing as the set of sets that are not members of themselves. (*Proof:* suppose there was such a set *R*. Is *R* a member of itself? If it is, it shouldn't be, since only sets that are not members of themselves should be members of *R*. If it isn't, it should be, since every set that is not a member of itself should be a member of *R*.) I'll say a bit more about Russell's Paradox in section 10.4.2.

Over the years, set theorists have identified ways of working with sets while steering clear of the paradoxes. Importantly, they have identified families of set-theoretic axioms that are strong enough to deliver interesting mathematics but are designed to retain consistency by ensuring that one can never introduce the set of ordinals or the set of sets that are not members of themselves. One such family of axioms is known as ZFC, after Ernst Zermelo and Abraham Fraenkel. (The *C* is for the Axiom of Choice.) ZFC is what I referred to earlier when I spoke of the "standard" set-theoretic axioms. Most set theorists believe that ZFC is a consistent axiom system and that it successfully steers clear of the paradoxes. But, as we will see in chapter 10, establishing the consistency of an axiom system is not as easy as one might have hoped.

Even if axiom systems like ZFC allow us to do set theory without lapsing into paradox, they do not immediately deliver a diagnosis of the paradoxes. They do not, for example, tell us what it is about the ordinals that makes it impossible to collect them into a set. A lot of interesting work has been done on the subject, but a proper discussion is beyond the scope of this text. (I recommend readings that provide more depth in section 2.7.)

My own view, in a nutshell, is that when we speak of "the ordinals," we do not succeed in singling out a definite totality of individuals. In other words, our concept of *ordinal* fails to settle the question of just how far the hierarchy of ordinals goes. And there are reasons of principle for such indeterminacy: any definite hypothesis about the length of the ordinal hierarchy must fail, since it could be used to characterize additional ordinals and therefore an ordinal hierarchy of even greater length.

## 2.6 Conclusion

This chapter has focused on generating sets of bigger and bigger cardinality by combining applications of the power set operation with applications of the union operation.

We gave a rigorous characterization of this procedure by introducing the notion of an ordinal (or a type of well-ordering). We saw, in particular, that an ordinal can be used as a "blueprint" for a sequence of applications of the power set and union operations. The farther up an ordinal is in the hierarchy of ordinals, the longer the sequence and the greater the cardinality of the end result.

This allows one to build sets of mind-boggling cardinality. But it is only the beginning. In order to describe cardinalities that a set theorist would describe as large, one needs to set forth hypotheses that are independent of the standard axioms of set theory.

## 2.7 Further Resources

- For a more formal treatment of ordinals, I recommend Drake and Singh's *Intermediate Set Theory*.
- If you'd like an informal description of some very large cardinals, I recommend Rucker's *Infinity and the Mind*.
- If you'd like to explore set theory more deeply, I recommend Kunen's *Set Theory*.
- If you'd like to know more about the problem of developing a principled and coherent conception of set that is strong enough to do interesting mathematics, I recommend George Boolos's "The Iterative Conception of Set" and "Iteration Again." Two interesting efforts to develop such conceptions are John Burgess's *Fixing Frege* and Øystein Linnebo's "The Potential Hierarchy of Sets."

## Appendix: Answers to Exercises

### Section 2.1.2

1. If $A = \wp(\mathbb{N})$, $\bigcup A = \mathbb{N}$. But $|\mathbb{N}| < |\wp(\mathbb{N})|$, so $|\bigcup A| < |A|$.

### Section 2.2.1

1. Suppose that $a < a$. Then by asymmetry, we also have not-$(a < a)$, which is a contradiction. So it can't be the case that $a < a$.
2. (a) Suppose $<$ orders $A$. Then $<$ is asymmetric and transitive. And we know from exercise 1 that it must also be irreflexive. So all we need to do to show that $<$ is a strict partial ordering on $A$ is to verify that it is anti-symmetric. But anti-symmetry follows immediately from asymmetry, since asymmetry guarantees that the antecedent of anti-symmetry will never be satisfied.

   Now suppose that $<$ is a strict partial ordering on $A$. Then $<$ is irreflexive, anti-symmetric, and transitive. So all we need to do to show that $<$ orders $A$ is to verify that it is asymmetric. Suppose, for *reductio*, that $<$ is not asymmetric. Then there are $a, b \in A$ such that $a < b$ and $b < a$. By anti-symmetry, this means that $a = b$ and therefore that $a < a$, contradicting irreflexivity.

   (b) We need to verify that $R'$ is irreflexive, anti-symmetric, and transitive. To see that $R'$ is irreflexive, suppose, for *reductio*, that $aR'a$ for some $a \in A$. Then, by the definition of $R'$, we must have $a \neq a$, which is impossible. To see that $R'$ is anti-symmetric, suppose that $aR'b$ and $bR'a$ for $a, b \in A$. By the definition of $R'$, it follows that $aRb$ and $bRa$. So by the symmetry of $R$ we have $a = b$. To see that $R'$ is transitive, suppose that $aR'b$ and $bR'c$ for $a, b, c \in A$. Then, by the definition of $R'$, we have $a \neq b$, $aRb$, and $aRc$. By the transitivity of $R$, it follows that we have $aRc$. Now suppose, for *reductio*, that $a = c$. Then it follows from $aR'b$ and $bR'c$ that $aR'b$ and $bR'a$. So anti-symmetry entails $a = b$, contradicting our earlier assertion that $a \neq b$.

   (c) We need to verify that $R'$ is reflexive, anti-symmetric, and transitive. To see that $R'$ is reflexive, note that $a = a$ (and therefore $aRa$ or $a = a$) for each $a \in A$. To see that $R'$ is anti-symmetric, suppose that $aR'b$ and $bR'a$ for $a, b \in A$. By the definition of $R'$, it follows that we must have either $a = b$ or both $aRb$ and $bRa$. But if we had both $aRb$ and $bRa$, $a = b$ would follow from the anti-symmetry of $R$. To see that $R'$ is transitive, suppose that $aR'b$ and $bR'c$ for $a, b, c \in A$. Then, by the definition of $R'$, it follows that we have either one of $a = b$ and $b = c$, or both $aRb$ and $bRc$. In the former case, the result follows immediately from the fact that we have both $aR'b$ and $bR'c$. In the latter case, the transitivity of $R$ entails $aRc$ and therefore $aR'c$.

### Section 2.2.2

1. Yes, $<^{-1}$ is a total ordering of $\mathbb{Z}$:

$$\ldots <^{-1} 2 <^{-1} 1 <^{-1} 0 <^{-1} -1 <^{-1} -2 <^{-1} -3 <^{-1} \ldots$$

2. No, $\subsetneq$ is not a total ordering of $\wp(\{0,1\})$. Totality fails because neither $\{0\} \subsetneq \{1\}$ nor $\{1\} \subsetneq \{0\}$ holds, even though $\{0\}$ and $\{1\}$ are distinct.
3. Yes, $<_{\mathbb{R}}$ is a total ordering of $[0,1]$. One indication of this is that the elements of $[0,1]$ can be represented as points in a line segment.
4. No, $<$ is not a total ordering of the relevant set of sequences. Totality fails because $0.5(0) = 0.4(9)$, so neither $\langle 5,0,0,0,\ldots\rangle < \langle 4,9,9,9,\ldots\rangle$ nor $\langle 4,9,9,9,\ldots\rangle < \langle 5,0,0,0,\ldots\rangle$ holds, even though $\langle 5,0,0,0,\ldots\rangle$ and $\langle 4,9,9,9,\ldots\rangle$ are distinct sequences.
5. A typical set of soldiers is not totally ordered by $<$, since it will include soldiers of the same rank, and soldiers such that neither one outranks the other. But one could, of course, imagine an army in which each rank is occupied by exactly one soldier. Such an army would be totally ordered by $<$.

### Section 2.2.3

1. No, $<_{\mathbb{Q}}$ is not a well-ordering of the positive rational numbers. There is no least element in, for instance, the set of all positive rational numbers. For every positive rational number, you can find a smaller one.
2. Yes, the natural numbers are well-ordered by $<_{\star}$.
3. Yes, the natural numbers are well-ordered by $<_0$.
4. Yes, a total ordering of a finite set is always a well-ordering.

### Section 2.2.4

1. These orderings do not have the same well-order type. One way to see this is to note that whereas the shape of $\{7,2,13,12\}$ is

   ||||

   the shape of $\{412,708,20,081\}$ is

   |||

   Generally speaking, two well-orderings can only be of the same well-order type if they order sets of the same cardinality, because the orderings can only be isomorphic to one another if there is a bijection between the sets they order.

2. They are of the same well-order type. Notice, in particular, that they both have the shape |||||.... Formally, the orderings are isomorphic to one another because the function $x+18$ is an order-preserving bijection from the set of natural numbers to the set of natural numbers greater than 17.

### Section 2.2.5

1. (a) Here is one such ordering: $3 < 4 < 5 < \ldots 0 < 1 < 2$
   (b) Here is one such ordering: $0 < 2 < 4 < \ldots 1 < 3 < 5 < \ldots$

### Section 2.2.8

1. To verify Stage Well-Ordering, we need to check that every non-empty set of ordinals has a smallest element. Let $S$ be such a set and let $S'$ be the set of stages that occur before every element in $S$. The Open-Endedness Principle guarantees that there is a "next" stage $s$; i.e., a first stage after every stage in $S'$. Since $s$ is not in $S'$, it must be in $S$ and therefore the smallest element of $S$.
2. Since $\alpha \in \beta$ if and only if $\alpha$ was introduced at an earlier stage than $\beta$, the membership relation among ordinals is, in fact, the relation of being-introduced-at-an-earlier-stage-than. So it follows from the fact that every set of stages is well-ordered by occurring-earlier-than that every set of ordinals must be well-ordered by $\in$.

### Section 2.2.9

1. Recall that every ordinal is built by pooling together the ordinals that were introduced at previous stages of the construction process. So suppose that $\alpha$ and $\beta$ are ordinals and that $\beta \in \alpha$. If $\beta$ has an element $\gamma$, $\gamma$ must be an ordinal that was introduced earlier than $\beta$. But since $\beta \in \alpha$, $\beta$ must have been introduced earlier than $\alpha$. So $\gamma$ was introduced earlier than $\alpha$. So $\gamma \in \alpha$.

### Section 2.3.1

1. The ordinal $(\omega + \omega)$ consists of two copies of the natural numbers appended to each other. If you put another $\omega$ on the end, you get three copies of the natural numbers, one after the other.

   Now suppose you start with $\omega$. That's one copy of the natural numbers. Then you append $(\omega + \omega)$ at the end. That's like appending two copies of the natural numbers after the initial copy of the natural numbers. So what you get at the end is three copies of the natural numbers, one after the other.

2. The result of appending three lines to one is the same as the result of appending one line to a row of three.
3. False. The left side of the equation equals $\omega+\omega+\omega$, and the right side equals $\omega+\omega+\omega+1$.

### Section 2.3.2

1. First consider $(\omega \times \omega)$. As we noted above, that is the result of replacing each position in a well-ordering of type $\omega$ with a copy of a well-ordering of type $\omega$. In other words:

   ||||...||||...||||... ...

   Accordingly $(\omega \times \omega) \times \omega$ is the result of replacing each position in a well-ordering of type $\omega$ with a copy of the order type above. Schematically:

   |||...|||...|||... ... |||...|||...|||... ... |||...|||...|||... ... • • •

   What about $\omega \times (\omega \times \omega)$? It is the result of replacing each position in a well-ordering of type $\omega \times \omega$ with a copy of a well-ordering of type $\omega$. So again, we get the following:

   |||...|||...|||... ... |||...|||...|||... ... |||...|||...|||... ... • • •

2. $(0'' \times 0''') = (0''' \times 0'')$ is true, because three copies of two units each yields the same result as two copies of three units each.
3. A well-ordering of type $(\omega + 0''') \times 0''$ is the result of replacing each position in a well-ordering of type $0''$ (i.e., ||) with a copy of a well-ordering of type of $\omega + 0'''$ (i.e., |||...|||). In other words, $(\omega + 0''') \times 0''$ is

   |||...||| |||...|||

   which is just:

   |||...|||...|||

   So $(\omega + 0''') \times 0'' = (\omega + \omega) + 0'''$

   In contrast, a well-ordering of type $(\omega \times 0'') + (0''' \times 0'')$ is the result of appending a well-ordering of type $0''' \times 0''$ (i.e., ||||||) to the end of a well-ordering of type $\omega \times 0''$ (i.e., "|||||...|||||..."). In other words, $(\omega \times 0'') + (0''' \times 0'')$ is:

   |||...|||...||||||

   which is just $(\omega + \omega) + 0''''''$. And, of course, $(\omega + \omega) + 0'''''' \neq (\omega + \omega) + 0'''$.
4. Let us start with $0'' \times (\omega + 0''')$. The result of replacing each position in a well-ordering of type $\omega + 0'''$ (i.e., "|||||...|||") with a well-ordering of type $0''$ (i.e., ||) is a well-ordering of type "|| || ||...|| || ||," which is just "|||||...||||||." So $0'' \times (\omega + 0''') = \omega + 0''''''$.

Now consider $(0'' \times \omega) + (0'' \times 0''')$. Since $(0'' \times \omega) = \omega$, and since $(0'' \times 0''') = 0''''''$, $(0'' \times \omega) + (0'' \times 0''') = \omega + 0''''''$.

## Section 2.3.4

1. The result follows immediately from the fact that $A \cup B = B \cup A$.
2. We need to verify $|A \times B| = |B \times A|$. We can do so by identifying a bijection $f$ from $|A \times B|$ to $|B \times A|$. For instance, $f(\langle a, b\rangle) = \langle b, a\rangle$.
3. The result follows immediately from the fact that $A \cup (B \cup C) = (A \cup B) \cup C$.
4. We need to verify $|A \times (B \times C)| = |(A \times B) \times C|$. We can do so by identifying a bijection $f$ from $A \times (B \times C)$ to $(A \times B) \times C$. For instance, $f(\langle a, \langle b, c\rangle\rangle) = \langle\langle a, b\rangle, c\rangle$.
5. The result follows immediately from the fact that $(A \cup B) \times C = (A \times C) \cup (B \times C)$.

# 3 Omega-Sequence Paradoxes

This chapter will be all about paradoxes, so I'd like to start by saying a few words about what philosophers take a paradox to be. A **paradox** is an argument that appears to be valid but goes from seemingly true premises to a seemingly false conclusion. Since it is impossible for valid reasoning to take us from true premises to a false conclusion, a paradox is a sure sign that we've made at least one of the following mistakes:

- Our premises aren't really true.
- Our conclusion isn't really false.
- Our reasoning isn't really valid.

Not every mistake of this kind makes for an interesting paradox, though. (You don't get a paradox by forgetting to carry a 1 while performing multiplication.) Interesting paradoxes are paradoxes that involve interesting mistakes: mistakes that teach us something important about the way the world is or the way our theories operate.

In his book *Paradoxes*, philosopher Mark Sainsbury talks about ranking paradoxes on a ten-point scale. A paradox that gets assigned a 1 is thoroughly boring: we made a mistake somewhere, but it's not a mistake that teaches us very much. A paradox that gets assigned a 10 is incredibly illuminating: it teaches us something extremely important. I like Sainsbury's idea and will implement it here. I'll assign a grade between 0 and 10 to each of the paradoxes in this chapter and tell you why I made my choice.

The paradoxes we'll be talking about in this chapter all involve omega-sequences or reverse omega-sequences. An **omega-sequence** (or $\omega$-sequence) is a sequence of items that is ordered like the natural numbers under the natural ordering: $0, 1, 2, 3, 4, \ldots$. In other words, it is an ordering with the shape ||||| .... An $\omega$-sequence has a first member and a second member and a third member and so forth—but no last member. A **reverse omega-sequence** (or reverse $\omega$-sequence) is a sequence of items that is ordered like the natural numbers but under a *reverse* ordering: $\ldots 4, 3, 2, 1, 0$. In other words, it is an ordering with the shape ... |||||. A reverse $\omega$-sequence has a last member and a next-to-last member and a next-to-next-to-last member and so forth—but no first member.

Of the paradoxes we'll be talking about in this chapter, I know how to solve some but not others. And the resolutions I do know are not all of the same kind (though they often are related). The complexity of the issues is part of what will make our discussion interesting.

## 3.1 Omega-Sequence Paradoxes

### 3.1.1 Zeno's Paradox (*Paradoxicality Grade:* 2)

The following is a variant of one of the paradoxes attributed to ancient philosopher Zeno of Elea, who lived in the fifth century BC.

> You wish to walk from point $A$ to point $B$:

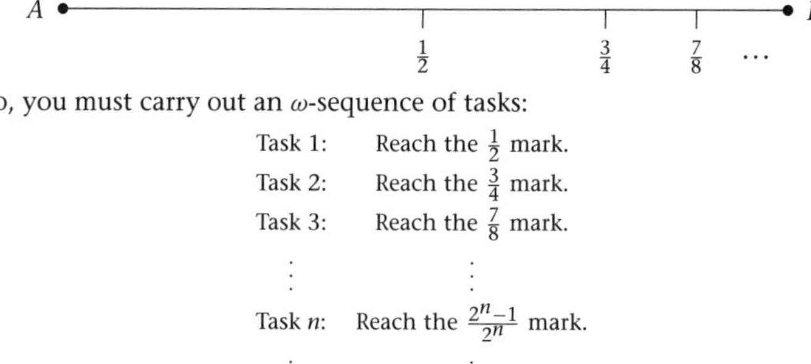


> To do so, you must carry out an $\omega$-sequence of tasks:
>
> Task 1: Reach the $\frac{1}{2}$ mark.
> Task 2: Reach the $\frac{3}{4}$ mark.
> Task 3: Reach the $\frac{7}{8}$ mark.
> ⋮
> Task $n$: Reach the $\frac{2^n-1}{2^n}$ mark.
> ⋮
>
> But nobody can complete infinitely many tasks in a finite amount of time. So it is impossible to get from point $A$ to point $B$. More generally: Movement is impossible.

### 3.1.2 Assessment

It seems to me that Zeno's Paradox deserves a lowish paradoxicality grade, from a contemporary perspective. This is because the mathematical notion of a limit is now commonplace, and limits can be used to show—contrary to the argument above—that it is possible to complete infinitely many tasks in a finite amount of time, as long as the time required to complete the tasks deceases quickly enough.

Here is an example: Suppose that Marty walks at a speed of 1 m/s and suppose that the distance from point $A$ to point $B$ is 1 m. This is how long it would take Marty to complete each of Zeno's infinitely many tasks:

| Task number | Task description | Time required, at 1m/s |
|---|---|---|
| Task 1: | Reach the $\frac{1}{2}$ mark. | $\frac{1}{2}$ s |
| Task 2: | Reach the $\frac{3}{4}$ mark. | $\frac{1}{4}$ s |
| Task 3: | Reach the $\frac{7}{8}$ mark. | $\frac{1}{8}$ s |
| ⋮ | ⋮ | ⋮ |
| Task $n$: | Reach the $\frac{2^n-1}{2^n}$ mark. | $\frac{1}{2^n}$ s |
| ⋮ | ⋮ | ⋮ |

By using the mathematical notion of a limit, one can show that even though Marty must complete infinitely many tasks to get from point $A$ to point $B$, he should be able to do so in 1 second, given the following:

$$\frac{1}{2}+\frac{1}{4}+\ldots+\frac{1}{2^n}+\ldots=\lim_{n\to\infty}\left(\frac{1}{2}+\frac{1}{4}+\ldots+\frac{1}{2^n}\right)=1$$

I certainly don't want to suggest that Zeno was a fool. He was not. For someone in Zeno's position, the paradox deserves a grade much higher than 2.

### Exercises

1. Show that $\lim_{n\to\infty}\left(\frac{1}{2}+\frac{1}{4}+\frac{1}{8}+\ldots+\frac{1}{2^n}\right)=1.$
2. Suppose Marty slows down as he approaches the 1m mark:

| Task number | Task description | Speed |
|---|---|---|
| Task 1: | Travel $\frac{1}{2}$m to reach the $\frac{1}{2}$m mark. | $\frac{1}{2}$m/s |
| Task 2: | Travel $\frac{1}{4}$m to reach the $\frac{3}{4}$m mark. | $\frac{1}{4}$m/s |
| Task 3: | Travel $\frac{1}{8}$m to reach the $\frac{7}{8}$m mark. | $\frac{1}{8}$m/s |
| ⋮ | ⋮ | ⋮ |
| Task $n$: | Travel $\frac{1}{2^n}$m to reach the $\frac{2^n-1}{2^n}$m mark. | $\frac{1}{2^n}$m/s |
| ⋮ | ⋮ | ⋮ |

Based on these revised assumptions, is it possible for Marty to reach the 1m mark in a finite amount of time?

### 3.1.3 Thomson's Lamp (*Paradoxicality Grade:* 3)

I would like to consider another example of running through an $\omega$-sequence in finite time with seemingly paradoxical results. It was devised by the late James Thomson, who was a professor of philosophy at MIT (and was married to the great philosopher Judith Jarvis Thomson).

Suppose you have a lamp with a toggle button: press the button once and the lamp goes on; press it again and the lamp goes off. It is 1 minute to midnight and the lamp is off; at 30 seconds before midnight, you press the button. At 15 seconds before midnight, you press it again. At 7.5 seconds before midnight, you press it again. And so forth.

| Time to midnight | Status of lamp shortly thereafter |
|---|---|
| 60s | off |
| 30s | on |
| 15s | off |
| 7.5s | on |

$$\begin{array}{cc} \vdots & \vdots \\ \frac{60}{2^{2n}}s & \text{off} \\ \frac{60}{2^{2n+1}}s & \text{on} \\ \vdots & \vdots \end{array}$$

At midnight, is the lamp on or off? Here are two arguments:

**Argument 1** For every time the lamp gets turned off before midnight, there is a later time before midnight when it gets turned on. So the lamp can't be off at midnight.

**Argument 2** For every time the lamp gets turned on before midnight, there is a later time before midnight when it gets turned off. So the lamp can't be on at midnight.

If both these arguments are right, the lamp is not on at midnight and not off at midnight. What would that even mean?

### 3.1.4 Assessment

It is natural to answer the paradox by arguing that there could be no such thing as a lamp that behaves in the way the argument demands. But I don't think such a response would take us to the heart of the matter.

It is certainly true that we are not in a position to carry out the experiment in practice. To use a phrase of Bertrand Russell's, it is "medically impossible" for a human to carry out each of the infinitely many required tasks. The experiment is also barred by the laws of physics. For example, the switch would eventually have to travel faster than the speed of light. If this is all we had to say about the paradox, however, we would fail to appreciate what is most interesting about it. For even if the relevant setup is not physically possible, Thomson's Lamp can teach us something interesting about the limits of *logical* possibility.

For a hypothesis to be **logically impossible** is for it to contain an absurdity. Consider, for example, the hypothesis that Smith raises her arm without moving it. This hypothesis contains an absurdity, because part of what it is to raise an arm is to move it. In contrast, the hypothesis that a button travels faster than the speed of light is not logically impossible. It involves a violation of the laws of physics, but there is no *absurdity* in hypothesizing different physical laws. It is for this reason that we can make sense of a world in which Newtonian mechanics is true (even though Newtonian mechanics is actually false). It's also why we can make sense of science fiction about faster-than-light travel (even though it's actually barred by the laws of physics). In contrast, there is no sense to be made of a story in which Smith raises her arm without moving it. The hypothesis that someone raises an arm without moving it isn't just contrary to physical law: it is absurd.

When we assess the paradoxes in this chapter, we will be interested in the limits of logical possibility. We will try to figure out whether various scenarios involve a genuine absurdity, and we will try to understand why the absurdity arises when it does.

Thompson's Lamp is interesting because it derives an apparent absurdity—that the lamp is neither on nor off at midnight—from seemingly innocuous premises. To defuse the paradox, we need to understand whether the absurdity is genuine; and if it is genuine, we need to understand how it arises by identifying either a flaw in our reasoning or a false premise.

I think the absurdity is genuine, at least on the assumption that the lamp exists at midnight. For there is no sense to be made of a lamp that is neither on nor off at a given time. It seems to me, however, that Arguments 1 and 2 are both invalid.

Let me be a little more specific. I do not deny that the setup of the case entails each of the following:

- For every time before midnight the lamp is on, there is a later time before midnight that the lamp is off.
- For every time before midnight the lamp is off, there is a later time before midnight that the lamp is on.

The crucial observation is that neither of these claims addresses the question of what happens at midnight. The sequence of button-pushings constrains what the Thomson Lamp must be like at every moment before midnight. But it does not give us any information about the status of the lamp at midnight. The story is consistent with the lamp's being on at midnight and consistent with the lamp's being off at midnight.

What makes the Thompson Lamp case strange is not that there is no consistent way of specifying what happens at midnight, but rather that there is no way of specifying what happens at midnight *without introducing a discontinuity*. Let me explain.

We tend to think of the world as a place in which macroscopic change is **continuous**. In other words, if a macroscopic object is in state $s$ at a given time $t$, then it must be in states that are arbitrarily similar to $s$ at times sufficiently close to $t$. For instance, if you walk to the grocery store and reach your destination at noon, then you must be arbitrarily close to the grocery store at times sufficiently close to noon. (Things work differently at quantum scales.)

The problem with Thomson's Lamp is that you get a discontinuity by assuming that the lamp is on at midnight and by assuming that the lamp is off at midnight. So you can't consistently believe that Thomson's Lamp exists if you think that there are no discontinuities.

Argument 1 and Argument 2 both tacitly assume that there are no discontinuities. So even if they constitute valid forms of reasoning in a world where one assumes continuity, they cannot be used to assess the question of whether a Thomson Lamp is logically consistent. What I think is interesting about Thomson's Lamp, and the reason I think it deserves a grade of 3, is that it shows that a situation can be logically consistent even if it is inconsistent with deeply held assumptions about the way the physical world works—assumptions such as continuity.

Princeton University philosopher Paul Benacerraf once suggested a clever modification of the case that does not require discontinuities. Suppose that each time you push the button, the lamp shrinks to half of its previous size. Suppose, moreover, that the lamp has size 0 at midnight (and therefore does not exist). Then, at midnight, it's not the case that the lamp is on and it's not the case that the lamp is off!

#### Exercises

1. Suppose the lamp's button must travel a fixed distance $d$ each time it is pressed. Then for each natural number $n$, the lamp's button must eventually travel faster than $n$ m/s. For, as we approach midnight, the time available for the button to traverse $d$ will approach 0. Is there some point before midnight at which the button must travel infinitely fast?

## 3.2 Rational Decision-Making

Infinite sequences can cause trouble for rational decision-making. In this section I'll describe a couple of paradoxes that help illustrate the point. I learned about them from philosophers Frank Arntzenius, Adam Elga, and John Hawthorne.

### 3.2.1 The Demon's Game (*Paradoxicality Grade:* 4)[1]

Imagine an infinite group of people $P_1, P_2, P_3, \ldots$. An evil demon suggests that they play a game. "I'll ask each of you to say *aye* or *nay*," he says. "As long as only finitely many of you say *aye*, there will be prizes for everyone: each of you will receive as many gold coins as there are people who said *aye*. Should infinitely many of you say *aye*, however, nobody will receive anything."

If the members of our group were in a position to agree on a strategy beforehand, they could end up with as many gold coins as they wanted. All they would need to do is pick a number $k$ and commit to a plan whereby persons $P_1, \ldots, P_k$ answer *aye* and everyone else answers *nay*. As long as everyone sticks to the plan, every member of the group will end up with $k$ gold coins.

But the demon is no fool. He knows that he could bankrupt himself if he allowed the group to agree on a strategy beforehand. So he isolates each member of the group as soon as the rules have been announced. As a result, each person must decide whether

1. An earlier version of this material appeared in Rayo, "El juego del diablo," *Investigación y Ciencia*, October 2010.

to say *aye* or *nay* without having any information about the decisions of other members of the group.

With such precautions in place, the demon has nothing to fear. To see this, imagine yourself as a member of the group. Isolated from your colleagues, you ponder your answer. Should you say *aye* or *nay*? In full knowledge that your decision will have no effect on other people's decisions, you reason as follows:

> If infinitely many of my colleagues answer *aye*, nobody will get any coins, regardless of what I decide. So my decision can only make a difference to the outcome on the assumption that, at most, finitely many of my colleagues answer *aye*. But in that case, I should definitely answer *aye*. For doing so will result in an additional gold coin for everyone, including myself! (If I were to answer *nay*, on the other hand, I wouldn't be helping anyone.)

So answering *aye* couldn't possibly make things worse and could very well make things better. The rational thing to do, therefore, is to answer *aye*!

Of course, other members of the group are in exactly the same situation as you are. So what is rational for you is also rational for them. That is why the demon has nothing to fear. As long as every member of the group behaves rationally, everyone will answer *aye*, and the demon won't have to cough up any money. (It goes without saying that the demon could lose a lot of money if only finitely many members of the group are fully rational. But the demon was careful to avoid selecting such a group for his game: nothing gives him more pleasure than torturing fully rational people.)

Our group is in a position to avail itself of as much money as it likes. And yet we can predict in advance that it will fail to do so, even when every member of the group behaves rationally. What has gone wrong?

### 3.2.2 Assessment

What I find interesting about this paradox is that it shows there can be a discontinuity between the rationality of a group and the rationality of its members. More precisely:

**Collective Tragedy** There are setups in which a group will predictably end up in a situation that is suboptimal for every member of the group, even though every member of the group behaves rationally.

What should we do about Collective Tragedy? My own view is that there is nothing to be done: we need to learn how to live with it.

(Collective Tragedy is in some ways reminiscent to the Prisoner's Dilemma, which we'll talk about in chapter 5.)

### 3.2.3 A One-Person Version of the Game (*Paradoxicality Grade:* 5)

There is a variation of the demon's evil game that can be applied to a single individual. Imagine that the demon whispers into your ear: "I am about to ask you an infinite sequence of questions $q_1, q_2, q_3, \ldots$. Each time I ask you a question, you must answer *aye* or *nay*. If you answer *aye* at most finitely many times, you will receive a prize: as many gold coins as *aye*-answers you give. If, however, you answer *aye* infinitely many times, you will leave empty-handed."

Note that for this exercise to make sense, there is no need to assume that you will lead an infinitely long life. We can assume, instead, that you will be able to answer the demon's questions faster and faster. Suppose, for example, that the second question is asked a half an hour after the first, that the third question is asked a quarter of an hour after the second, and so forth. (In general, question $q_{n+1}$ is asked $1/2^n$ hours after question $q_n$.) This entails that the demon will have asked all his questions within the span of an hour. So as long as you're able to answer each question before he asks the next one, you will answer all of his questions within the span of an hour.

If you were able to commit once and for all to a plan about how to answer the demon's questions, you could end up with as many gold coins as you wanted. Just as in the multiperson case, you could simply select a number $k$ and commit to answering *aye* to questions $q_1, \ldots, q_k$ and no further.

But would you be able to stick to such a plan? Imagine that you've responded *aye* to the first $k$ questions and that the demon asks question $q_{k+1}$. If you respected the original plan, you would answer *nay*. But why respect the original plan? It seems clear that a better plan is available to you: answer *aye* to questions $q_1, \ldots, q_{k+1}$ and no further. This new plan will bring you all the benefits of the old plan, plus 1 gold coin!

The problem, of course, is that this process could be iterated. Each time your current plan committed you to answering *nay*, you would be tempted to switch plans. And if you switched plans every time, you'd end up empty-handed, since you would end up answering *aye* to every single question.

There are those who possess an unflappable will: people who are able to make a plan and stick to it even if a better plan presents itself further along the line. Others need a trick to ensure that they stick to their plans. There is a famous example of this in Homer's *Odyssey*. Here is a paraphrase of the relevant passage:

> So enchanting were the songs of the sirens that no sailor could resist them. Bewitched, he would follow the music into the sea and drown.
>
> Odysseus wanted to hear the song of the sirens, but he didn't want to die. So he ordered his men to bind him to the mast of his ship and made clear that he was not to be released until the deadly sirens had been left behind. He then ordered his men to cover their own ears with wax.
>
> When the sirens began to sing, it seemed to Odysseus that his original plan ought to be replaced with a different one: the plan of following the sirens' songs. But his men forced him to stick to the original plan and refused to release him until they were safely out of the sirens' reach.

There is an interesting analogy between Odysseus's predicament and the Demon's Game. In both cases, things are set up so that one is tempted to abandon one's original plan at the crucial time. As in the case of Odysseus, the new plan seems like a good idea only to someone who has fallen under the sirens' spell. When it comes to the Demon's Game, however, everyone should agree that switching plans is the rational thing to do. Notice, for example, that even before the start of the game, everyone should agree that answering *aye* to the first $k+1$ questions (and *nay* to the rest) is better than answering *aye* to the first $k$ questions (and *nay* to the rest).

You will be able to beat the evil demon only if you're able to make decisions on the basis of a plan that can be seen to be inferior from the start, and then stick to that plan. You'll have to somehow find a way of "binding yourself to the mast." But, for most people, that is not easy to do.

### 3.2.4 Assessment

Think of an individual person as a *collective*: the collective consisting of her different "time slices." The reason I like the one-person version of the Demon's Game, and the reason I think it deserves a higher paradoxicality grade than its multiperson counterpart, is that it suggests there are scenarios in which you predictably will end up in a suboptimal situation, even though each of your time slices behaves rationally. In other words, it suggests that when an individual is thought of as a collective of time slices, he is just as subject to Collective Tragedy as a collective consisting of many people.

How should we respond to this? As before, my own view is that there's nothing to be done: we need to learn to live with it. If your decisions are the decisions of your time slices, you may end up doing things that are predictably suboptimal, even though none of your time slices ever decides irrationally.

#### Exercises

1. The Devil offers you a deal: spend a day in hell, and you will get two in heaven. You value a day in heaven just as much as you disvalue a day in hell. And you value two days in heaven twice as much as one day in heaven. Also, you do not value immediate pleasures over future pleasures or disvalue immediate pains over future pains. Given your values, should you accept the Devil's deal?[2]
2. Suppose that you do accept the deal above. Toward the end of your horrible day in hell, the Devil sidles up to your telemarketing cubicle and offers you another deal: stay in hell tomorrow and you get two more days in heaven, for a total of four days in heaven. (You value four days in heaven twice as much as you value two days in heaven.) Given your values, should you accept this deal? (As you consider your answer, note that by iterating this strategy, the Devil could get you to spend an eternity in hell....)

2. Both of these exercises are based on text by Damien Rochford.

## 3.3 Reverse Omega-Sequence Paradoxes

In this section we'll talk about some paradoxes based on reverse $\omega$-sequences.

### 3.3.1 The Bomber's Paradox (*Paradoxicality Grade*: 6)

I heard about the following puzzle from Josh Parsons, who was a fellow at Oxford until shortly before his untimely death in 2017. (The puzzle is a version of Bernadete's Paradox.)

There are infinitely many electronic bombs, $B_0, B_1, B_2, \ldots$, one for each natural number. They are set to go off on the following schedule:

| Bomb | Time set to go off |
|---|---|
| $B_0$ | 12:00 p.m. |
| $B_1$ | 11:30 a.m. |
| $B_2$ | 11:15 a.m. |
| $\vdots$ | $\vdots$ |
| $B_k$ | $\frac{1}{2^k}$ hours after 11:00 a.m. |
| $\vdots$ | $\vdots$ |

Our bombs are a special kind: they target electronics. Should one of the bombs go off, it will instantaneously disable all nearby electronic devices, *including other bombs*. This means that a bomb goes off if and only if no bombs have gone off before it. More specifically:

**(0)** $B_0$ goes off if and only if, for each $n > 0$, $B_n$ fails to go off.

**(1)** $B_1$ goes off if and only if, for each $n > 1$, $B_n$ fails to go off.

**(2)** $B_2$ goes off if and only if, for each $n > 2$, $B_n$ fails to go off.

$\vdots$

**($k$)** $B_k$ goes off if and only if, for each $n > k$, $B_n$ fails to go off.

**($k+1$)** $B_{k+1}$ goes off if and only if, for each $n > k+1$, $B_n$ fails to go off.

$\vdots$

Will any bombs go off? If so, which ones? Here's a proof that bomb $B_k$ can't go off, for arbitrary $k$:

> Suppose that bomb $B_k$ goes off. It follows from statement ($k$) above that $B_n$ must fail to go off for each $n > k$ (and therefore for each $n > k+1$). So it follows from statement ($k+1$) that $B_{k+1}$ will go off, contradicting our assumption that $B_k$ goes off.

We have seen that $B_k$ must fail to go off for arbitrary $k$. But wait! Here's a proof that bomb $B_k$ must go off for arbitrary $k$:

> Suppose that $B_k$ fails to go off. It follows from statement $(k)$ above that $B_n$ must go off for some $n > k$, which is shown to be impossible by the previous argument.

What's going on?

### 3.3.2 Assessment

I have a preferred response to the Bomber's Paradox, but I should concede from the start that it is somewhat unintuitive. In fact, that is part of the reason I give the Bomber's Paradox a high-ish paradoxicality grade. I think the paradox teaches us about an interesting respect in which our intuitions cannot be trusted.

My response starts with the observation that the argument involved in the Bomber's Paradox is, in fact, valid. So it must be impossible for there to be a sequence of bombs set up as the case describes. And it is not just medical or physical impossibility. I think the paradox's setup entails an absurdity and is therefore *logically* impossible.

There is no denying that there is something disconcerting about this claim. One way to bring this out is to imagine an infinite being who attempts to set up the relevant configuration of bombs. If the setup is logically impossible, something will go wrong. But what?

I'm inclined to think that this question deserves a flat-footed answer. The best way to see this is to consider an analogy. Suppose that our infinite being tries to bring about a more straightforward absurdity. Perhaps she wishes to build three objects—$A$, $B$, and $C$—such that:

$$\text{Mass}(A) < \text{Mass}(B) < \text{Mass}(C) < \text{Mass}(A)$$

We know that something will go wrong and that she won't succeed. But what? The answer is not determined by the story so far, but note that it could turn out to be something rather mundane. Perhaps she creates objects with masses of 1kg, 2kg, and 3kg, respectively, and thereby fails to conform to the required setup. Or perhaps she discovers the futility of the project and loses heart.

The reason our infinite being fails to put together the relevant configuration of bombs could turn out to be similarly mundane. Perhaps she messes up the timers, thereby failing to conform to the paradoxical setup. Or perhaps she discovers the futility of the project and loses heart.

There is, of course, an important disanalogy between the two cases. When we consider the case of the three objects, we get a strong feeling that the constraints of the problem are at odds with one another: that the only way to satisfy some of the constraints is to fail to satisfy others. In contrast, when we think about the bomber's case, it feels like the constraints of the case are not at odds with one another and that our infinite being should be able to build a suitable sequence of bombs. ("She first builds bomb $B_0$," I think. "She then builds bomb $B_1$....") Why the disanalogy?

Part of the answer, I think, is that whereas we're smart enough to see at a glance that the infinite being is trying to do something impossible in the case of the three objects, it's harder to see at a glance that an infinite being would be trying to do something impossible if she were trying to set up infinitely many bombs. But perhaps there is also something deeper going on. As my friend Øystein Linnebo pointed out, it is possible to have a *potentially infinite* sequence of bombs meet the requirements of the Bomber's Paradox. (More precisely, we know that any finite sequence that meets the requirements can be extended to an even longer finite sequence that meets the requirements.) But, as we saw above, it is impossible for there to be an *actual infinity* of bombs satisfying the problem's requirements. So part of the reason the two cases seem disanalogous may be due to a slip between potential infinity and actual infinity. Perhaps we're prone to assume that whenever a potential infinity could exist to satisfy a particular constraint, it must be possible for there to be an actual infinity to exist that satisfies that constraint. The Bomber's Paradox shows that such an assumption would be incorrect.

### 3.3.3 Yablo's Paradox (*Paradoxicality Grade*: 8)

Let me tell you about a second reverse-omega-sequence paradox. It does not involve time but is structurally very similar to the Bomber's Paradox. It was discovered by Steve Yablo, who is a famous philosophy professor at MIT (and was a member of my dissertation committee many years ago).

Suppose you have infinitely many sentences, one for each natural number. We label the sentences $S_0, S_1, S_2 \ldots$, and characterize them as follows:

| Label | Sentence |
|---|---|
| $S_0$ | "For each $i > 0$, sentence $S_i$ is false." |
| $S_1$ | "For each $i > 1$, sentence $S_i$ is false." |
| $S_2$ | "For each $i > 2$, sentence $S_i$ is false." |
| ⋮ | ⋮ |
| $S_k$ | "For each $i > k$, sentence $S_i$ is false." |
| ⋮ | ⋮ |

We therefore know that each of the following statements must be true:

**(0)** $S_0$ is true if and only if, for every $n > 0$, $S_n$ is false.

**(1)** $S_1$ is true if and only if, for every $n > 1$, $S_n$ is false.

**(2)** $S_2$ is true if and only if, for every $n > 2$, $S_n$ is false.

⋮

(*k*) $S_k$ is true if and only if, for every $n > k$, $S_n$ is false.

**$(k+1)$** $S_{k+1}$ is true if and only if, for every $n > k+1$, $S_n$ is false.

⋮

Which sentences are true and which ones are false? Here is a proof that $S_k$ can't be true:

> Suppose that sentence $S_k$ is true. It follows from statement $(k)$ above that $S_n$ must be false for each $n > k$ (and therefore for each $n > k+1$). So it follows from statement $(k+1)$ that $S_{k+1}$ must be true, contradicting our assumption that $S_k$ is true.

But wait! Here's a proof that sentence $S_k$ must be true:

> The previous argument shows that sentence $S_m$ is false for arbitrary $m$. This means, in particular, that sentence $S_m$ is false for every $m > k$. So it follows from statement $(k)$ above that $S_k$ must be true.

What's going on?

#### Exercises

1. Suppose we modify the setup of Yablo's Paradox, so that $S_0, S_1, S_2, \ldots$ are instead characterized as follows:

| Label | Sentence |
|---|---|
| $S_0$ | "For some $i > 0$, sentence $S_i$ is false." |
| $S_1$ | "For some $i > 1$, sentence $S_i$ is false." |
| $S_2$ | "For some $i > 2$, sentence $S_i$ is false." |
| ⋮ | ⋮ |
| $S_k$ | "For some $i > k$, sentence $S_i$ is false." |
| ⋮ | ⋮ |

Is it possible to come up with a stable assignment of truth and falsity to $S_0, S_1, S_2 \ldots$?

### 3.3.4 Yablo and the Bomber

Yablo's Paradox is closely analogous to the Bomber's Paradox, since one paradox can be turned into the other by substituting talk of bombs going off for talk of sentences being true. But there is also a respect in which the two paradoxes might seem to come apart, at least initially.

My response to the Bomber's Paradox was based on the claim that no bombs could ever be configured in the way that the paradox requires. It is not initially clear, however, that an analogous response is available in the case of Yablo's Paradox. For whereas in the case of the Bomber's Paradox we were speaking of hypothetical bombs, in the case of Yablo's Paradox we are talking about sentences of *English* (or, more precisely, sentences of the language that results from enriching English with unproblematic logical notation). And one might think that, unlike the bombs, which don't actually exist, the relevant sentences do exist—not in the sense of being represented by actual blobs of ink or spoken sounds, but in the sense of being well-defined and available for use by English speakers.

That is one reason Yablo's Paradox is so interesting. But there is another...

### 3.3.5 The Liar Paradox (*Paradoxicality Grade*: 10)

Yablo's Paradox teaches us something important about the most famous of all the semantic paradoxes:

**The Liar Paradox** Consider the starred sentence:

(⋆) The starred sentence is false.

Is the starred sentence true or false? If it is true, what it says is true, so it is false. If it is false, what it says is false, so it is true.

It is tempting to think that the Liar Paradox is, at its root, a puzzle about self-reference and to think that the way out of the paradox is to find a way to disallow sentences that make reference to themselves in "vicious" ways. But Yablo's Paradox suggests that self-reference isn't really at the root of the problem, because no sentence involved in Yablo's Paradox makes reference to itself. So any solution to the Liar Paradox that is general enough to apply to Yablo's Paradox must go beyond self-reference.

In this section I'll describe my own answer to the Liar Paradox, but please keep in mind that my view is far from orthodox. The paradox has generated an enormous literature over the years, and there is no real consensus about how to address it. (If you'd like to learn more, have a look at the readings I recommend in section 3.6.)

I think linguistic communication is best thought of as a *coordination game* between Speaker and Hearer. The goal of the game is for Hearer to interpret Speaker's assertion as Speaker intends. When Speaker and Hearer successfully coordinate on an interpretation of Speaker's assertion, the asserted sentence is thereby rendered meaningful in the relevant context. When they fail to coordinate, the sentence remains meaningless. As a result, I think that sentences have no meanings independent of the interpretations that are agreed upon by Speaker and Hearer in the context of particular assertions.

Of what resources do Speaker and Hearer avail themselves in their effort to coordinate on an interpretation of Speaker's assertion? They use what they know (and assume each other to know) about past linguistic usage. But they also use what they know (and assume each other to know) about the *context* in which the assertion takes place.

Consider an example. Suppose that I am Speaker and you are Hearer. I make an assertion. Your knowledge of past linguistic usage is enough to establish that what I said is either "the last word in Zoyman's assertion is obscene" or "the last word in Zoyman's assertion is 'obscene.'" But you cannot immediately decide between these two interpretations because quotation marks are not pronounced in English. (Note that in the first case, I am referring to an obscenity-laced assertion; in the second, I am referring to an assertion that ends with the word *obscene*.)

You turn to context for help. Suppose, first, that Zoyman is a drunken sailor who speaks no English and is known to be remarkably vulgar. You and I are in a bar with him, and we both know that you don't speak his language and I do. When my assertion is made in this context, you can reasonably interpret it to mean "the last word in Zoyman's assertion is obscene." I can reasonably assume that my assertion will be so interpreted, since I wouldn't have made it in that context unless I had intended it to be so interpreted. So we have coordinated successfully on an interpretation of my assertion.

Next, suppose that Zoyman is not a drunken sailor but an Oxford don. We are all sitting at a high table having a conversation about Victorian standards of lewdness. Someone sneezes, and you and I both see that you couldn't quite catch the last word of Zoyman's previous statement. When my assertion is made in this context, you can reasonably interpret it to mean "the last word in Zoyman's assertion is 'obscene.'" I can reasonably assume that my assertion will be so interpreted, since I wouldn't have made it in that context unless I had intended it to be so interpreted. So, again, we have coordinated successfully on an interpretation of my assertion.

Now consider a third scenario. Suppose that *I* am Zoyman and that it is common knowledge that this is so. It is also common knowledge that the only relevant assertion I have made is the very assertion you are trying to interpret. Then your ability to find a stable interpretation for my assertion will break down. For if you assume that the last word of my assertion is *obscene* then you will be forced to interpret the assertion as an obvious falsehood (since *obscene* is not an obscene word). And if you assume that the last word in my assertion is the result of my putting quotation marks around *obscene*, then again you will be forced to interpret the assertion as an obvious falsehood (since "obscene" does not refer to itself but to the word *obscene*).

When it is common knowledge that I am Zoyman, you lose the ability to find a stable interpretation of my assertion. But note that there is nothing mysterious or paradoxical going on. Here is an analogy: We are playing a game in which one wins if one does something Martha is not doing. Most people are in a position to do well in such a game, but poor Martha is not, since in her case the strategy of doing something Martha is not doing becomes unstable. A game has been set up, and it is built into the rules that Martha can't win. Something similar is going on when it comes to my assertion about Zoyman. A coordination game has been set up. In most ordinary contexts, the assertion is readily interpreted, so the game can be won. But when things are set up so that I am Zoyman and the only assertion I have made is the very assertion you are trying to interpret, ordinary interpretative strategies become unstable.

I claim that something similar is going on in the Liar Paradox. In most ordinary contexts, an assertion of "the starred sentence is false" is easily interpretable. For example, in a context in which the starred sentence is "walruses are tuskless," one would have no difficulty interpreting the assertion of "the starred sentence is false." But a context in which the starred sentence is "the starred sentence is false" is like the context in which I am Zoyman: efforts to interpret the relevant assertion become unstable, and the possibility of coordination between Speaker and Hearer breaks down. As before, nothing mysterious or paradoxical is going on: a coordination game has been set up, and our ordinary interpretative strategies are such that the game cannot be won in the relevant context.

I have argued that an assertion of "the starred sentence is false" cannot be interpreted in a context in which the starred sentence is "the starred sentence is false." But recall that, in my view, it is a mistake to think that sentences have meanings independent of particular contexts of use. Instead, Speaker and Hearer render a sentence meaningful by coordinating on a suitable interpretation in a particular conversational context. Since "the starred sentence is false" resists such coordination in the relevant context, it fails to be rendered meaningful and is neither true nor false. (This is all exceedingly rough, but I develop these ideas further in one of the articles listed in section 3.6.)

If I had to assign the Liar Paradox a paradoxicality grade, I'd assign it a perfect 10. This is partly to do with the tenacity of the problem: even though versions of the paradox have been known since antiquity, philosophers have yet to settle on an answer that generates broad agreement. It is also to do with my own assessment of the problem: I think the Liar Paradox calls for a fairly radical rethinking of the way language works.

### 3.3.6 Assessment of Yablo's Paradox

It seems to me that Yablo's Paradox ought to be addressed in the same sort of way as the Liar Paradox. Rather than assuming that $S_0, S_1, \ldots$ have meanings independent of particular contexts of use, one should think that Speaker and Hearer must attempt to render these sentences meaningful in conversation by coordinating on suitable interpretations. But as in the case of the Liar Paradox, there is no stable way of doing so. So the sentences must remain uninterpreted and are neither true nor false.

Note that this response to Yablo's Paradox is in some ways analogous to my response to the Bomber's Paradox. For in saying that $S_0, S_1, \ldots$ must remain uninterpreted, I am claiming that one couldn't really set up a configuration of (meaningful) sentences that gives rise to the paradox, which is analogous to saying that one couldn't really set up a configuration of bombs that gives rise to the Bomber's Paradox.

## 3.4 Hat Problems

### 3.4.1 The Easy Hat Problem (*Paradoxicality Grade*: 0)

You may have heard of the following brain teaser:

> *Setup:* Ten prisoners, $P_1, P_2, \dots, P_{10}$, are standing in line. $P_1$ is at the end of the line. In front of her is $P_2$, in front of him is $P_3$, and so forth, with $P_{10}$ standing at the front of the line. Each of them is wearing a hat. The color of each prisoner's hat is determined by the toss of a fair coin: each person whose coin lands Heads gets a blue hat, and each person whose coin lands Tails gets a red hat. Nobody knows the color of her own hat, but each prisoner can see the colors of the hats of everyone standing in front of her. (For instance, $P_6$ can see the colors of the hats worn by $P_7, P_8, P_9$, and $P_{10}$.) Now assume that a guard will start at the end of the line and ask each prisoner to guess the color of her own hat. Prisoners answer in turn, with everyone being in a position to hear the answers of everyone else. Prisoners who correctly call out the color of their own hats will be spared. Everyone else will be shot.
>
> *Problem:* If the prisoners are allowed to confer with one another beforehand, is there a strategy they can agree upon to improve their chances of surviving?

The answer is yes. In fact, the prisoners can guarantee that $P_2, \dots, P_{10}$ survive and that $P_1$ has a 50% chance of survival. (I'll ask you to verify this in an exercise below.)

That's a fun puzzle. It's a bit tricky, but it's certainly not paradoxical. Here we will be concerned with a much harder version of the hat problem, one which arguably *is* paradoxical. Read on.

#### Exercises

1. Identify a strategy that guarantees $P_2, \dots, P_{10}$ survive and gives $P_1$ a 50% chance of survival.

### 3.4.2 Bacon's Puzzle (*Paradoxicality Grade*: 7)[3]

Let me now tell you about a hard version of the hat problem, for which credit is due to University of Southern California philosopher Andrew Bacon.

> *Setup:* As before, we have a line of prisoners. They are all wearing hats, and for each prisoner, a coin was flipped to determine whether she would wear a red hat or a blue hat. This time, however, the prisoners form an $\omega$-sequence, with one prisoner for each positive integer, $P_1, P_2, P_3, \dots$. As before, $P_1$ is at the very end of the line; in front of her is $P_2$, in front of him is $P_3$, and so forth. As before, each person in the sequence can see the hats of the prisoners

3. An earlier version of this material appeared in Rayo, "Sombreros e infinitos," *Investigación y Ciencia*, April 2009.

> in front of her, but cannot see her own hat (or the hat of anyone behind her). This time, however, the prisoners will not be called on one at a time to guess her hat color; instead, at a set time, everyone has to guess the color of her own hat by crying out "Red!" or "Blue!" Everyone is to speak simultaneously, so that nobody's guess can be informed by what others say. People who correctly call out the color of their own hats will be spared. Everyone else will be shot.
>
> *Problem:* Find a strategy that $P_1, P_2, P_3, \ldots$ could agree upon in advance that would guarantee that at most finitely many people are shot.

Astonishingly, it is possible to find such a strategy. In fact, describing the strategy is relatively straightforward, and we shall do so shortly. The hard part, and the reason I give this puzzle a paradoxicality grade of 7, is making sense of the fact that the strategy exists.

### 3.4.3 The Strategy: Preliminaries

Let us represent an assignment of hats to individuals as an $\omega$-sequence of 0s and 1s. A 0 in the $k$th position means that $P_k$'s hat is red and a 1 in the $k$th position means that $P_k$'s hat is blue. (So, for instance, the sequence $\langle 0, 1, 1, 1, 0, \ldots \rangle$ represents a scenario in which $P_1$'s hat is red, $P_2$'s hat is blue, and so forth.)

Let $S$ be the set of all $\omega$-sequence 0s and 1s. We'll start by **partitioning** $S$. In other words, we'll divide $S$ into a family of non-overlapping "cells," whose union is $S$. Cells are defined in accordance with the following principle:

> Sequences $s$ and $s'$ in $S$ are members of the same cell if and only if there are at most finitely many numbers $k$ such that the $s$ and $s'$ differ in the $k$th position.

For instance, $\langle 0, 0, 0, 0, \ldots \rangle$ and $\langle 1, 0, 0, 0, \ldots \rangle$ are in the same cell because they differ only in the first position. But $\langle 0, 0, 0, 0, \ldots \rangle$ and the forever alternating sequence $\langle 1, 0, 1, 0, \ldots \rangle$ are in different cells because they differ in infinitely many positions.

#### Exercises

1. Show that our definition of *cell* succeeds in partitioning $S$. More specifically, for $a, b, c \in S$, verify each of the following:
   a) $a$ is in the same cell as itself.
   b) If $a$ is in the same cell as $b$, then $b$ is in the same cell as $a$.
   c) If $a$ is in the same cell as $b$ and $b$ is in the same cell as $c$, then $a$ is in the same cell as $c$.
2. How many sequences does a given cell contain?
3. How many cells are there?

### 3.4.4 The Strategy

Our strategy relies on getting $P_1, P_2, \ldots$ to agree in advance on a "representative" from each cell. So, for instance, they might agree that $\langle 0, 1, 0, 0, \ldots \rangle$ is to be the representative for the cell that contains $\langle 0, 0, 0, 0, \ldots \rangle$. (The ability to pick a representative from each cell presupposes the Axiom of Choice, which I mentioned in chapter 2 and will discuss further in chapter 7.)

Imagine that $P_1, P_2, \ldots$ are all lined up and that they've agreed on a representative from each cell. Each of $P_1, P_2, \ldots$ can see the hat colors of everyone ahead of her. Now consider person $P_k$. She can see the colors of the hats of $P_{k+1}, P_{k+2}, P_{k+3}, \ldots$, but not of $P_1, \ldots, P_k$. So she doesn't know exactly what the actual assignment of hats to persons is. But, crucially, she is in a position to determine which *cell* contains the sequence of 0s and 1s corresponding to that assignment. To see this, suppose that $P_k$ sees the following:

| $P_1$ | $P_2$ | ... | $P_k$ | $P_{k+1}$ | $P_{k+2}$ | $P_{k+3}$ | $P_{k+4}$ | ... |
|---|---|---|---|---|---|---|---|---|
| $\downarrow$ | $\downarrow$ | ... | $\downarrow$ | $\downarrow$ | $\downarrow$ | $\downarrow$ | $\downarrow$ | ... |
| ? | ? | ... | ? | red | blue | red | red | ... |

Let her fill in the missing information arbitrarily—using "blue," for example:

| $P_1$ | $P_2$ | ... | $P_k$ | $P_{k+1}$ | $P_{k+2}$ | $P_{k+3}$ | $P_{k+4}$ | ... |
|---|---|---|---|---|---|---|---|---|
| $\downarrow$ | $\downarrow$ | ... | $\downarrow$ | $\downarrow$ | $\downarrow$ | $\downarrow$ | $\downarrow$ | ... |
| *blue* | *blue* | ... | *blue* | red | blue | red | red | ... |

Since the resulting assignment of hat colors differs from the actual assignment at most in the first $k$ positions, the corresponding sequences of 0s and 1s must belong to the same cell. So $P_k$ has all the information she needs to identify the relevant cell. And, of course, everyone else is in a position to use a similar technique. So even though each of $P_1, P_2, \ldots$ has incomplete information about the distribution of hat colors, each is in a position to know which cell contains the $\omega$-sequence representing the actual hat distribution. Call this cell $O$, and let $r_O$ be the representative that was previously agreed upon for $O$. Let the group agree on the following strategy:

> Everyone is to answer their question on the assumption that the actual sequence of hats is correctly described by $r_O$.

In other words, $P_k$ will cry out "Red!" if $r_O$ contains a 0 in its $k$th position, and she will cry out "Blue!" if $r_O$ contains a 1 in its $k$th position. Because $r_O$ and the sequence corresponding to the actual hat distribution are members of the same cell, we know they differ in at most finitely many positions. So, as long as everyone conforms to the agreed-upon strategy, at most finitely many people will guess incorrectly.

(It goes without saying that this strategy presupposes that every prisoner has superhuman capabilities. For instance, they must be able to absorb information about infinitely many hat colors in a finite amount of time, and they must be able to pick representatives from sets with no natural ordering. No actual human could implement this strategy. But what matters for present purposes is that the strategy exists. Recall that we are interested in logical possibility, not medical possibility.)

### 3.4.5 Puzzle Solved?

We have succeeded in identifying a strategy that guarantees that at most finitely many people will guess incorrectly. Unfortunately, our success leads to paradox. For there is a seemingly compelling argument for the conclusion that such a strategy should be *impossible*. That argument is as follows:

> We know that a random process will be used to decide what kind of hat to place on each person. Let us imagine that it works as follows: Each person has an assistant. While the person closes his or her eyes, the assistant flips a coin. If the coin lands Heads, the assistant places a red hat on the person's head. If the coin lands Tails, the assistant places a blue hat on the person's head.
>
> We shall assume that assistants always use fair coins. Accordingly, the probability that a coin toss lands Heads is exactly 50%, independent of how other coin tosses might have landed. This means, in particular, that knowing the colors of the hats ahead of you gives you *no information whatsoever* about the color of your own hat. So even after you've seen the colors of the hats of everyone in front of you, you should assign a probability of 50% to the proposition that your own hat is red.
>
> This seems to entail that none of the prisoners could have a better than 50% chance of correctly guessing the color of her hat. But if this was so, it should be impossible for there to be a strategy that guarantees that the vast majority of $P_1, P_2, P_3, \ldots$ answers correctly.

Since we have found such a strategy, we know that this argument must go wrong somewhere. To solve the paradox, we must understand where the argument goes wrong.

### 3.4.6 Toward an Answer

Let me begin with a preliminary observation. There is a difference between saying that the group increases its probability of *collective* success and saying that each member of the group increases her own probability of success. To see this, imagine that the group has two choices:

**Strategy** Each member makes her decision by following the strategy outlined above.

**Coin Toss** Each member of the group makes her decision by tossing a coin.

One should expect that Strategy would lead to a better *collective* outcome than a Coin Toss. This is because a coin toss allows for uncountably many different options and only countably many of them are such that at most finitely many people answer incorrectly.

On the other hand, it is not at all clear that choosing Strategy over Coin Toss would increase the probability that any member of the group answered correctly. To see this, suppose that the group decided to follow Strategy and imagine yourself in $P_k$'s position. Although you could be confident that the number of people who would die is finite, you would have no idea how large that finite number would be. Notice, moreover, that the vast majority of values that number could take—all except for the first $k$—are too large to offer you much comfort. So it's not clear that you should expect to be better off following Strategy than following Coin Toss.

The core of the paradox we are considering is that it seems hard to reconcile two facts. On the one hand, there are reasons for thinking that none of the $P_1, P_2, P_3, \ldots$ could have a better than 50% chance of correctly guessing the color of her own hat. On the other, we have found a strategy that guarantees that the vast majority of $P_1, P_2, P_3, \ldots$ answers correctly. This would seem very puzzling indeed if the upshot of the strategy was that each member was able to bring her probability of success over 50%. But we have now seen that it is not clear that this is so. All we know is that the group increases its probability of *collective* success.

To fully answer the paradox, we need an understanding of how the group is able to increase its probability of collective success. To this we now turn.

### 3.4.7 The Three Prisoners[4]

Let me start by telling you about a different puzzle: the problem of the Three Prisoners. I don't know who invented it, but I learned about it thanks to philosopher and computer scientist Rohit Parikh, from the City University of New York.

*Setup:* Suppose there are three prisoners in a room. They all close their eyes, and each of them is approached by a guard. Each guard flips a fair coin. If the coin lands Heads, he gives his prisoner a red hat; if it lands Tails, he gives his prisoner a blue hat.

Once all three prisoners have been assigned hats, they are allowed to open their eyes. Each can see the colors of the others' hats but has no idea about the color of his own hat.

As soon as everyone knows the color of everyone else's hat, the prisoners are taken into separate cells, so that they are unable to communicate with each other. At that point, each is asked about the color of his own hat. He is free to offer an answer or remain silent. The guards will then proceed as follows:

- If all three prisoners remain silent, all three will be killed.
- If one of them answers incorrectly, all three will be killed.
- If at least one prisoner offers an answer, and if everyone who offers an answer answers correctly, then all three prisoners will be set free.

4. An earlier version of this material appeared in Rayo, "Los prisoneros y María," *Investigación y Ciencia*, August 2009.

| Prisoner *A* | Prisoner *B* | Prisoner *C* | Result of following strategy |
|---|---|---|---|
| Red | Red | Red | Everyone answers incorrectly |
| Red | Red | Blue | *C* answers correctly |
| Red | Blue | Red | *B* answers correctly |
| Red | Blue | Blue | *A* answers correctly |
| Blue | Red | Red | *A* answers correctly |
| Blue | Red | Blue | *B* answers correctly |
| Blue | Blue | Red | *C* answers correctly |
| Blue | Blue | Blue | Everyone answers incorrectly |

**Figure 3.1**
The eight possible hat distributions, along with the result of applying the suggested strategy.

> *Problem:* Find a strategy that the prisoners could have agreed upon before being assigned hats that would guarantee that their chance of survival is above 50%.

Before attempting to solve the puzzle, notice that there is a strategy that gives the prisoners a chance of survival of *exactly* 50%. The prisoners can select one among themselves to serve as their captain, and agree that only the captain is to offer an answer. Since the captain has no idea of the color of his own hat, he must answer at random. And because the color of his hat was chosen by the toss of a fair coin, he has a 50% chance of answering correctly.

### 3.4.8 A Solution to the Three Prisoners Puzzle

One strategy is for each prisoner to follow these instructions:

- If you see that the other two prisoners have hats of the same color, answer the guard's question on the assumption that your hat is of the opposite color.
- If you see that the other two prisoners have hats of different colors, refrain from answering the guard's question.

If all three prisoners follow this procedure, their chance of survival will be 75%. For (as shown in figure 3.1) there are eight possible hat distributions, all equally likely, and as long as they follow the strategy, the prisoners will be set free in six of those eight possibilities ($6/8 = 75\%$).

Note that although the prisoners increase their chance of collective success to 75%, the chance that any given prisoner answers correctly (assuming he offers an answer) remains fixed at 50%. Suppose, for example, that you are prisoner *A* and that you see two red hats before you. The agreed upon strategy asks you to answer "Blue!" But because the color of each hat was chosen on the basis of an independent coin toss, a ⟨Blue, Red, Red⟩ hat distribution is exactly as likely as a ⟨Red, Red, Red⟩ hat distribution. So the probability that you will answer correctly is precisely 50%.

The group improves its chance of collective success not because it increases the chance of success of individual answers but because it manages to coordinate individual successes and failures in a certain kind of way. To see this, go back to figure 3.1 and consider the eight different hat distributions that are possible. Each prisoner answers correctly in two of these eight hat distributions, answers incorrectly in two, and remains silent in four. So each prisoner has as many successes as failures (which is why his chances of answering correctly, given that he answers, are 50%). But the prisoners' individual successes and failures are distributed unevenly across the space of possible hat distributions. They are coordinated in such a way that only two of the eight possibilities involve failures (which is why the strategy gives the group a 75% chance of success).

### Exercises

1. You and I each toss a fair coin, and then each of us is asked to guess how the other person's coin landed. (Neither of us has any information about how the other's coin landed.) If at least one of us guesses correctly, we both get a prize; otherwise we get nothing. Show that there is a strategy that we could agree upon ahead of time that would guarantee that we win the prize. (Thanks to Cosmo Grant for telling me about this puzzle!)

### 3.4.9 Back to Bacon's Puzzle

Now that we have discussed the Three Prisoners puzzle, I would like to return to the problem of understanding how the prisoners are able to increase their probability of collective success in Bacon's Puzzle, even though it is not clear that any individual member of the group is able bring her probability of success over 50%.

A useful way to address this point is to draw on an analogy between the Three Prisoners and Bacon's Puzzle. In each case, the group improves its chances of collective success not by increasing the chance of success of individual answers but by coordinating successes and failures in a certain kind of way.

Recall that in Bacon's Puzzle, $P_1, P_2, P_3, \ldots$ are able to identify the cell containing the actual hat distribution, and everyone agrees to answer on the assumption that her actual hat color is as described by the representative of that cell. Because everyone knows which cell contains the actual hat distribution, and because they're all agreed on a representative for that cell, they are able to *coordinate* their individual successes and failures so as to make sure their answers form a sequence that is included in the same cell as the actual distribution of hat colors. And since members of the same cell differ at most finitely, this guarantees that at most finitely many people answer incorrectly, even if none of the $P_1, P_2, P_3, \ldots$ increases her chances of guessing correctly beyond 50%.

This concludes my response to Bacon's Puzzle. I have proceeded in two steps. The first step is to note that, although it would be very surprising if someone could increase her probability of individual success by following the proposed strategy, the proposed strategy does not presuppose such an increase. All it requires is that the group increase its probability of *collective* success by following the strategy. The second step is to explain how the group is able to increase its probability of collective success even if no member of the group increases her probability of individual success. I have suggested that this is possible because, as in the Three Prisoners, individual members of the group are able to *coordinate* their successes and failures in the right sort of way.

One final question: What is the probability that an individual who follows the strategy will answer correctly? I don't know the answer to this question, but I suspect that when one follows the strategy, one's probability of success is best thought of as ill-defined. A little more specifically, I doubt there is a reasonable way of assigning a probability to the proposition that the representative of a given cell has a 1 in its $k$th position and therefore to the proposition that $P_k$ guesses correctly, given that she guesses in accordance with the strategy.

If it is true that there is no reasonable way of assigning a probability to the proposition that someone acting in accordance with the hat strategy will answer correctly, then Bacon's Puzzle brings out an important limitation of standard ways of thinking about probability theory. That is why I think Bacon's Puzzle deserves a paradoxicality grade of 7. (I'll return to the problem of ill-defined probabilities when we talk about non-measurable sets in chapter 6.)

## 3.5 Conclusion

In this chapter we discussed four families of $\omega$-sequence (or reverse $\omega$-sequence) paradoxes.

The first family consists of Zeno's Paradox and Thomson's Lamp. Each of these paradoxes teaches us something important about logical consistency. Zeno's Paradox teaches us that there is nothing logically inconsistent about an infinite series of tasks that is completed in a finite amount of time, and Thomson's Lamp teaches us that a scenario can be logically consistent even if it does not satisfy continuity principles of the sort we're accustomed to in everyday life.

The second family consists of the multiperson and one-person versions of the Demon's Game. Both of these paradoxes teach us something important about rationality. The multiperson version teaches us that there are situations in which a group will predictably end up in a situation that is suboptimal for every member, even though everyone behaves rationally. The one-person version teaches us that there are scenarios in which you will predictably end up in a suboptimal situation, even though each of your time slices behaves rationally.

The third family consists of the Bomber's Paradox and Yablo's Paradox. I think these are hard paradoxes, which don't have obvious resolutions. But I suggested that, in each case, one should respond by rejecting the assumption that the scenario described by the paradox is coherent.

The fourth family consists of Bacon's Puzzle. I suggested that this paradox may help bring out an important limitation of standard ways of thinking about probability theory.

## 3.6 Further Resources

- A. C. Paseau has taken the idea of assigning paradoxicality grades one step further, by suggesting a family of formulas that assigns paradoxes a precise measure of paradoxicality. See his "An Exact Measure of Paradox."
- For Thomson's official description of his famous lamp, and for Benacerraf's response, see Benacerraf's "Tasks, Super-Tasks, and the Modern Eleatics."
- Bernadete's Paradox, on which the Bomber's Paradox is based, appears in his *Infinity: An Essay in Metaphysics*. For further discussion of this kind of case, see John Hawthorne's "Before-Effect and Zeno Causality."
- Yablo's Paradox first appeared in his "Paradox without Self-Reference." Yablo's "A Reply to New Zeno" is worth reading too: it's a response to Bernadete's Paradox.
- The classic treatment of the Liar Paradox is Alfred Tarski's "The Concept of Truth in Formalized Languages." Among the more contemporary treatments of the paradox, my favorites are Saul Kripke's "Outline of a Theory of Truth," Vann McGee's *Truth, Vagueness and Paradox*, and Hartry Field's *Saving Truth from Paradox*.
- My own views about the Liar Paradox, which I mention in section 3.3.5, are developed in my article "A Plea for Semantic Localism." (The "obscene" example is inspired by the questionnaire at the end of Cartwright's *Philosophical Essays*.)
- Some philosophers think that the best way to deal with some of the paradoxes in this section is to grant that certain contradictions are true. They think that, for some propositions $p$, it is both the case that $p$ and not-$p$. This view is sometimes called Dialetheism. If you'd like to learn more about Dialetheism, I recommend Graham Priest's "What Is So Bad about Contradictions?"
- The examples in section 3.2 are drawn from Frank Arntzenius, Adam Elga, and John Hawthorne's "Bayesianism, Infinite Decisions, and Binding."
- The hard version of the hat problem is drawn from Andrew Bacon's "A Paradox for Supertask Decision Makers." (I first learned about it from Dan Greco.)

## Appendix: Answers to Exercises

### Section 3.1.2

1. Let $f(n) = \frac{1}{2^1} + \frac{1}{2^2} + \ldots + \frac{1}{2^n}$. Then it is easy to verify that for each $n$, $f(n) = 1 - \frac{1}{2^n}$. (*Proof:* It is obvious that $f(1) = \frac{1}{2^1}$. For $n > 1$, let $n = m + 1$ and assume, as an inductive hypothesis, that $f(m) = 1 - \frac{1}{2^m}$. It follows that $f(n) = 1 - \frac{1}{2^m} + \frac{1}{2^n}$. But $\frac{1}{2^m} = 2\frac{1}{2^n}$. So $f(n) = 1 - 2\frac{1}{2^n} + \frac{1}{2^n} = 1 - \frac{1}{2^n}$.)

   Since $f(n) = 1 - \frac{1}{2^n}$ for each $n$, $\lim_{n\to\infty} f(n)$ must be 1. To see this, note that for any $\epsilon > 0$, we can find $\delta$ such that for any $k > \delta$, $|1 - f(k)| < \epsilon$: we simply let $\delta$ be such that $\frac{1}{2^\delta} < \epsilon$.
2. No. Under our new assumptions, Marty would require an infinite amount of time to make it to the end point. For he would have to complete each of the following tasks:

| Task number | Task description | Speed | Time required |
|---|---|---|---|
| Task 1: | Travel $\frac{1}{2}$m to reach the $\frac{1}{2}$m mark. | $\frac{1}{2}$m/s | 1s |
| Task 2: | Travel $\frac{1}{4}$m to reach the $\frac{3}{4}$m mark. | $\frac{1}{4}$m/s | 1s |
| Task 3: | Travel $\frac{1}{8}$m to reach the $\frac{7}{8}$m mark. | $\frac{1}{8}$m/s | 1s |
| ⋮ | ⋮ | ⋮ | ⋮ |
| Task $n$: | Travel $\frac{1}{2^n}$m to reach the $\frac{2^n-1}{2^n}$m mark. | $\frac{1}{2^n}$m/s | 1s |
| ⋮ | ⋮ | ⋮ | ⋮ |

   So the number of seconds that Marty would need to complete her infinitely many tasks is $1 + 1 + 1 + \ldots = \infty$.

### Section 3.1.4

1. No, there is no point before midnight at which the button must travel infinitely fast. At each time before midnight, the button will have a finite amount of time to travel a finite distance. And one can always use a finite speed to travel a finite distance in a finite amount of time.

### Section 3.2.4

1. Here is an argument for thinking that you ought to take the deal. Suppose you value the status quo at 0 and that you value a day in heaven at 10. It follows from the setup that you value a day in hell at $-10$ and two days in heaven at 20. So taking the deal has a positive net value: $-10 + 20 = 10$. So if you think the rational thing to do is to choose the option with highest net value, you should take the deal.

2. As in the previous question, taking the deal has a positive net value. So if you think the rational thing to do is to choose the option with highest net value, you should take the deal.

### Section 3.3.3

1. There is no stable assignment of truth and falsity to $S_0, S_1, S_2, \ldots$. Here is a proof that $S_k$ can't be false:

    Suppose that sentence $S_k$ is false. Then it follows from the definition of $S_k$ that $S_n$ must be true for each $n > k$. This means, in particular, that $S_{k+1}$ must be true. But it follows from the definition of $S_{k+1}$ that the only way for that to happen is for $S_m$ to be false for some $m > k+1$. And that's impossible: we concluded earlier that $S_n$ must be true for each $n > k$.

    Here's a proof that sentence $S_k$ must be false:

    The previous argument shows that sentence $S_m$ is true for arbitrary $m$. This means, in particular, that sentence $S_m$ is true for every $m > k$. So it follows from the definition of $S_k$ that $S_k$ must be false.

### Section 3.4.1

1. Let $P_1$ say "Red!" if she sees an odd number of red hats in front of her, and "Blue!" otherwise. Since the outcomes of fair coin tosses are independent of one another, this gives $P_1$ a 50% chance of survival.

    Now consider things from $P_2$'s perspective. If $P_1$ said "Red!," $P_2$ knows that the number of red hats among $P_2, \ldots, P_{10}$ is odd. But $P_2$ can see the colors of the hats on $P_3, \ldots, P_{10}$, so she has all the information she needs to figure out whether her own hat is red. (The case in which $P_1$ says "Blue!" is similar.)

    $P_2$ now says the color of her hat. So $P_3$ knows (i) that the number of red hats among $P_2, \ldots, P_{10}$ is odd, (ii) the color of $P_2$'s hat, and (iii) the colors of the hats on $P_4, \ldots, P_{10}$. So she has all the information she needs to figure out the color of her own hat.

    By iterating this process, $P_2, \ldots, P_{10}$ can each figure out the color of her own hat.

### Section 3.4.3

1. (a) $a$ is in the same cell as itself because there are no numbers $k$ such that $a$ differs from itself in the $k$th position.

    (b) Suppose $a$ is in the same cell as $b$. Then, for some finite $n$, there are $n$ numbers $k$ such that $a$ differs from $b$ in the $k$th position. But difference is symmetric. So it follows that there are $n$ numbers $k$ such that $b$ differs from $a$ in the $k$th position, which means that $b$ is in the same cell as $a$.

(c) Suppose $a$ is in the same cell as $b$ and $b$ is in the same cell as $c$. Then there are finite numbers $n$ and $m$ such that there are $n$ numbers $k$ such that $a$ differs from $b$ in the $k$th position and there are $m$ numbers $k$ such that $b$ differs from $c$ in the $k$th position. Since $n+m$ is finite if $n$ and $m$ are, this means that there are at most $n+m$ numbers $k$ such that $a$ differs from $c$ in the $k$th position. So $a$ is in the same cell as $c$.

2. A cell $C$ contains as many sequences as there are natural numbers. Here is one way to see this. Let $\langle a_1, a_2, a_3, \ldots \rangle$ be an arbitrary sequence in $C$ (with each $a_i$ in $\{0, 1\}$). Since, by the definition of *cell*, the sequences in $C$ differ only finitely, we can use $\langle a_1, a_2, a_3, \ldots \rangle$ to list the elements of $C$:

   - The first element of our list is the sequence that nowhere differs from $\langle a_1, a_2, a_3, \ldots \rangle$. In other words, it is just $\langle a_1, a_2, a_3, \ldots \rangle$ itself.
   - The next element of our list is the sequence that differs from $\langle a_1, a_2, a_3, \ldots \rangle$ in at most the first position (and that has not already been counted). In other words, it is the sequence $\langle (1-a_1), a_2, a_3, \ldots \rangle$.
   - The next two elements of our list are the two sequences that differ from $\langle a_1, a_2, a_3, \ldots \rangle$ in at most the first two positions (and that have not already been counted). In other words, they are the sequences $\langle (1-a_1), (1-a_2), a_3, \ldots \rangle$ and $\langle a_1, (1-a_2), a_3, \ldots \rangle$.
   - The next four elements of our list are the four sequences that differ from $\langle a_1, a_2, a_3, \ldots \rangle$ in at most the first three positions (and that have not already been counted). In other words, they are the sequences $\langle (1-a_1), (1-a_2), (1-a_3), a_4, \ldots \rangle$, $\langle a_1, (1-a_2), (1-a_3), a_4, \ldots \rangle$, $\langle (1-a_1), a_2, (1-a_3), a_4, \ldots \rangle$, and $\langle a_1, a_2, (1-a_3), a_4, \ldots \rangle$.
   - And so forth.

   We can then use this enumeration of $C$ to define a bijection between $C$ and the natural numbers.

3. There are more cells than there are natural numbers. (In fact, there are just as many cells as there are real numbers.)

   To see this, suppose otherwise: suppose that there are no more cells than there are natural numbers. Then there can be no more sequences than there are natural numbers. For we have just seen that each cell is countable, and we saw in chapter 1 that the union of countably many countable sets must be countable. But we also showed in chapter 1 that the set of sequences of 0s and 1s has as many elements as the set of real numbers. So there must be more cells than there are natural numbers.

### Section 3.4.8

1. Here is one strategy: you guess that my coin landed on the same side as yours, and I guess that your coin landed on the opposite side of mine. If the coins landed alike, you are guessing correctly and we win; if the coins landed differently, I am guessing correctly and we win.

# II Decisions, Probabilities, and Measures

# 4 Time Travel

In this chapter we'll talk about time travel and about an interesting connection between time travel and free will.

We'll first consider the question of whether time travel is consistent by focusing on the Grandfather Paradox. We'll then talk about what a system of laws might look like if it is to allow for time travel while excluding paradox. After that, we'll turn to free will. We'll consider the question of what it is to act freely and discuss whether the Grandfather Paradox entails that a time traveler cannot act freely.

Many of the ideas in this chapter are drawn from the work of David Lewis, who was a professor of philosophy at Princeton University until his untimely death in 2001.

## 4.1 Time Travel[1]

Everyone travels in time: we travel one second into the future every second! Here, however, we will talk about extraordinary cases of time travel in which there is a discrepancy between, on the one hand, the start-time and end-time of the journey, and on the other hand, the duration of the journey as measured from the perspective of the traveler.

Suppose, for example, that it takes you ten seconds to travel one year into the future (or one year into the past). Then there is a discrepancy between the length of your journey as measured by its start-time and end-time, and the length of the journey as measured from your perspective. So you will be counted as a time traveler. (Note that to measure time "from your perspective" is not to measure time using your *experience* of time—otherwise, being a time traveler would require nothing more than sitting through a boring lecture. The relevant measurement is what would be recorded by a clock accompanying the time traveler.)

---

1. An earlier version of this material appeared in Rayo, "Viajes a través del tiempo," *Investigación y Ciencia*, October 2009. The article was inspired by a lecture given by my MIT colleague Brad Skow.

This characterization of time travel is not perfect: all sorts of things go wrong when we bring in relativistic scenarios. (If you'd like to know more, have a look at the recommended readings in section 4.6.) It is not perfect, but it is good enough for present purposes.

### 4.1.1 An Inconsistent Time Travel Story

Telling a consistent time travel story is harder than you might think. The Hollywood film *Back to the Future* is an example of a fun but inconsistent time travel story. Here is a quick summary. In 1985, George McFly is unhappy. But George's son, Marty, travels back in time to 1955 and meets his father as a young man. As a result of that meeting, the father conquers his fears and blossoms into a bold and courageous man. When Marty returns to 1985, he is faced with a very different situation from the one that was described at the beginning of the film. George leads a happy life.

It is not hard to see why this story is inconsistent. What we are told at the beginning of the film contradicts what we are told at the end of the film:

| What we're told | When we're told |
|---|---|
| In 1985, George is unhappy. | Beginning of the film |
| In 1985, George is happy. | End of the film |

The fact that these contradictory statements are made at different times in the film doesn't matter. Notice, for example, that we wouldn't count a witness as consistent if she said one thing at the beginning of her testimony and the opposite at the end of her testimony.

For a time travel story to be consistent, it must never make conflicting statements about what the world of the story is like at a given time. If at some point in the story we are told that, at a given time $t$, George is unhappy, then the story is never allowed to renege on this commitment: it cannot go on to tell us that, at time $t$, George is happy. And this is so regardless of whether there is any time travel in the story. If, according to the story, Marty travels back in time and tries to improve George's situation at $t$, then the story must have it turn out that somehow or other George is unhappy at $t$ in spite of the attempted interference.

**4.1.1.1 A changing-timeline interpretation** One way of restoring consistency to *Back to the Future* is to interpret the story as involving a "change" in George's timeline. (See figure 4.1.) Before the "change," George's timeline is such that he is born, grows up, and ends up unhappy. After the "change," George's timeline is such that he is born, grows up, and ends up happy.

That, at any rate, is supposed to be the view, but the truth is that I find it somewhat obscure. Notice, in particular, that it relies on two different senses of time. It relies on

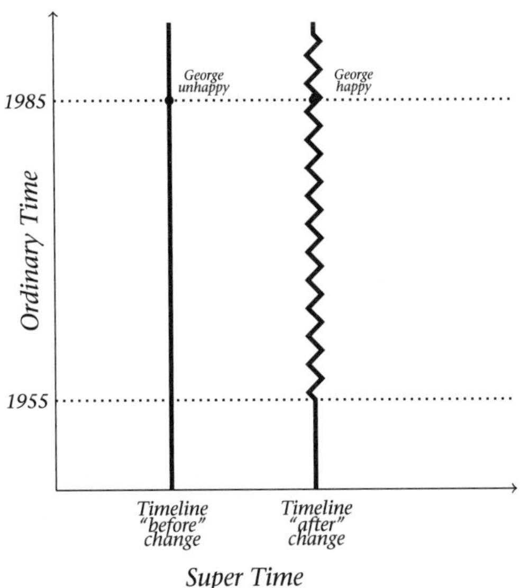


**Figure 4.1**
A change in George's timeline. The straight lines represent events as they "originally" occurred. The jagged line represents events as they occur "after" the change.

an ordinary notion of time when it describes changes *within* George's timeline: when it says that he is born, grows up, and ends up happy or unhappy. It uses a different, non-ordinary sense of time when it says that the timeline itself "changes": that George's development through ordinary time, as it is described at the beginning of the story, is different from his development through ordinary time, as it is described later in the story.

Let us use "super time" to talk about time in this second sense. The view is then supposed to be that facts about how George evolves in ordinary time "change" in super time. At an initial super time, George lives his life within ordinary time in a way that ends badly; at a later super time, he lives his life within ordinary time in a way that ends happily. The problem with this way of thinking about the story is that it is not clear that there is good sense to be made of super time. Notice, in particular, that none of us has ever experienced changes in super time. When we experience change in everyday life, it is always by experiencing ordinary change: by observing sand trickling through an hourglass or listening to the leaves rustling in the wind. And ordinary change involves differences across ordinary time, not differences across super time.

Whether the notion of super time can ultimately be made sense of, however, will not be our concern here. Whenever I speak about time in this chapter, I am concerned with ordinary time. And whenever I speak about time travel, I mean extraordinary travel within ordinary time—not changes in the world's timeline across super time.

**4.1.1.2 A world-travel interpretation** There is a different way to try to make *Back to the Future* consistent. One could think of the film as a story about travel *between* worlds rather than a story of time travel within a single world. The idea is that, when Marty travels back to 1955, he travels to the 1955 of a *different universe* and then comes back to the 1985 of that other universe. Accordingly, our George remains miserable in the 1985 of our universe, while alternative-universe George is happy in his own otherworldly 1985.

A "world travel" story of this kind is consistent. Notice, however, that based on this interpretation of *Back to the Future,* we should feel very differently about events in the film than the film itself encourages us to feel. For instance, the film is supposed to have a happy ending. But in the world travel interpretation, the ending is tragic, since it has Marty traveling to the 1985 of the alternate universe and abandoning his home in our universe. His original family is still miserable. They are probably even more miserable now that Marty has left.

### 4.1.2 A Consistent Time Travel Story

I have argued that *Back to the Future* is inconsistent, assuming it is interpreted as an ordinary time travel story rather than a super time story or a world travel story. Not every time travel story is inconsistent, however. Good examples of consistent time travel stories include the films *Twelve Monkeys* and *The Time Traveler's Wife.* Here is another example of a consistent time travel story:

> On the eve of your 21st birthday, an elderly stranger shows up at your doorstep and explains how to build a time machine. It takes you several years of hard work, but you eventually manage to construct a working version of the machine. You use it to lead a fascinating life of time travel. Toward the end of your life you use your time machine to travel to the eve of your 21st birthday and explain to your younger self how to build a time machine.

According to this narrative, there are causal loops. (For instance: your building a time machine late in life is a cause of your traveling back in time to give your younger self information about how to build a time machine, which is a cause of your building a time machine late in life.) That means that the scenario depicted by the story is weird. But weirdness is not the same as logical inconsistency and what's at issue here is whether the story is logically consistent. In spite of its causal loops, there is nothing logically inconsistent about our story: there is no time $t$ and proposition $p$ such that at one point in the narrative we are told that $p$ is the case at $t$ and at a different point in the narrative we are told that $p$ is not the case at $t$.

### Exercises

Determine whether each of the following stories can be interpreted as consistent (ordinary time, one-world) time travel stories:

1. You are an avid gambler. You are at the races, and decide to bet all your savings on Easy Goer. Tragically, Master Charlie wins the race by a nose. Because of your recklessness, you end up leading a life of hardship. Toward the end of your life, however, you catch a lucky break: an old aunt bequeathes you a time machine. You travel back to the time you had placed the disastrous bet and head straight for the tracks. You see your younger self standing in line, eagerly awaiting the chance to bet on Easy Goer. You approach him and whisper into his ear, "Master Charlie." Your younger self is receptive, and bets his money on Master Charlie. He wins big and spends the rest of his life in luxury and style.
2. *Pixit*, the innovative pig-sitting app, is the latest start-up to disrupt the pet-sitting industry. You travel back in time to a time when *Pixit* shares were cheap and buy as many as you can. You place the share certificates in a box and bury the box in an empty lot—the empty lot where you know your home will be built a few years later. You then travel back to the present day, run to your back yard, and dig up the chest. The certificates are a little moldy, but they're still there!

## 4.2 The Grandfather Paradox

Bruno hates his grandfather. Grandfather was a gambler and a drunk, and he orchestrated a large-scale fraud that left hundreds of people in poverty. Nothing would please Bruno more than killing Grandfather. But Grandfather died many years ago, long before Bruno was born.

Bruno is undeterred. He builds a time machine and travels to September 13, 1937, a time before Grandfather had children and committed the fraud. Grandfather is on his morning walk. Bruno has climbed up a church steeple and positioned himself at the belfry with a sniper's rifle. He aims the rifle with the precision of an expert gunman. Grandfather stops to tie his shoelaces. The church bells toll. Noon has arrived. Bruno caresses the trigger. Grandfather stands still for a moment. A breeze ruffles some nearby leaves. Bruno prepares to shoot....What happens next?

Suppose Bruno kills Grandfather. Then Grandfather will never have children. (Assume no funny business: no rising from the dead, no frozen sperm, etc.) So Bruno's mother will never be born. So Bruno will never be born. So he won't go back in time. So he won't kill Grandfather after all. Contradiction! But if Bruno doesn't kill Grandfather, what stops him? There is Bruno on the belfry with his rifle. Killing Grandfather is his lifelong ambition. He has a clear shot....

We seem to have been left with a paradox. On the one hand, we know that Bruno won't succeed, on pain of contradiction. On the other, we're at a loss to explain what could possibly stop him.

### 4.2.1 Assessment

What should we conclude from the Grandfather Paradox?

One possible conclusion is that time travel is an inherently contradictory idea, like a figure that is both a circle and a square. I think that would be a mistake. The key observation is that there are perfectly consistent ways of filling in the details in Bruno's story. Suppose, for example, that we continue the story by saying that Bruno loses his nerve at the last minute and puts down the gun. Or by saying that Bruno gets distracted by a barking dog and aims a little too far to the right. In each case we have a consistent time travel story and in each case we have a perfectly acceptable explanation of why Bruno's assassination attempt fails.

You might think that this misses the point. Perhaps the real problem is not to explain why the assassination attempt fails in particular versions of the story. Perhaps the real problem is to explain why it *must* fail: why it fails in *any* consistent way to fill out the details. It seems to me, however, that there is a fairly flat-footed explanation to be given.

Let me begin with an analogy. Suppose you are allowed to paint each of the points on a sheet of paper any color you like. Could you draw a figure that is both a circle and a square? You cannot. And there is no reason to expect an interesting explanation of *why* you cannot. Presumably all there is to be said is that no distribution of ink-blots on a page yields a figure that is both a circle and a square, and therefore any attempt to draw such a figure will fail. Now consider a second analogy: imagine that God makes the world by starting with a grid of spacetime points and "drawing" on the grid by deciding which physical properties to fill each spacetime point with. In some regions, God "draws" positive charge; in some, She "draws" mass. Could God "draw" a world in which Bruno kills Grandfather before Grandfather has children? She cannot. And, as before, there is no reason to expect an interesting explanation of why She cannot. Perhaps all there is to be said is that no distribution of physical properties across spacetime yields a scenario in which a future grandfather is killed before having children, and so any attempt to "draw" such a scenario will fail.

With this as our background, let us return to our question: Why does Bruno's assassination attempt fail in *any* consistent way of filling out the details? I see no reason to expect a more interesting explanation than the following: We have seen that there is no coherent scenario—no world God could "draw"—in which Bruno kills Grandfather before Grandfather has children. So any coherent way of filling out the details of the story must be such that the assassination attempt fails.

I do not mean to suggest that these considerations set the Grandfather Paradox to rest, but I do think they show that explaining why the paradox is supposed to be paradoxical is not as easy as one might think. We have seen that there is no reason to think that a fully spelled-out version of the story will fail to explain why Bruno's assassination attempt goes wrong. And we have seen that there is a (fairly uninteresting) explanation of why the assassination attempt will go wrong on any way of spelling out the details. So what is the problem supposed to be? I will use the remainder of this chapter to explore two possible answers:

**Possible answer 1** The Grandfather Paradox is problematic because it raises questions about whether the laws of physics could rule out paradoxical time travel in a principled way, without banning it altogether.

**Possible answer 2** The Grandfather Paradox is problematic because it shows that time travel is incompatible with free will.

I'll discuss each of them in turn.

## 4.3 Time Travel and Physical Law

What might a system of laws look like if it were to allow for time travel and do so in a way that steers clear of paradox?

It is easy to imagine a system of laws that avoids paradox by banning time travel altogether. It is also easy to imagine a system of laws that avoids paradox in an unprincipled way: the system could, for example, postulate an "anti-paradox force" to deflect Bruno's bullet before it hits Grandfather; or it could disallow initial conditions that lead to problematic instances of time travel. That's not what we're after here. What we'd like to understand is whether there could be a system of laws that allowed for interesting forms of time travel but managed to avoid paradox in a *principled* way.

We will explore these issues by considering a toy model: a system of laws that allows for interesting forms of time travel but is much simpler than the actual laws of physics. This will allow us to bring some of the key issues into focus while avoiding the complexities of contemporary physics. (As it turns out, there are solutions to Einstein's equations that allow for time travel, in the sense that there are paths through spacetime that form a closed loop and do not require traveling faster than the speed of light. If you'd like to know more, have a look at the recommended readings in section 4.6.)

### 4.3.1 The Toy Model[2]

The system of laws I would like to discuss comes from philosophers Frank Arntzenius and Tim Maudlin. It governs a world with two dimensions: a temporal dimension and a single spatial dimension.

2. An earlier version of this material appeared in Rayo, "¿Cómo sería el mundo si pudiéramos viajar al pasado?" *Investigación y Ciencia*, February 2016.

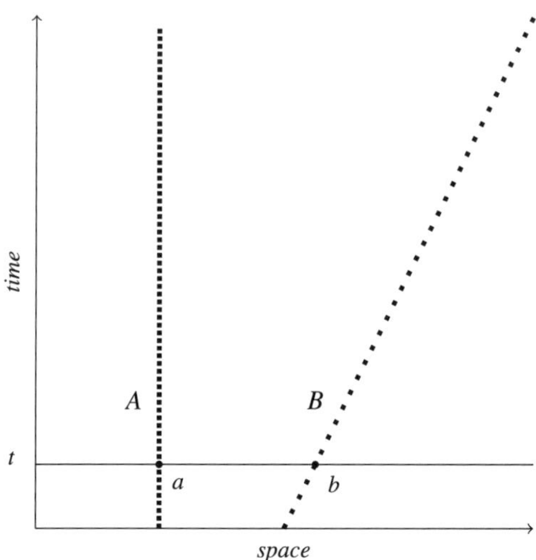


**Figure 4.2**
*A* is stationary; *B* is moving rightward at a constant speed.

We will represent our world's spacetime using two-dimensional diagrams like the one in figure 4.2. The *x*-axis represents a spatial dimension. The *y*-axis represents a temporal dimension, with earlier times closer to the origin. Events that take place at time *t* correspond to points on the dotted line. Particle *A*'s location at *t* is labeled *a*, and particle *B*'s position at time *t* is labeled *b*. *A* is at rest, so its spatiotemporal trajectory is represented as a perfectly vertical line; *B* is moving rightward at a constant speed, so its spatiotemporal trajectory is represented by a diagonal line. (Slower speeds are represented by steeper diagonals.)

We shall assume that our particles obey two exceedingly simple dynamical laws:

**Law 1** In the absence of collisions, a particle's velocity remains constant.

(This is our version of Newton's First Law of Motion. We represent it in our diagrams by ensuring that the spatiotemporal trajectory of a freely moving particle is always a straight line, as in figure 4.2. Although this will not be represented in our diagrams, we will assume that the relevant trajectories are infinite in both directions.)

**Law 2** When two particles collide, they exchange velocities. (There are no collisions involving more than two particles.)

(This is our version of Newton's Third Law of Motion. We represent it in our diagrams by ensuring that the spatiotemporal trajectories of colliding particles always form an *X*, as in figure 4.3.)

Now imagine that our world contains a **wormhole**, which causes different regions of our diagram to represent the same region of spacetime. To see what this means, consider

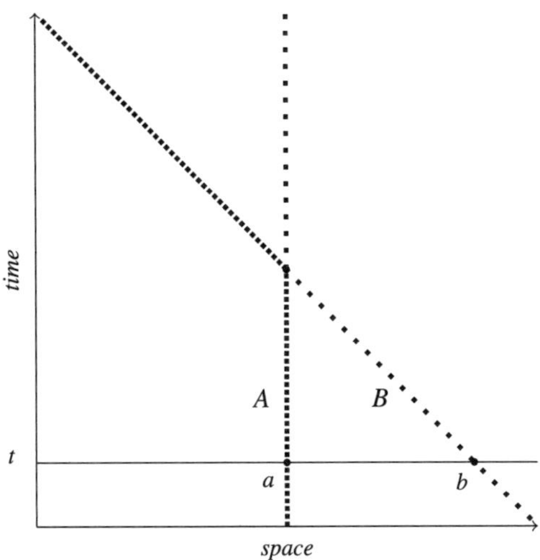


**Figure 4.3**
*A* and *B* collide, exchanging velocities.

the wormhole of figure 4.4, in which the spacetime points on line *W*- are identified with the spacetime points on line *W*+. This identification has two consequences:

1. When a particle approaches line *W*- from below, it continues its upward trajectory from line *W*+ with no change in velocity. (See object *A* in figure 4.4.)
   In other words, when an object enters *W*- from below, it leaps forward in time by exiting the wormhole at the corresponding point on *W*+.
2. When a particle approaches line *W*+ from below, it continues its upward trajectory from line *W*- with no change in velocity. (See object *B* in figure 4.4.)
   In other words, when an object enters *W*+ from below, it leaps backwards in time by exiting the wormhole at the corresponding point on *W*-.

Our toy physical model is now in place. Let us use it to illustrate some surprising consequences of time travel as it occurs in this setting.

**4.3.1.1 Indeterminacy** In the absence of wormholes, our world is fully deterministic. In other words, one can use a specification of the positions and velocities of the world's particles at any given time to determine the positions and velocities of the world's particles at any other time. This is illustrated in figure 4.3. At time *t*, *A* is at rest and occupies spacetime point *a*, and *B* is traveling leftward with speed 1 and occupies spacetime point *b*. To determine the position and velocity of *A* and *B* at other times, we draw a straight line through each of *a* and *b*, at an angle corresponding to the velocity of the object at *t*. As long as no collisions occur, the spacetime trajectory of a

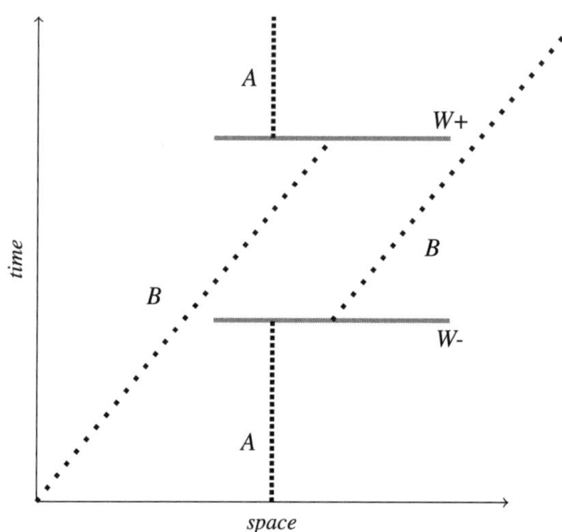


**Figure 4.4**
A wormhole in which the points represented by *W-* are identified with the points represented by *W+*. *A* jumps to the future when its spacetime trajectory reaches a point at which the wormhole is active; *B* jumps to the past when its spacetime trajectory reaches a spacetime point at which the wormhole is active.

particle is given by the straight line that intersects its position at *t*. When a collision takes place, the spacetime trajectories of the two particles swap, with each continuing along a straight line corresponding to the trajectory of the other prior to the collision. The spacetime trajectory of a particle can then be used to determine its position and velocity at any given time. The particle's position is given by the point at which the particle's spacetime trajectory intersects the horizontal line corresponding to the relevant time; the velocity is given by the angle of the particle's spacetime trajectory at that point.

Now for the interesting part: when a wormhole is introduced, determinism is lost. Consider, for example, the wormhole in figure 4.5. At time *t*, no particles exist. If we had determinism, this would be all the information we would need to determine how things stand at every other time. In particular, we should be able to figure out how many particles exist in the "wormhole region" (i.e., the spacetime region between *W-* and *W+*). But our laws do not determine an answer to this question. As far as the laws go, there could be no particles in the wormhole region, or there could be one or two or any number whatsoever, as long as there weren't too many to fit into spacetime. (Figure 4.5 depicts a case in which there are exactly two particles in the wormhole region.)

**4.3.1.2 Paradox** Now suppose that our world contains a "mirror": a stationary object that reflects particles by inverting their velocity.

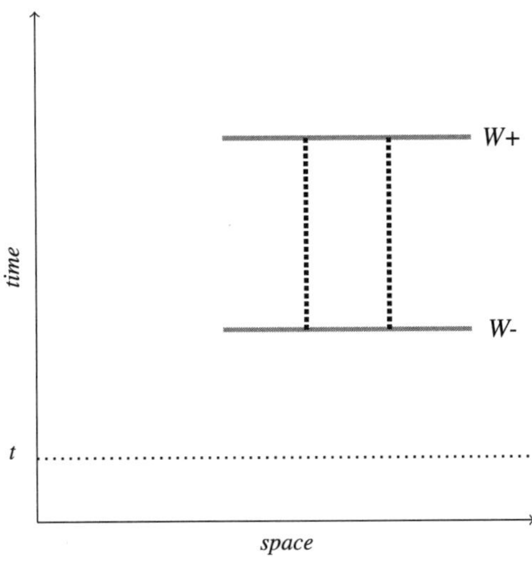


**Figure 4.5**
There is no way of predicting how many particles will exist between *W*- and *W*+ using only information about time *t*.

We will use the mirror to construct a version of the Grandfather Paradox in our toy world. The relevant scenario is depicted in figure 4.6. Particle *A* is on a "paradoxical path." It travels rightward, passes through spacetime point *a*, and enters the wormhole at spacetime point *b*, jumping to the past. It exits the wormhole and continues its rightward trajectory until it reaches the mirror at spacetime point *c*. But what happens next? It is not clear that we could draw out *A*'s trajectory in full without violating the dynamical laws of our toy model. Notice, in particular, that if we attempted to complete *A*'s trajectory as depicted in figure 4.7, particle *A* would be blocked from entering the wormhole region at point *a* by a future version of itself. (Not just that: it would be blocked from crossing point *b* on its way to the mirror by a later version of itself, and it would get blocked from crossing point *b* on its way back from the mirror by an earlier version of itself.)

So does particle *A* of figure 4.6 succeed in entering the wormhole region or not? If it does enter, it shouldn't have, since its entry should have been blocked by a future version of itself. And if it doesn't, it should have, since there is nothing to block it. In other words, the scenario depicted by figure 4.6 is paradoxical, given the laws of our toy model.

Notice, however, that the laws do allow for a situation in which particle *A* starts out traveling on the "paradoxical path" of figure 4.6 but is prevented from entering the wormhole region by a *different* particle. One way in which this can happen is depicted in figure 4.8. Note that figure 4.8 is very much like figure 4.7, but it represents a situation

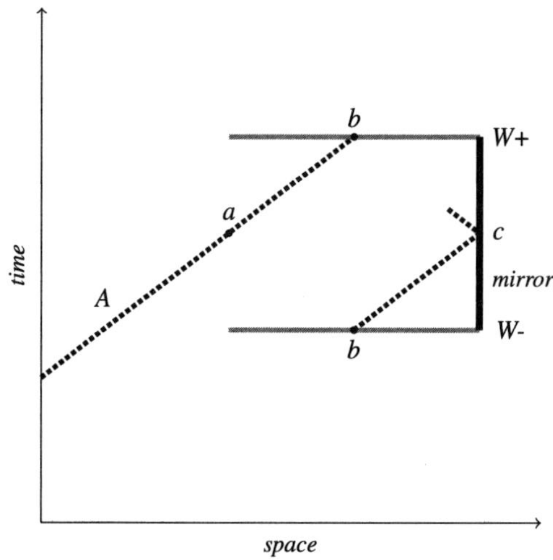


**Figure 4.6**
After passing through point *a* and crossing through the wormhole at point *b*, particle *A* is reflected by a mirror at point *c*.

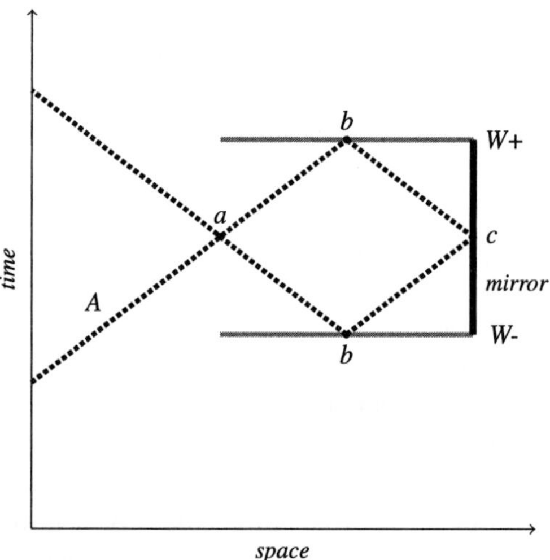


**Figure 4.7**
Particle *A* is prevented from entering the wormhole region by a future version of itself.

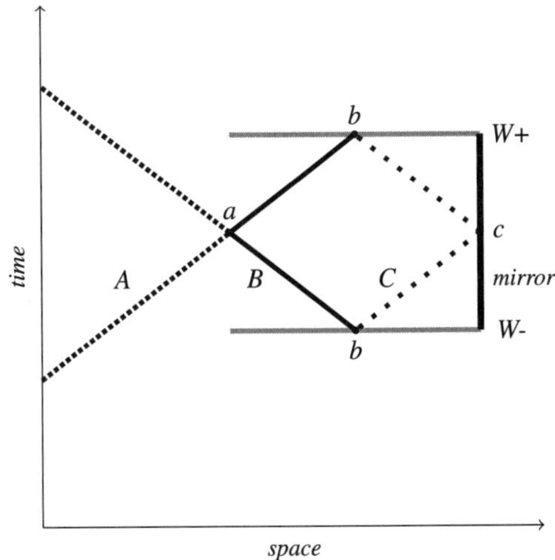


**Figure 4.8**
Particle *A* fails to enter the wormhole region after colliding with particle *B* at spacetime point *a*. Particles *B* and *C* are each caught in a loop within the wormhole.

in which there are two additional particles living within the wormhole region and one of them blocks particle *A* from entering. (Think of this as analogous to a version of Bruno's story, in which consistency is achieved because something blocks Bruno from carrying out his assassination attempt.)

Here is a more detailed description of the behavior of each of the particles in figure 4.8:

- Particle *A* moves rightward until it reaches spacetime point *a*, where it collides with *B* and bounces off, moving leftward.
- Particle *B* is trapped in a loop. It departs spacetime point *b* at the time of *W*-, moving leftward until it collides with particle *A* at spacetime point *a*. It then bounces rightward until it reaches spacetime point *b* at the time of *W*+. Two things happen next. First, particle *B* collides with particle *C* and bounces off leftward. Second, *B* enters the wormhole and jumps back in time.
- Particle *C* is also trapped in a loop. It departs spacetime point *b* at the time of *W*-, moving rightward until it collides with the mirror at spacetime point *c*. It then bounces leftward until it reaches spacetime point *b* at the time of *W*+. Two things happen next. First, particle *C* collides with particle *B* and bounces off rightward. Second, *C* enters the wormhole and jumps back in time.

Figure 4.8 demonstrates that our toy model allows for consistent scenarios in which particle *A* starts out on the "paradoxical path" of figure 4.6. But one might worry that there is something artificial about this way of escaping inconsistency. For one might think it amounts to the postulation of a "pre-established harmony," wherein there happens to be an additional particle in just the right place to avert paradox. Such harmony could certainly be imposed by restricting the range of admissible "initial conditions" of our toy model so as to allow for situations in which particle *A* starts out on a paradoxical path only when another particle is ready to block its entry to the wormhole region. But the resulting theory would be horribly unprincipled. (It would be a bit like saying that we should allow for "initial conditions" wherein Bruno goes back in time to kill Grandfather but only if the circumstances are such as to ensure that the assassination attempt would be derailed.)

When one thinks of the toy model in the right kind of way, however, there is no need for a postulation of pre-established harmony. The trick is to be careful about how one characterizes one's worlds. One does not characterize a world by *first* deciding how many particles the world is to contain (assigning each a position and velocity, one at a time) and *then* using the dynamical laws to calculate the particles' spacetime trajectories. Instead, one characterizes a world by *first* drawing a family of spacetime trajectories that conform to the dynamical laws and *then* using the laws to determine how many particles the resulting world must contain. In this way of thinking, it is a mistake to suppose that one can characterize a world by stipulating that it is to contain a single particle traveling as in figure 4.6 and then asking what happens when the dynamical laws are used to calculate the particle's spacetime trajectory. Instead, one draws a family of spacetime trajectories, as depicted in figures 4.7 and 4.8, and then uses the laws to conclude that a world with those spacetime trajectories must contain three different objects (as shown in figure 4.8).

This delivers a very satisfying result. We get a system of laws that allows for interesting forms of time travel and yet has a principled way of avoiding paradox. Notice, moreover, that one is able to explain why there is no world in which particle *A* completes a paradoxical path. The explanation is analogous to our earlier explanation of why one cannot draw a figure that is both a circle and a square. Just as there is no distribution of ink-blots on a page that yields a figure that is both a circle and a square, so there is no lawful distribution of spacetime trajectories in our toy model such that both a particle is on a paradoxical path and there isn't another particle there to block it.

There is a final point I would like to emphasize before bringing our discussion of the toy model to a close. We have seen that our laws allow for the situation depicted in figure 4.8, in which particle *A* starts out on a paradoxical path but is prevented from entering the wormhole region by another particle. It is essential to keep in mind that the presence of the additional particles in the wormhole is not *caused* by particle *A*'s paradoxical path. The additional particles are not an "anti-paradox" mechanism that is activated by particles on paradoxical paths. What is going on is simply that there is no lawful distribution of spacetime trajectories such that a particle is on a paradoxical path and there isn't another particle there to block it.

**4.3.1.3 Conclusion** We have been trying to get clear on why the Grandfather Paradox is supposed to be problematic. We considered whether the problem is that the Grandfather Paradox makes it hard to see how a system of laws could rule out paradoxical time travel in a principled way, without banning it altogether. We tackled the question in a simplified case, identifying a toy model that appears to allow for interesting forms of time travel while avoiding paradox in a principled way.

### Exercises

1. I noted earlier that the number of particles living in the wormhole region between *W*- and *W*+ is not settled by facts prior to the appearance of the wormhole. For this reason, the scenario depicted in figure 4.8 is not the only way to restore consistency to a situation in which particle *A* is the only particle that exists prior to the appearance of the wormhole and is traveling at constant speed on a path that would (in the absence of collisions) lead to *b* via *a*. Describe an additional resolution.

## 4.4 Free Will

I would like to end this chapter by considering the question of whether the Grandfather Paradox shows that time travel is incompatible with free will.

Let us start with a preliminary question: *What is it to act freely?* The Control Hypothesis is the hypothesis that to act freely is to be such that one *is in a position to act otherwise.* A little more precisely:

**Control Hypothesis** An agent acts freely in doing *X* if and only if (i) she does *X* by making a certain decision, and (ii) she is in a position to do something other than *X* by making a different decision.

As its name suggests, the Control Hypothesis is meant to capture the idea that someone who acts freely has *control* over the action she performs. We'll see below that there are good reasons for thinking that the Control Hypothesis is incorrect. But it is a good starting point for elucidating the connection between time travel and free will, so we'll treat it as our working hypothesis for now.

With the Control Hypothesis as our background, I will consider a couple of arguments purporting to show that Bruno fails to act freely because he is not in a position to make a different decision about how to take his shot.

### 4.4.1 The First Argument

Suppose that Bruno's assassination attempt was well-documented in the local press. It is known that the would-be assassin fired a single shot. The bullet missed Grandfather—but only barely: it burnt a scar on his left cheek.

Bruno has seen that scar in photographs on countless occasions. He can remember it clearly. "A hideous scar on a hideous face," he thinks, as he takes his position at the belfry and loads a single bullet into his rifle. He prepares to take aim and remembers one of the old newspaper articles about the shooting: "The assassination attempt occurred at noon. The church bells were tolling. A single shot was fired. The bullet missed but only barely, burning a scar on the victim's left cheek." Bruno's train of thought is interrupted by the tolling bells. Noon has arrived. Bruno caresses the trigger. Grandfather stands still for a moment as a breeze ruffles some nearby leaves. Bruno prepares his shot....

In this version of the story, we know exactly how Bruno will act: he will take a single shot and miss—but only barely, burning a scar on Grandfather's left cheek. This means that *what we know* about the rest of the story is incompatible with Bruno's killing Grandfather. It is tempting to conclude from this that Bruno was not in a position to do otherwise and to use this conclusion to argue—via the Control Hypothesis—that he did not act freely. It seems to me, however, that that would be a mistake.

Let me begin with an analogy. You meet your friend Susan for breakfast in New York. She tells you that she decided to leave on a train trip to Alaska last night and set off to the train station. Susan has yet to end her story but you already know that her attempt to get to Alaska by train was unsuccessful. How do you know this? Because here she is in New York the next morning, telling you her story. And you know that it takes several days to get to Alaska by train. So *what you know* about the actual situation is incompatible with her making a train trip to Alaska.

Notice, however, that so far, nothing entails that Susan did not act freely when she failed to leave town. Although Susan's story could go on to reveal that she remained in New York against her will—perhaps she was prevented from getting on the train by an anxious friend—it could also reveal that she acted freely: perhaps she changed her mind at the last minute. The key feature of the case is that, even before Susan gets to the end of her story, you have information about the story's future: you know that Susan will somehow fail to make a train trip to Alaska. But what is relevant to free will, according to the Control Hypothesis, is whether Susan was in a position to go to Alaska by deciding differently, regardless whether you—who are with her in New York the next day—happen to know that things won't turn out that way.

Now return to the case of Bruno and his Grandfather. We—who live in the present day—have information about Grandfather's future: we know that he will, in fact, live long enough to have children and, therefore, that Bruno's assassination attempt will fail. But, according to the Control Hypothesis, this information is not relevant to the question of whether Bruno acted freely. What matters is whether Bruno was in a position to kill Grandfather, regardless of whether we—who live in the present day—happen to know that things won't turn out that way.

## Exercises

1. Suppose that Susan interrupts her story to take a phone call. She's told you that she went to the train station and bought her ticket, but she has not yet revealed whether she set foot on the train. Should you accept the following conditional?

   If Susan set foot on the train, she managed to make the trip to Alaska.

2. You later learn that Susan did not, in fact, set foot on the train. Would this be enough to justify accepting the following conditional?

   Had Susan set foot on the train, she would have made the trip to Alaska.

3. Here is the story of Death in Damascus:[3]

   One day a traveling merchant met Death in the souks of Damascus. Death appeared surprised, but she quickly recovered her characteristic cool and intoned with cadaverous solemnity, "Prepare yourself; I am coming for you tomorrow."

   The merchant was terrified and fled that very night to Aleppo.

   The next day, the merchant woke up and—horror of horrors!—found Death at her bedside. Her voice quaking, she managed to squeak, "I thought you were looking for me in Damascus!"

   "No, I was merely shopping in Damascus," said Death. "That's why I was surprised to see you: it is written that our final meeting is in Aleppo."

   Could the merchant have stayed in Damascus and avoided her date with Death?

   The answer is not fully specified by the story so far, but let me tell you about a way of interpreting the story so that, yes, the merchant could have avoided her date with Death; it's just that she wasn't going to, and Death knew this.

   Think of the merchant as analogous to your friend Susan, who was thinking about taking a train trip to Alaska, and think of Death as analogous to you. Death knows that she will meet the merchant on the appointed date the way that you know your friend will fail to make the trip to Alaska: you both have information about the future. But it doesn't follow that the merchant wasn't in a position to avoid meeting Death in Aleppo, just as it doesn't follow that Susan wasn't in a position to take the train to Alaska.

   Let us suppose that if the merchant had decided to stay in Damascus, she would have succeeded in staying. Does the Control Hypothesis entail that she acted freely in traveling to Aleppo?

---

3. Based on text by Damien Rochford, which draws from Gibbard and Harper (1978).

### 4.4.2 The Second Argument

I would now like to consider a second strategy for arguing that Bruno does not act freely when he fails to kill Grandfather. Let us start by going back to Susan and her unsuccessful attempt to travel by train to Alaska, but now let's add a new element to the story. Suppose Susan was prevented from leaving by an external influence. She was being monitored by a powerful enemy, who is committed to keeping her in New York. As Susan was about to board the train, the enemy arranged for the train to be canceled. In fact, the enemy wouldn't have stopped there: he would have thwarted any other means by which Susan could leave New York.

In this version of the story, Susan is not in a position to make a train trip to Alaska. This has nothing to do whether whether we—who are having breakfast with Susan in New York the next morning—happen to have information about how Susan's story will, in fact, turn out. Instead, it has to do with whether Susan was *in control* of the outcome of the story. And, of course, Susan was not in control of the outcome in this case: the enemy would have made sure that any effort to leave New York was thwarted. So the Control Hypothesis entails that Susan did not act freely when she stayed in New York.

The argument I consider is based on the idea that something similar should be said of Bruno and his unsuccessful attempt to kill Grandfather. The claim is that Bruno fails to act freely because he faces an analogue of Susan's enemy. If Bruno were ever on track to kill Grandfather, the laws of physics would intervene to derail him, much like the enemy intervened to derail Susan's attempt to leave New York.

To assess this argument, start by thinking about physical law.

**4.4.2.1 Two conceptions of physical law** For a system of laws to be **determinisitic** is for it to entail, on the basis of a full specification of the state of the world at any given time, a full specification of the state of the world at any later time.

There is a temptation to think that possessing free will is simply a matter of having a decision-making process that is not subject to deterministic laws. As it turns out, however, the relationship between determinism and free will is not as straightforward as that. To see this, imagine that your brain is equipped with a "quantum module," which uses indeterministic quantum phenomena to issue "yes" or "no" outputs entirely at random. Imagine, moreover, that you make your choices by consulting your quantum module and acting in accordance with its randomly generated output. Under such circumstances, would you be acting freely? Certainly not. Far from being free, you would be a slave to the random outputs of your quantum module.

What then is the relationship between determinism and free will? This is a complex question, the answer to which philosophers disagree on. (For recommended readings over this topic, see section 4.6.) I won't attempt to reproduce the debate here, except to point out that whether or not one thinks determinism is compatible with free will

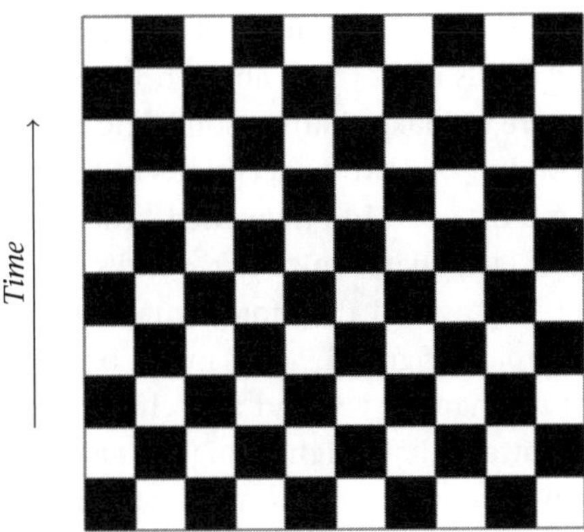


**Figure 4.9**
In the above mosaic, neighboring cells are never colored alike. This pattern can determine, based on the color and location of any given cell, the color distribution of the rest of the mosaic. This means, in particular, that when the $y$-axis is thought of as a temporal dimension, with each row of the mosaic corresponding to one instant of time, the pattern can be used to characterize a deterministic law that determines, based on a specification of the mosaic at any given time, a full specification of the mosaic at every later time.

might depend on one's conception of physical law. Broadly speaking, there are two different conceptions to consider:

1. The laws tell us what *will* happen, on the basis of what has happened.
2. The laws tell us what *must* happen, on the basis of what has happened.

To get a sense of the difference, think of the world as a "mosaic" of events distributed across spacetime. In an orderly world, such as our own, the distribution of properties across the mosaic is subject to patterns. These patterns can be used to make projections. Patterns can be used to determine, based on information about one region of the mosaic, what other regions of the mosaic are like. Consider, for example, the mosaic in figure 4.9. It consists of square cells, each painted black or white. The coloring of the mosaic satisfies a pattern: neighboring cells are never colored alike. This pattern allows for very strong projections. For instance, it can determine, based on the color and location of any given cell, the color distribution of the rest of the mosaic.

In the first of our two conceptions of the physical laws—a law tells us what *will* happen based on what has happened—a system of laws is nothing more than a description of the patterns that, as a matter of fact, characterize our world's mosaic. In other words, a law is just a device for inferring what one region of the mosaic is actually like based on information about what another region is actually like. It tells us nothing about what a given region of the mosaic *would* look like if were to make changes somewhere else.

(For example, learning that the mosaic in figure 4.9 is, in fact, such that neighboring cells are never colored alike tells us nothing about how the cells in one region of the grid would change if we were to make changes somewhere else.)

In contrast, in the second of our two conceptions of the physical laws—a law tells us what *must* happen based on what has happened—the laws don't just describe the patterns that, as a matter of fact, characterize our world's mosaic; they also tell us that those patterns *must* hold, and would therefore remain in place if we were to modify the mosaic. (In the example of figure 4.9, this might be cashed out as the claim that the color of a cell cannot be changed without also changing the colors of other cells, so as to keep in place the pattern that neighboring cells are never colored alike.)

In this second conception of physical law, it is natural to think that determinism is incompatible with free will. For a deterministic world will be a world in which the laws make it *impossible* to do something other than what we actually do, assuming the initial conditions of the universe remain fixed. As far as I can tell, however, there is nothing in contemporary physics that encourages such a conception of physical law. Physics is in the business of uncovering the patterns that characterize the world as we find it. But, as far as I can tell, it remains silent on the more philosophical question of whether the relevant patterns *must* hold.

In our first conception of physical law, in contrast, it is not obvious that we should think of determinism as incompatible with free will. Assume that the Control Hypothesis is correct. The fact that the world as it actually is instantiates a particular pattern does not obviously entail that we aren't in a position to decide differently than we actually did. And this is so even if the pattern in question is highly projectable and can be used to determine, based on a full description of the mosaic at one particular time, what the mosaic is like at any future time. So there is no obvious tension between determinism and free will.

**4.4.2.2 Back to the argument** Let us now return to our second argument for the conclusion that Bruno fails to act freely when he misses his shot. The claim, recall, is that Bruno faces an analogue of Susan's enemy in the laws of physics. He fails to act freely because the laws would intervene to derail him if he were ever on track to kill Grandfather.

I would like to suggest that the lesson of the preceding discussion is that whether or not it is right to think of the physical laws as an analogue of Susan's enemy depends on one's conception of physical law. In the second conception, a law tells us what *must* happen on the basis of what has happened so it is indeed the case that the laws make it impossible for Bruno to act otherwise (assuming the initial conditions of the universe remain fixed), much the way Susan's enemy makes it impossible for Susan to leave town. But in the first conception, a law tells us what *will* happen on the basis of what has happened so the laws are *not* analogous to Susan's enemy: they are simply descriptions of the patterns that, as a matter of fact, characterize our world's mosaic.

Before bringing this section to a close, I'd like to ask you to think back to the toy model of section 4.3.1. First, note that we were given no reason to think of the laws of the toy model as telling us what *must* happen on the basis of what has happened, as opposed to simply telling us what *will* happen on the basis of what has happened. Next, recall that the laws of the model entail that any situation in which particle *A* is on a "paradoxical path" is also a situation in which there are additional particles living inside the wormhole region, and it is a situation in which one of these particles collides with particle *A* before it is able to enter the wormhole, averting the paradox. Notice, however, that the additional particles in this scenario are not exactly analogous to the enemy in Susan's story. The enemy's actions are partly *caused* by Susan's decisions: the enemy interferes with the train *because* he sees that Susan is about to board. But, as noted in section 4.3.1, we have been given no reason to think that the presence of the additional particles in the wormhole are caused by particle *A*'s paradoxical path. The laws tell us that the additional particles will be in the wormhole if particle *A* is on the paradoxical path, but they do not tell us that their presence in the wormhole is *caused* by particle *A*'s paradoxical path.

This difference is significant because it affects our judgements about counterfactual scenarios. Suppose that, as a matter of fact, particle *A* is not on a paradoxical path, and that, as a matter of fact, there are no particles within the wormhole region. What would have happened if we had altered particle *A*'s velocity slightly, so as to put it in a paradoxical path? Don't say that additional particles would have appeared in the wormhole. That assumes that the presence of the additional particles is caused by particle *A*'s trajectory, and we have been given no reason to think that it is. What we should say instead is that the laws of the toy model entail that we will never be in a situation in which particle *A* is in a paradoxical path without anything to stop it within the wormhole region. So if we were to alter particle *A*'s velocity in the relevant way, we would end up in a situation that cannot be accounted for within our model.

When we think of the laws of physics as telling us what *will* happen, as opposed to what *must* happen, it is natural to think that Bruno's situation is akin to particle *A*'s. We know that any scenario consistent with physical law in which Bruno travels back in time in an attempt to kill his childless Grandfather is also a scenario in which his efforts are defeated somehow. Suppose that, in fact, his efforts are defeated because he aims a little too far to the right. What would have happened if Bruno had aimed just right? Don't say that some additional defeater would have appeared and saved Grandfather. That assumes that aiming just right would have *caused* the additional defeater to come about, and we have been given no reason to think that such a causal structure is in place. What we should say instead is that if Bruno had managed to aim just right, we would have ended up in a situation that cannot be accounted for while keeping the rest of the story fixed.

The upshot of our discussion is that it is not clear that we should think of Bruno's situation as analogous to Susan's, when she is the victim of an enemy bent on preventing her departure. So even if it is clear that Susan does not act freely when she is thwarted by the enemy, it is a mistake to conclude on that basis that Bruno does not act freely when he fails to kill Grandfather.

### 4.4.3 Reasons to Doubt the Control Hypothesis

To assess whether Bruno is acting freely, we have been working with the Control Hypothesis: the view that to act freely is to be in a position to act otherwise.

As noted earlier, however, it is not clear that the Control Hypothesis is correct. The best way to see this is to consider a slight variation of Susan's story, which is adapted from a famous example provided by Princeton philosopher Harry Frankfurt. As before, we will assume that Susan is on her way to board a train to Alaska. As before, we will assume that Susan has a powerful enemy, who is committed to keeping her in New York by any means necessary. And, as before, we will assume that Susan ends up staying in New York. This time, however, we will assume that the enemy never interferes. He doesn't have to. Susan changes her mind at the last minute and heads home on her own volition. So the enemy doesn't arrange for the train to be canceled. He would have certainly interfered had Susan gotten any closer. But, as a matter of fact, he did nothing at all.

In this scenario, Susan acts freely when she fails to leave New York, since she heads home on her own volition, with no outside interference. But the Control Hypothesis entails, incorrectly, that Susan does not act freely. For Susan was not in a position to leave New York by deciding differently: had she decided differently, the enemy would have interfered, and her efforts would have been thwarted.

Following Frankfurt, we can even imagine a scenario in which Susan not only fails to be in a position to *act* differently, but also to *decide* differently. Suppose the enemy interferes not by canceling the train, but by meddling with her decision-making. He is able to keep track of Susan's deliberations. As long as Susan makes no decision to leave New York, the enemy does nothing. But should Susan ever come close to such a decision, the enemy is ready to interfere by administering a drug that will cause Susan to decide to stay in New York. Now suppose that, as it happens, Susan considers whether to leave New York, and she decides to stay. Then she will have acted freely in staying, even though she won't have been in a position to make a different decision, since the enemy would have interfered. But the Control Hypothesis entails, incorrectly, that Susan does not act freely, since she was not in a position to decide differently.

## 4.5 Conclusion

If I were to assign the Grandfather Paradox a paradoxicality grade of the kind we discussed in chapter 3, I would assign it a 7. I don't think the paradox shows that there is anything inherently inconsistent about time travel, but I do think it helps highlight some of its hidden complexities.

First, the Grandfather Paradox brings out just how tricky it is to explain how a consistent system of physical laws might allow for time travel while excluding incoherence. The toy model of section 4.3.1 illustrates the problem. The laws of this world guarantee that any scenario in which a particle is traveling on a paradoxical path is also a scenario in which the paradox will be blocked by other particles. But we are given no reason to think that the presence of the additional particles is *caused* by a paradoxical path.

Second, the Grandfather Paradox raises the question of whether time travel can bring about failures of free will. We did not resolve this question here, but we made some progress. For it is tempting to think that when Bruno travels back in time to kill Grandfather and fails, he does not act freely. But we have seen that arguing for such a claim is not as easy as one might have thought.

## 4.6 Further Resources

As I mentioned earlier, many of the ideas discussed in this chapter were originally set forth by David Lewis. If you'd like to go into these subjects in more depth, I highly recommend the articles in volume II of his *Philosophical Papers*. Three articles in this collection are of special relevance to the topics we have discussed here: "The Paradoxes of Time Travel," "Are We Free to Break the Laws?," and "Counterfactual Dependence and Time's Arrow."

### 4.6.1 Additional Texts

- For more on why the characterization of time travel I offer above breaks down in relativistic scenarios, see Frank Arntzenius's "Time Travel: Double Your Fun."
- The toy model of section 4.3.1 is drawn from Arntzenius and Tim Maudlin's "Time Travel and Modern Physics" in the *Stanford Encyclopedia of Philosophy*, which is available online. This is also a good place to look if you're interested in ways in which Einstein's equations allow for time travel or in systems of laws that that exclude paradoxical time travel scenarios in interesting ways.
- If you'd like to know more about free will, a good place to start is Michael McKenna and Justin Coates's "Compatibilism," in the *Stanford Encyclopedia of Philosophy*. It's available online and contains an excellent list of references.

- The Death in Damascus story of sectoin 4.4.1, in one version or another, has existed for a long time, but it was introduced to the philosophical literature in a paper on decision theory by Allan Gibbard and William Harper called "Counterfactuals and Two Kinds of Expected Utility."
- For a discussion of the connection between determinism and free will, see Derk Pereboom's "Determinism *Al Dente*."
- The Frankfurt Cases were introduced by Harry Frankfurt in "Alternate Possibilities and Moral Responsibility." Frankfurt's first presentation of his theory of free will is in "Freedom of the Will and the Concept of a Person."

## Appendix: Answers to Exercises

### Section 4.1.2

1. The narrative cannot be interpreted as a consistent time travel story because it makes contradictory claims about your life after the race. Toward the beginning of the narrative we are told that you live in poverty after the race. Toward the end of the narrative we are told that you live in luxury after the race.
2. The narrative is consistent: there is no time $t$ and proposition $p$ such that we are told that $p$ is the case at $t$ at one point in the narrative, and that $p$ is not the case at $t$ at a different point in the narrative.

   One way to bring this out is to add some detail to the narrative. Perhaps your interest in *Pixit* was originally piqued by a news story describing an "angel investor" who seemed to appear out of nowhere and who believed in the company when nobody else would. The investor bought shares by selling a collection of gold coins—a collection not unlike the one you see sitting on your bookshelf.

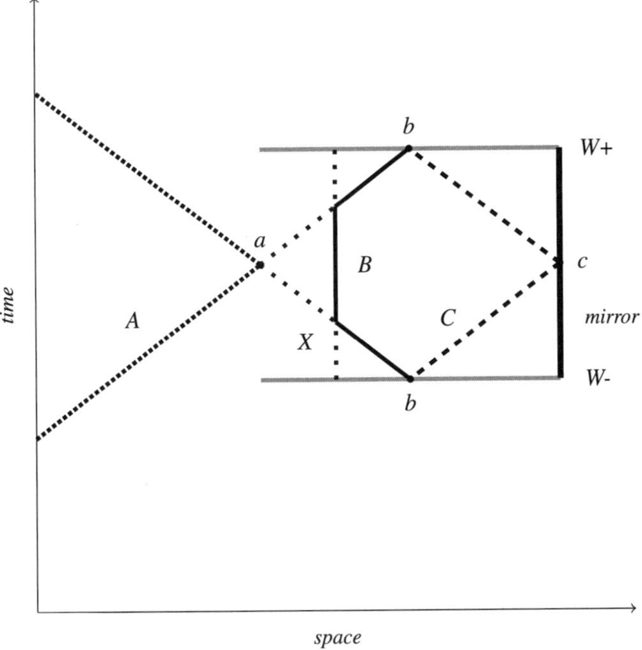


**Figure A.4.1**
A resolution of the paradox with an extra particle, $X$.

### Section 4.3.1.3

1. A resolution with one additional particle is depicted in figure A.1.1.

### Section 4.4.1

1. You should not accept the conditional. You know that your friend is in New York the next morning, because she is right there with you. So if she did set foot on the train, she must have gotten off at some point.
2. It is not clear that you would be justified in accepting the conditional just by learning that she set foot on the train. It is compatible with the story so far that no trains left New York because of a storm. In that scenario, it is not the case that she would have made the trip to Alaska if she had set foot on the train.
3. Yes. When we interpret the story in such a way that the merchant could have decided to stay in Damascus (and succeeded), the Control Hypothesis entails that she acted freely in traveling to Aleppo.

# 5 Newcomb's Problem

In this chapter we'll talk about Newcomb's Problem and show that it has profound implications for the foundations of decision theory. Many of the ideas we'll discuss are drawn from David Lewis.

## 5.1 The Problem[1]

You are led into a room and shown two boxes; a large one and a small one. You're told that the small box contains a thousand dollars. You're not told what the large one contains, but you're told that it either contains a million dollars or is completely empty. You are then offered two choices:

**Two-Box** Keep both boxes.

**One-Box** Keep the large box; leave the small box behind.

The boxes were sealed before you entered the room, and your choice will not cause their contents to change. How should you proceed? Should you one-box or two-box? The answer seems totally obvious. You should take both boxes! How could you possibly benefit from leaving a thousand dollars behind? Whether or not the large box contains a million dollars, you'll end up with more money if you take the small box as well.

### 5.1.1 A Twist

Wait! Let me tell you what happened *before* you entered the room. A couple of weeks ago, a personality expert was handed as much information about you as could be gathered. She was then asked to predict whether you would one-box or two-box. If she concluded that you would one-box, the large box would be filled with a million dollars.

1. An earlier version of this material appeared in Rayo, "El Problema de Newcomb," *Investigación y Ciencia*, September 2008.

If she concluded that you would two-box, the large box would be left empty. In other words:

| Expert's prediction | Contents of large box | Contents of small Box |
|---|---|---|
| One-box | $1,000,000 | $1,000 |
| Two-box | $0 | $1,000 |

The expert issued her prediction last night, and the boxes have been sealed since. If the large box was filled with a million dollars last night, it will continue to hold a million dollars regardless of what you decide. And if it was left empty last night, it will remain empty regardless of what you decide.

One final point: the predictor is known to be highly reliable. She has participated in thousands of experiments of this kind and has made accurate predictions 99% of the time. There is nothing special about your case, so you should think that the expert is 99% likely to issue an accurate prediction about whether you will one-box or two-box.

How should you proceed now that you know about the procedure that was used to fill the boxes? Is it still obvious that you should two-box?

(Keep in mind that the predictor knows that you'll be told how the experiment works and about the method used to decide how to fill the boxes. So she knows that you'll be engaging in just the kind of reasoning that you are engaging in right now!)

### 5.1.2 The Predicament

The fact that the predictor is 99% likely to issue an accurate prediction gives you the following two pieces of information:

1. If you decide to two-box, it is almost certain (99%) that the large box will be empty.
2. If you decide to one-box, it is almost certain (99%) that the large box will contain a million dollars.

Before learning about the expert, it seemed clear that two-boxing was the right thing to do. But now you know that if you one-box you are almost certain to end up with a million dollars, and that if you two-box you are almost certain to end up with just a thousand. Should you be a one-boxer after all?

This problem is often referred to as "Newcomb's Problem" after Lawrence Livermore National Laboratory physicist William Newcomb. It captured the imagination of the philosophical community in the last third of the twentieth century, partly because of the writings of Harvard philosopher Robert Nozick and the legendary *Scientific American* columnist Martin Gardner.

### 5.1.3 Could a Newcomb Predictor Really Exist?

When people are introduced to Newcomb's Problem, they often worry about whether there could really be an expert of the kind the story requires.

It is certainly *logically possible* for there to be such an expert. Consider, for example, a super scientist who creates a molecule-by-molecule duplicate of your body and makes her prediction by allowing your duplicate to experience a Newcomb case and observing the results. As long as you and your duplicate are subjected to identical stimuli, the two of you should carry out the same reasoning and reach the same decision. Even taking quantum effects into account, there is no reason such a predictor couldn't be 99% reliable or more.

Of course, none of this shows that perfect (or near-perfect) predictions are possible in practice. But for our purposes, it doesn't matter if the Newcomb case could be carried out in practice or not. We will be using the Newcomb case as a *thought experiment* to help us understand what rational decision-making is all about. So all we really need is for Newcomb cases to be logically possible.

It is also worth keeping in mind that the discussion below won't require perfect, or even near-perfect, predictions. As you'll be asked to show in exercise 3 of section 5.2.3, anything better than chance will do. And it is certainly possible to imagine real-life predictors who do better than chance. Consider, for example, a predictor who works with identical twins. She starts by testing the Newcomb setup on one of the twins and observing the results. She then predicts that the second twin will choose in the same way as the first. Even if such a predictor isn't perfectly accurate, one might expect her to do better than chance.

#### Exercises

1. I just suggested that it might be possible to generate better-than-chance Newcomb predictions using identical twins. Describe a different method for doing so.

## 5.2 Maximizing Expected Value

In the previous section I tried to give you an intuitive sense of why the Newcomb case might seem puzzling. Now I will give you some mathematical machinery that will help make the puzzle a little more precise. I will introduce you to *decision theory.*

Suppose that you have different options to choose from. According to the most straightforward form of decision theory, you should make your decision in accordance with the following principle:

**Expected Value Maximization** Choose an option whose expected value is at least as high as that of any rival option.

The best way to get a handle on the notion of expected value is to consider an example, and we'll do so shortly. But the intuitive idea is that the expected value of an option is a measure of how desirable the world is expected to be if you choose that option.

### 5.2.1 An Example

Suppose you're wondering what to do tonight. You have two **options**: *go out for drinks* and *study*. There are two relevant **states of the world**: *easy exam* and *hard exam*. With two options and two relevant states of the world, your choice has four possible **outcomes**, which are represented as the empty cells in the following matrix:

| | Easy exam | Hard exam |
|---|---|---|
| Drinks | | |
| Study | | |

How desirable is each of these outcomes? Let us suppose that you only care about two things: having fun and passing the exam. You'll only have fun if you go out for drinks, but you'll only pass a hard exam if you stay home and study. (You'll pass an easy exam regardless of whether you study.) So the desirability of each outcome is as follows:

| | Easy exam | Hard exam |
|---|---|---|
| Drinks | Fun + pass | Fun + no pass |
| Study | No fun + pass | No fun + pass |

Let us now try to put numerical values on the desirability of each outcome. Suppose you think that failing the exam would be twice as bad as passing the exam would be good. So if we assign an arbitrary value of 20 to passing the exam, we should assign a value of −40 to failing the exam. Having fun would be almost as good as passing the exam, so we'll assign it a value of 15. The absence of fun, in contrast, would be only somewhat undesirable, so we'll assign it a value of −2. This allows us to assign a desirability value to each of our four possible outcomes:

| | Easy exam | Hard exam |
|---|---|---|
| Drinks | $15+20$ | $15+(-40)$ |
| Study | $-2+20$ | $-2+20$ |

The next thing we need to do is associate a *probability* with each of the two states of the world that's relevant to our outcome: easy exam and hard exam. Let us suppose that the probability of an easy exam is 80% and, therefore, that the probability of a hard exam is 20%.

We've now assigned a value to each outcome and a probability to each relevant state of the world. We can use this to assign expected values to each of the two options you can choose from: going out for drinks and staying home to study. We do so by taking the *weighted average* of the values of the different possible outcomes, with weights given by the probabilities of the relevant states of the world. The expected value of going out for drinks ($D$) is:

$$EV(D) = \underbrace{35}_{\substack{\text{Value of drinks}\\\text{and an easy}\\\text{exam}}} \cdot \underbrace{0.8}_{\substack{\text{Probability}\\\text{of an easy}\\\text{exam}}} + \underbrace{-25}_{\substack{\text{Value of drinks}\\\text{and a hard}\\\text{exam}}} \cdot \underbrace{0.2}_{\substack{\text{Probability}\\\text{of a hard}\\\text{exam}}} = 23$$

Similarly, the expected value of studying ($S$) is:

$$EV(S) = \underbrace{18}_{\substack{\text{Value of study-}\\\text{ing and an easy}\\\text{exam}}} \cdot \underbrace{0.8}_{\substack{\text{Probability}\\\text{of an easy}\\\text{exam}}} + \underbrace{18}_{\substack{\text{Value of study-}\\\text{ing and a hard}\\\text{exam}}} \cdot \underbrace{0.2}_{\substack{\text{Probability}\\\text{of a hard}\\\text{exam}}} = 18$$

We can now look to Expected Value Maximization to decide what to do. This principle tells us we ought to choose an option whose expected value is at least as high as that of any rival. Since $EV(D) > EV(S)$, that means you ought to go out for drinks rather than stay home and study.

**5.2.1.1 An evil teacher** An important feature of our example is that the probability of a given state of the world (easy exam or hard exam) is independent of which option you select (going out for drinks or staying home to study). But this needn't be true in general. Suppose, for example, that you have an evil teacher who knows whether you go out for drinks and is more likely to assign a hard exam if you do. Whereas the probability of an easy exam *given that you go out for drinks* is 30%, the probability of an easy exam *given that you stay home to study* is 90%. The resulting probabilities might be represented as follows:

| | Easy exam | Hard exam |
|---|---|---|
| Drinks | 30% | 70% |
| Study | 90% | 10% |

Note that each cell in this matrix corresponds to a *conditional probability*. For example, the probability in the lower right corner corresponds to the probability of a hard exam, given that you study.

In the evil teacher scenario, we need to use these conditional probabilities to calculate expected value. For instance, the expected value of going out for drinks ($D$) should be calculated as follows:

$$EV(D) = \underbrace{35}_{\substack{\text{Value of drinks}\\\text{with an easy}\\\text{exam}}} \cdot \underbrace{0.3}_{\substack{\text{Probability of an}\\\text{easy exam, given}\\\text{drinks}}} + \underbrace{-25}_{\substack{\text{Value of drinks}\\\text{with a hard}\\\text{exam}}} \cdot \underbrace{0.7}_{\substack{\text{Probability of}\\\text{a hard exam,}\\\text{given drinks}}} = -7$$

Similarly, the expected value of studying ($S$) is now calculated as follows:

$$EV(S) = \underbrace{18}_{\substack{\text{Value of study-}\\\text{ing with an easy}\\\text{exam}}} \cdot \underbrace{0.9}_{\substack{\text{Probability}\\\text{of an easy}\\\text{exam}}} + \underbrace{18}_{\substack{\text{Value of study-}\\\text{ing with a hard}\\\text{exam}}} \cdot \underbrace{0.1}_{\substack{\text{Probability}\\\text{of a hard}\\\text{exam}}} = 18$$

Since $EV(S) > EV(D)$, Expected Value Maximization entails that when you have an evil teacher, you ought to stay home and study rather than go out for drinks.

### 5.2.2 Definition of Expected Value

Now that you know how expected value works, I can give you a more precise definition. The **expected value** of an option $A$ is the weighted average of the value of the outcomes that $A$ might lead to, with weights determined by the probability of the relevant state of affairs, given that you choose $A$. Formally:

$$EV(A) = v(AS_1) \cdot p(S_1|A) + v(AS_2) \cdot p(S_2|A) + \ldots + v(AS_n) \cdot p(S_n|A)$$

where $S_1, S_2, \ldots, S_n$ is any list of (exhaustive and mutually exclusive) states of the world; $v(AS_i)$ is the value of being in a situation in which you've chosen $A$ and $S_i$ is the case; and $p(S|A)$ is the probability of $S$, given that you choose $A$.

Here is how to apply this formula in the case of the evil teacher. Recall that there are two relevant states of affairs: easy exam ($E$) and hard exam ($H$). So the expected value of going out for drinks ($D$) and staying home to study ($S$) should be calculated as follows:

$$\begin{aligned} EV(D) &= v(DE) \cdot p(E|D) + v(DH) \cdot p(H|D) \\ &= 35 \cdot 0.3 + -25 \cdot 0.7 = -7 \\ EV(S) &= v(SE) \cdot p(E|S) + v(SH) \cdot p(H|S) \\ &= 18 \cdot 0.9 + 18 \cdot 0.1 = 18 \end{aligned}$$

#### Exercises

1. A fair coin will be tossed, and you must choose between the following two bets:

**$B_1$:** \$1,000 if Heads; −\$200 if Tails.

**$B_2$:** \$100 if Heads; \$50 if Tails.

What is the expected value of accepting $B_1$? What is the expected value of accepting $B_2$? Which of the two bets should you accept according to Expected Value Maximization? (Assume that the degree to which you value a given outcome corresponds to the amount of money you will end up with. So, for example, assign value 1,000 to an outcome in which you receive \$1,000 and value −200 to an outcome in which you pay \$200.)

### 5.2.3 Back to Newcomb

What does Expected Value Maximization say we should do in a Newcomb situation? Recall that you can choose between two options: one-box and two-box. Recall, moreover, that there are two relevant states of the world: the large box is full and the large box is empty.

This means that there are four possible outcomes, depending on whether you one-box or two-box and on whether the large box is full or empty. The value of each outcome is as follows:

| | Large box full ($F$) | Large box empty ($E$) |
|---|---|---|
| One-box ($1B$) | \$1,000,000 | \$0 |
| Two-box ($2B$) | \$1,001,000 | \$1,000 |

When the predictor is assumed to be 99% accurate, we work with the following probabilities:

| | Large box full ($F$) | Large box empty ($E$) |
|---|---|---|
| One-box ($1B$) | 99% | 1% |
| Two-box ($2B$) | 1% | 99% |

As before, each cell in this matrix corresponds to a conditional probability. For example, the probability in the lower right corner corresponds to the probability of finding the large box empty, given that you two-box. The expected value of one-boxing and two-boxing can then calculated as follows:

$$\begin{aligned} EV(1B) &= v(1B\,F) \cdot p(F|1B) + v(1B\,E) \cdot p(E|1B) \\ &= 1,000,000 \cdot 0.99 + 0 \cdot 0.01 = 990,000 \end{aligned}$$

$$\begin{aligned} EV(2B) &= v(2B\,F) \cdot p(F|2B) + v(2B\,E) \cdot p(E|2B) \\ &= 1,001,000 \cdot 0.01 + 1,000 \cdot 0.99 = 11,000 \end{aligned}$$

The expected value of one-boxing (\$990,000) is much greater than the expected value of two-boxing (\$11,000). So according to Expected Value Maximization, we should one-box. Is that the right result? We'll try to tackle that question in the next section.

#### Exercises

1. Assume the predictor has an 80% chance of making the right prediction. What is the expected value of one-boxing? What is the expected value of two-boxing?
2. How accurate does the predictor need to be in order for Expected Value Maximization to entail that one should one-box? In other words, identify the smallest number $x$ such that, as long as the predictor has an accuracy of more than $x$, Expected Value

Maximization entails that one should one-box. (Assume that the small box contains a hundred dollars and the large box contains a million dollars or nothing.)

3. Show that as long as one has a predictor who does better than chance, it is possible to find payoffs that generate a Newcomb Problem. More precisely, show that, for any small-box value $s$ and any positive value $\epsilon$ $(0 < \epsilon \leq 0.5)$, one can find a large-box value $l$ such that when the predictor has an accuracy of $0.5 + \epsilon$, Expected Value Maximization entails that one should one-box in a Newcomb scenario based on $s$ and $l$.

4. Consider a variant of the Newcomb case that works as follows: You can choose either the large box or the small box, but not both. If the predictor predicted that you would choose the large box, then she left the large box empty and placed a hundred dollars in the small box. If the predictor predicted that you would choose the small box, then she left the small box empty and placed a thousand dollars in the large box. According to Expected Value Maximization, which of the two boxes should you choose? (Assume the predictor is not perfectly accurate.)

## 5.3 In Defense of Two-Boxing

Many philosophers believe that one-boxing is irrational and that the form of decision theory I described above should therefore be rejected, however natural it may seem. So let me tell you about one of the rationales that has been set forth in defense of two-boxing.

### 5.3.1 Dominance

An important argument for two-boxing is based on the following observation: In a Newcomb case, there is one key matter you have no control over—whether or not the large box is full. And regardless of how this matter turns out, you will be better off if you two-box than if you one box. In particular:

- If the large box is empty, you'll be better off if you two-box than if you one-box.
- If the large box is full, you'll be better off if you two-box than if you one-box.

Decision theorists sometimes summarize this by saying that two-boxing **dominates** one-boxing. (In general, one says that option $A$ (strictly) dominates option $B$ if however matters you have no control over turn out, you'll be better off choosing $A$ than choosing $B$.)

The fact that two-boxing dominates one-boxing is a powerful argument for thinking that you ought to two-box. One way to bring this idea out is to imagine that you have a friend who has your best interests at heart, and who knows whether the large box is empty or full. How would your friend want you to choose? If the large box is empty, your friend will be hoping that you two-box so that you at least get the one thousand dollars. If the large box is full, your friend will be absolutely delighted, because she knows you'll be rich whatever you decide. But she'll still be hoping that you two-box, so you get the extra one thousand dollars.

This means that you needn't bother asking your friend for advice: you know from the start that she'll recommend two-boxing. Actually, you don't even need the friend; you know from the start that if you did have a friend who knew what was in the boxes and had your best interests at heart, she would recommend two-boxing!

### 5.3.2 Common Causes[2]

I have given you an argument for two-boxing. Let me now say a little more about how the two-boxer thinks of Newcomb scenarios.

We will begin with an analogy. You're more likely to see wet sidewalks when there are people using umbrellas. But that's not because umbrella use causes wet sidewalks (or because wet sidewalks cause umbrella use); it is because wet sidewalks and umbrella use have a *common cause*: they are both caused by rain. Two-boxers tend to think that something similar is going on in the case of Newcomb's Problem. You're more likely to find money in the large box when you one-box, but that's not because one-boxing causes the large box to contain money (or because the contents of the large box cause your decision). It is because one-boxing and a full large box have a *common cause*: they are both caused by your psychological constitution.

We can bring home the point by considering a different example in which your constitution—in this case your physical constitution—is a common cause. Suppose there is a gene with the following two features:

1. Having the gene increases the probability of doing mathematics. (Perhaps the gene causes a certain hormone to be released, which causes one to enjoy doing mathematics.)
2. Having the gene increases the probability of suffering a terrible disease: *mathematosis*. (The symptoms of mathematosis are too horrible to describe here.)

Mathematosis is much more prevalent among people who do mathematics than in the population at large. But this is *not* because doing mathematics causes the disease (or because having the disease causes one to do mathematics). It is because having the disease and doing mathematics have a common cause: they are both caused by the gene.

Now suppose that you very much enjoy doing mathematics. What should you do? Should you refrain from doing mathematics even though you enjoy it? Of course not! For there is a dominance argument. If you carry the gene, you're likely to get

2. Partly based on text by Damien Rochford.

the disease, but there is nothing you can do about it. Better to do mathematics and enjoy life while you're still healthy. And if you don't carry the gene, there is no need to worry: you won't get the disease, regardless of whether you do mathematics. So, again, better to enjoy yourself doing mathematics. Either way, you should do mathematics!

*Compare:* Suppose you don't want the sidewalks to be wet, but you'd very much like to bring along your favorite umbrella. Should you refrain from bringing your umbrella, even though you'd enjoy it? Of course not! For, again, there is a dominance argument. If rain is on the way, the sidewalks will get wet regardless of whether you bring your umbrella. So better to bring it along and get both the pleasure of holding your favorite umbrella and the ability to stay dry. And if rain is not on its way, there is no need to worry: the sidewalks will remain dry regardless of whether you bring your umbrella. So you'll be better off bringing it if it would give you pleasure.

Two-boxers think the dominance argument of the preceding section is analogous. If you want to be rich, it makes no sense to one-box. For if the large box is full, it will be full regardless of whether you one-box or two-box. So better to two-box and get the extra thousand dollars. And if the large box is empty, there's nothing you can do about it now: better to keep the extra thousand dollars as a consolation prize.

### 5.3.3 The Pull of One-Boxing

Even if you are a committed two-boxer, it can be difficult to escape the temptation to one-box. Imagine that a billionaire comes to town and offers everyone the chance to participate in a Newcomb case. After the first day, your one-boxer friends are absolutely delighted. They have each found a million dollars in the large box, and they have spent the night celebrating with caviar and champagne. Your two-boxer friends, on the other hand, are all crestfallen. They all found the large box empty. There's nothing wrong with getting a thousand dollars, but it's not quite the same as a million.

You get your chance to participate the next day. When the time finally comes, you decide to two-box, and, predictably, you find nothing in the large box. Your one-boxing friends can't stop laughing. "What a fool!" they cry. "What on earth possessed you to two-box?"

Are they right? Did you make a mistake? Although philosophers are not all in agreement about this, my own view is that you did not make a mistake when you two-boxed. The billionaire came to town with the intention of rewarding irrational behavior (i.e., one-boxing). So it would have certainly been in your interest to somehow make her *believe* that you are irrational, since that would cause her to reward you. But by the time you entered the room, it had already been decided that you would not be rewarded: your large box had been empty from the start, and there's nothing you could have done to change that. Leaving the thousand dollars behind would certainly not have helped.

### 5.3.4 Exceptions

I have been arguing that one ought to be a two-boxer. But I'd like to mention a couple of special cases in which I think one-boxing would be rational.

First, consider a Newcomb scenario in which the predictor has a time machine. She travels to the future, observes whether you one-box or two-box, travels back to the past, and makes her prediction in accordance with what she observed. In ordinary Newcomb scenarios, your choice has no causal effect on the contents of the large box. But in this case your choice has a causal effect on what the predictor observes in the future and, therefore, on the contents of the large box. It seems to me that one-boxing would be rational under such circumstances.

Next, consider a predictor who is *necessarily accurate*: there is *no possibility* of her issuing the wrong prediction. With such a predictor in place, there are really only two possible outcomes, rather than the original four:

| | Large box full | Large box empty |
|---|---|---|
| One-Box 1*B* | | [not available] |
| Two-Box 2*B* | [not available] | |

Accordingly, one-boxing is *equivalent* to one-boxing with a full large box, and two-boxing is *equivalent* to two-boxing with an empty large box. So your options are actually these:

| Option 1 | Option 2 |
|---|---|
| Take the contents of a full large box | Take the contents of the small box (plus the non-existent contents of an empty large box) |

When these are your options, you should certainly choose option 1!

It is worth keeping in mind, however, that describing a predictor who is necessarily accurate is not as easy as one might think. Consider, for example, a deterministic world and a predictor who knows the laws and the initial conditions and is able to use them to determine whether you will one-box or two-box. Such a predictor would certainly have an accuracy rate of 100%. It is not clear, however, that she would count as necessarily accurate. To see this, recall our discussion of determinism from section 4.4.2.1, where we considered two different conceptions of physical law:

1. The laws tell us what *will* happen, on the basis of what has happened.
2. The laws tell us what *must* happen, on the basis of what has happened.

The second of these conceptions entails that, in a deterministic world, you cannot choose differently than you actually will (assuming the initial conditions are fixed).

So it is indeed true that there is no possibility that our predictor will issue the wrong prediction. Recall, however, that we found little reason to endorse such a conception of physical law.

What about the first of the above conceptions of physical law? When we assume that the laws tell us what *will* happen on the basis of what has happened, but they don't tell us what *must* happen, it remains true that the predictor has an accuracy rate of 100%. For she is able to use the laws to calculate what you *will* decide on the basis of the initial conditions of the universe, and do so with unfailing accuracy. And yet it is not clear that she should be counted as necessarily accurate. For recall that on the conception of physical law we are considering, determinism may well be compatible with your being in a position to act otherwise. And if you are in a position to act otherwise, then the predictor is not necessarily accurate, since her prediction would be falsified were you to act otherwise. (We know, of course, that you will not, in fact, act differently, so we know that her prediction will not, in fact, be falsified. But it *might* have been falsified, and that's what matters in the present context.)

It is not clear to me how a necessarily accurate predictor could work, unless it relied on some form of time travel—in which case we would have independent reasons for thinking that one-boxing is rational.

## 5.4 Causal Decision Theory

A decision theory based on the definition of expected value I gave in section 5.2.2 tells us that the right thing to do in a Newcomb case is to one-box. So if you are a two-boxer, you must think that that form of decision theory is incorrect. In this section I'll tell you how to develop an alternative.

### 5.4.1 Dependence, Causal and Probabilistic

It is useful to start by distinguishing between *causal* dependence and *probabilistic* dependence.

- Two events are **causally independent** of each other if neither of them is a cause of the other; otherwise, the effect is **causally dependent** on the cause.
- Two events are **probabilistically independent** of each other if the assumption that one of them occurring does not affect the probability that the other one will occur; otherwise, each of them is **probabilistically dependent** on the other.

Our umbrella example can be used to illustrate the difference between these two types of dependence:

- There being wet sidewalks is *probabilistically dependent* on umbrella use, because the assumption that people are using umbrellas increases the probability of wet sidewalks.

- There being wet sidewalks is *causally independent* of umbrella use, because umbrella use does not *cause* sidewalks to be wet and wet sidewalks do not cause umbrella use. (What we have instead is a *common cause*: rain causes both wet sidewalks and umbrella use.)

Something similar happens in the case of Newcomb's Problem:

- Whether or not the large box contains a million dollars is *probabilistically dependent* on your choice to one-box or two-box, because the assumption that you one-box increases the probability that the large box contains the money.
- Whether or not the large box contains a million dollars is *causally independent* from your choice to one-box or two-box, because your action doesn't cause the box to have the money in it and the amount of money in the box does not cause your action. (Here, too, we have a *common cause*: your psychological constitution causes both your decision and the predictor's prediction.)

#### Exercises

1. Alice and Bob make independent decisions to go for a walk in the park on Tuesday afternoon, and neither brings an umbrella. Let $A$ be the event of Alice's getting soaked in rain and $B$ be the event of Bob's getting soaked in rain. Are $A$ and $B$ probabilistically independent? Are they causally independent?
2. Let $S$ be the event of Jones's smoking and $C$ be the event of Jones's being diagnosed with lung cancer. Are $S$ and $C$ probabilistically independent? Are they causally independent?

### 5.4.2 Two Types of Conditionals

Like many other natural languages, English uses different kinds of conditionals to track probabilistic and causal dependence.

To see this, let us start by considering a famous pair of conditionals that can be credited to philosopher Ernest Adams:

**[Didn't]** If Oswald didn't kill Kennedy, somebody else did.

**[Hadn't]** If Oswald hadn't killed Kennedy, somebody else would have.

Interestingly, these conditionals have different truth values. [Didn't] is true: Kennedy was definitely assassinated, so if Oswald didn't do it, it must have been someone else. In contrast, we have every reason to think that [Hadn't] is false. For—on the reasonable assumption that there was no conspiracy—nobody was waiting in the wings to kill Kennedy had Oswald failed to act.

- [Didn't] is an **indicative conditional**. (In general, an indicative conditional is a conditional of the form *if A, then B*.) Indicative conditionals are useful because they track probabilistic dependence. More specifically, the probability of *if A, then B* is always the probability of *B* conditional on *A*. So, for example, the probability of [Didn't] is the probability that someone else killed Kennedy conditional on the assumption that Oswald didn't do it, which is very high. Or consider another indicative conditional:

  *If people are using their umbrellas, it's raining.*

  Since the probability of rain conditional on people using their umbrellas is high, we are correspondingly confident that the conditional is true.
- [Hadn't] is a **subjunctive conditional**. (Subjunctive conditionals take many different forms; here, we will mostly focus on those of the form *if it were that A, then it would be that B*.) Unlike indicative conditionals, subjunctive conditionals do not track probabilistic dependence. We have seen, for example, that [Hadn't] is false, even though the probability that someone else killed Kennedy conditional on the assumption that Oswald didn't do it is extremely high.

  Here is another example: Suppose that it's sunny outside. Were people to start opening their umbrellas, it would remain just as sunny. So the following subjunctive conditional is false, even though rain and umbrella use depend probabilistically on each other:

  *If people were to use their umbrellas, there would be rain.*

  Subjunctive conditionals don't track probabilistic dependence, but they can be used to track something in the vicinity of causal dependence. The precise connection between causation and subjunctive conditionals is fairly subtle, and we won't discuss it here. (See section 5.6 for recommended readings.) But, to a rough first approximation, if *A* causes *B*, then the subjunctive conditional *if it were that A, then it would be that B* is true. For example, the fact that rain causes umbrella use means that the following subjunctive conditional is true:

  *If it were to rain, people would use their umbrellas.*

  Subjunctive conditionals can also be used to keep track of causal independence. For example, one can capture the fact that umbrella use has no causal effect on rain by noting that the following subjunctive conditionals are both true, assuming it's sunny outside:

  *If people were to use their umbrellas, it would (still) be sunny outside.*

  *If people were to refrain from using their umbrellas, it would (still) be sunny outside.*

**Exercises**

1. Recall the "Death in Damascus" story from section 4.4.1. Imagine the merchant's situation just after Death has warned her that they will meet tomorrow, and suppose she says to herself, "If I stay in Damascus, Death will meet me in Damascus." Is this indicative conditional true? (Assume that Death is completely reliable.)[3]
2. Now consider the merchant's situation after she wakes up to find Death in Aleppo, and suppose she says to herself, "Had I stayed in Damascus, Death would have met me in Damascus." Is this counterfactual conditional true?

### 5.4.3 An Alternate Characterization of Expected Value

Causal Decision Theory is a variant of decision theory. It agrees with Expected Value Maximization and holds onto the idea that the expected value of an option is a weighted average of the values of the different outcomes that the option might lead to. But it does the weighting differently and so works with a different characterization of expected value.

The characterization of expected value from section 5.2.2 is sometimes called Evidential Decision Theory. It uses *probabilistic dependence* to determine how much weight to give each of the possible outcomes of one's actions, because it weighs an outcome using the conditional probability of the outcome given the relevant action.

Let us review how this plays out in a Newcomb scenario. In calculating the expected value of one-boxing, one weighs an outcome in which you one-box ($1B$) and in which the large box is full ($F$) using $p(F|1B)$; and one weighs an outcome in which you one-box and in which the large box is empty ($E$) using $p(E|1B)$. Calculating the expected value of two-boxing ($2B$) is similar. This yields the following familiar results:

$$EV(1B) = v(1B\,F) \cdot p(F|1B) + v(1B\,E) \cdot p(E|1B)$$
$$EV(2B) = v(2B\,F) \cdot p(F|2B) + v(1B\,E) \cdot p(E|2B)$$

Because the probability of an indicative conditional *if A, then B* is just the conditional probability of *B given A*, these equations could also be written as follows, where $A \rightarrow B$ is shorthand for if *A, then B*:

$$EV(1B) = v(1B\,F) \cdot p(1B \rightarrow F) + v(1B\,E) \cdot p(1B \rightarrow E)$$
$$EV(2B) = v(2B\,F) \cdot p(2B \rightarrow F) + v(1B\,E) \cdot p(2B \rightarrow E)$$

In either version of the equations, Expected Value Maximization delivers the conclusion that one ought to one-box.

Causal decision theorists see things differently. They think that one should use *causal dependence* to determine how much weight to give each of the possible outcomes of

3. The exercises in this section are based on text by Damien Rochford.

one's actions. Accordingly, they think one shouldn't weigh an outcome by using the probability of an indicative conditional. One should instead use the probability of a subjunctive conditional: *were you to perform the relevant action, the outcome would come about.*

(Nerdy observation: I am oversimplifying a bit because subjunctive conditionals have different kinds of readings and not every reading will serve the causal decision theorist's purposes. Roughly, we need to avoid "backtracking" readings of the conditionals, which are evaluated by revising events that occured prior to the action referred to in the conditional's antecedent. If you'd like to know more, have a look at the recommended readings in section 5.6.)

Here is an example. Suppose you wish to calculate the expected value of one-boxing. How should you weigh an outcome in which you one-box and the large box is full? According to Causal Decision Theory, you should use the probability of the following subjunctive conditional: *Were you to one-box, a full-box outcome would come about.* Using $A \Box\!\!\rightarrow B$ as shorthand for *were it that A, it would be that B*, we have:

$$EV(1B) = v(1B\,F) \cdot p(1B \Box\!\!\rightarrow F) + v(1B\,E) \cdot p(1B \Box\!\!\rightarrow E)$$

And, similarly, for two-boxing, we have:

$$EV(2B) = v(2B\,F) \cdot p(2B \Box\!\!\rightarrow F) + v(2B\,E) \cdot p(2B \Box\!\!\rightarrow E)$$

Notice that these are exactly like the equations endorsed by the evidential decision theorists, except that we use subjunctive conditionals instead of indicative conditionals.

As it turns out, the causal decision theorist's equations can be simplified a bit in the present case. For recall that you have no control over the contents of the large box. Now consider this question: Under what circumstances would it be the case that were you to one-box, you would find a full large box? Answer: Exactly when the large box is full to begin with. In other words, $p(1B \Box\!\!\rightarrow F)$ is just $p(F)$, which is the (unconditional) probability of $F$. (And it would go similarly for other cases.) So the equations above reduce to:

$$EV(1B) = v(1B\,F) \cdot p(F) + v(1B\,E) \cdot p(E)$$
$$EV(2B) = v(2B\,F) \cdot p(F) + v(1B\,E) \cdot p(E)$$

Since the payoff structure of a Newcomb situation guarantees that $v(2B\,F) > v(1B\,F)$ and $v(2B\,E) > v(1B\,E)$, our equations guarantee that the expected value of two-boxing is higher than the expected value of one-boxing, regardless of the values of $p(F)$ and $p(E)$. So Causal Decision Theory vindicates the idea that the rational thing to do in a Newcomb scenario is two-box.

As the Newcomb case makes clear, Causal Decision Theory can come apart from Evidential Decision Theory when causal and probabilistic dependence come apart. But, importantly, they always deliver the same results in cases where causal and probabilistic dependence don't come apart.

### Exercises

1. Consider a variant of the Newcomb scenario in which the predictor decides whether to fill the large box by flipping a coin. Do Causal Decision Theory and Evidential Decision Theory deliver different results?

### 5.4.4 A Connection to Free Will

There is an interesting connection between Causal Decision Theory and some of the ideas we discussed in chapter 4, when we talked about free will. Recall your friend Susan. You are having breakfast with her in New York, and she tells you that last night she was thinking of taking a train trip to Alaska. This puts you in a position to know that the following indicative conditional is true:

*If Susan decided to make a train trip to Alaska last night, she failed to do so.*

The reason is that you can see that Susan is in New York the next morning and you know that a train trip from New York to Alaska cannot be completed in a single night. In contrast, it is not clear whether you should think that the following subjunctive conditional was true at the time when Susan was still deciding what to do:

*Were Susan to decide to make the trip, she would succeed.*

If all trains out of New York were canceled last night because of a storm, this conditional is false. But if a suitable train left New York last night, and if there was nothing to stop Susan from taking it had she decided to do so, the conditional is true.

Recall that our discussion of free will in chapter 4 focused on the Control Hypothesis: the hypothesis that to act freely is to be in a position to do otherwise. According to the Control Hypothesis, the truth of the indicative conditional above is irrelevant to the question of whether Susan acted freely in staying in New York. But the subjunctive conditional is extremely relevant. For the Control Hypothesis sees acting freely as a matter of being in a position to *cause* a different outcome, and it is the subjunctive conditional, rather than its indicative counterpart, that tracks causal dependence.

Something similar is true of Causal Decision Theory. Notice, in particular, that the causal decision theorist agrees that the following two indicative conditionals are both true of a Newcomb scenario:

1. If you one-box, you'll almost certainly be rich.
2. If you two-box, you'll almost certainly be poor.

The causal decision theorist thinks, however, that their truth is irrelevant to the question of whether one should one-box or two-box. What matters is the fact that the following subjunctive conditionals are both true:

1. Were you to one-box, you would fail to bring about a situation in which you end up with as much money as is presently available to you.

2. Were you to two-box, you would succeed in bringing about a situation in which you end up with as much money as is presently available to you.

This is because Causal Decision Theory thinks that the question of what to do turns on causal, rather than probabilistic, dependence, and it is the subjunctive conditionals above, rather than their indicative counterparts, that track causal dependence.

The moral of our discussion is that the Control Hypothesis and Causal Decision Theory have something important in common: they are both *causalist* accounts of their subject matters. Since indicative conditionals track probabilistic dependence rather than causal dependence, this means that when probabilistic dependence and causal dependence come apart, the Control Hypothesis and the Causal Decision Theory will both count indicative conditionals relating actions to outcomes as irrelevant to their subject matters.

## 5.5 Conclusion

The focus of this chapter has been Newcomb's Problem. We have seen that even though there are reasons for thinking that one-boxing is irrational, a natural formulation of decision theory yields the result that you should choose one-boxing. We then considered Causal Decision Theory, a variant of decision theory that is designed to deliver the right result in Newcomb cases.

I have included two appendices in this chapter. The first discusses a connection between Newcomb's Problem and the Prisoner's Dilemma. The second discusses the Tickle Defense, a strategy that has sometimes been used to defend one-boxing. They are both based on text by Damien Rochford.

## 5.6 Further Resources

Many of the ideas discussed in this chapter were originally set forth by David Lewis. If you'd like to go into these subjects more in-depth, I highly recommend the articles in volume II of his *Philosophical Papers*. Three articles in this collection are of special relevance: "Causation," "Causal Decision Theory," and "Prisoner's Dilemma Is a Newcomb Problem."

### 5.6.1 Additional Texts

- For a defense of one-boxing and an example of the Tickle Defense, see Terry Horgan's "Counterfactuals and Newcomb's Problem."
- For a comprehensive treatment of Expected Value Maximization and the ensuing decision theory, see Richard Jeffrey's *The Logic of Decision*.

- Ernest Adams discusses his famous pair of conditionals in "Subjunctive and Indicative Conditionals."
- For a comprehensive treatment of Causal Decision Theory, see Jim Joyce's *The Foundations of Causal Decision Theory*.
- Causal Decision Theory is controversial. For purported counterexamples, see Bostrom's "The Meta-Newcomb Problem," Egan's "Some Counterexamples to Causal Decision Theory," Ahmed's "Dicing with Death," and Spencer and Wells's "Why Take Both Boxes?"
- If you'd like to know more about Newcomb's Problem in the context of Causal Decision Theory, a good place to start is Paul Weirich's article on Causal Decision Theory in the *Stanford Encyclopedia of Philosophy*. It's available online and contains an excellent list of references.
- For a comprehensive discussion of conditionals, I recommend Jonathan Bennett's *A Philosophical Guide to Conditionals*.

## Appendix A: The Prisoner's Dilemma

Newcomb's Problem is closely connected to a classic problem in Game Theory: the Prisoner's Dilemma. Here is the dilemma: Imagine you and Jones committed a crime together, and you've both been arrested. The police don't have much evidence against either of you though. If both of you keep quiet, you'll each be convicted of a relatively minor offense and spend only 1,000 days in prison (a payoff of −1,000). But there is an offer on the table. If either of you agrees to sign a statement accusing the other, the defector will be allowed to leave scot-free (a payoff of 0), and the accused will be convicted of a felony offense and sentenced to 10,000 days in prison (a payoff of −10,000). Unfortunately, there is a catch. Should *both* you and Jones defect, you will both be convicted with felony offenses but be given the lesser sentence of 9,000 days in prison because of your cooperation with the police (a payoff of −9,000). Here is a summary of the payoffs:

| | You defect | You keep quiet |
|---|---|---|
| Jones defects | Jones → −9,000<br>You → −9,000 | Jones → 0<br>You → −10,000 |
| Jones keeps quiet | Jones → −10,000<br>You → 0 | Jones → −1,000<br>You → −1,000 |

Before the offer is made, you and Jones are placed in separate interrogation rooms. You're then each told about the offer, and each of you is also told that the other has been told about the offer. But, you're not allowed to communicate with each other. So neither of you can do anything to affect the other's decision. Finally, assume that all you care about is minimizing your jail time. So, in particular, you do not care about Jones or about consequences beyond those already mentioned. (No need to worry about Jones exacting revenge, or about feeling guilt, or about acquiring a bad reputation, or anything like that.)

Should you defect or should you keep quiet?

### A.5.1 The Standard Answer

Philosophers and economists tend to think that the right thing to do in the Prisoner's Dilemma is to defect. Why? Because defecting **dominates** not defecting, in the sense discussed in section 5.3.1:

- If Jones defects, you'll be better off if you defect than if you don't.
- If Jones doesn't defect, you'll be better off if you defect than if you don't.

So, regardless of what Jones does, you'll be better off if you defect.

#### Exercises

1. Suppose that Jones will, in fact, defect. Why will you be better off if you defect? (Remember that you are in separate interrogation rooms, so there is nothing you can do to affect Jones's decision.)
2. Suppose that Jones will, in fact, keep quiet. Why will you be better off if you defect?

### A.5.2 The Newcomb Problem as a Prisoner's Dilemma

An interesting observation by philosopher David Lewis is that the Newcomb Problem can be thought of as a special case of the Prisoner's Dilemma, in which you and Jones are similar enough that you're highly likely to end up making the same decisions, even though you haven't communicated with one another.

Suppose, for example, that Jones is your clone. The two of you are identical, molecule-for-molecule and have grown up in parallel environments. The probability that you'll make the same decision in this case is 99%. Now imagine yourself facing 10,000 days of prison. You have two boxes in front of you. The small box contains a voucher that reduces your prison time by 1,000 days; the large box contains a voucher that will either reduce your prison time by 9,000 days or be declared void. You are offered two choices: take both boxes or take just the large box.

How is the value of the voucher in the large box determined? Jones is in a situation analogous to yours. If he decides to two-box, the voucher will be declared void; and if he decides to one-box, the voucher will give you a 9,000-day reduction. (Similarly, if you decide to two-box, the voucher in Jones's large box will be declared void, and if you decide to one-box, it will be counted as giving him a 9,000-day reduction.) Since there is no communication between you, your choices are causally independent. But since you are clones, the choices are not probabilistically independent: it is 99% likely that you will both end up making the same choice.

We have constructed a version of the Prisoner's Dilemma in which defecting is, in effect, two-boxing, while keeping quiet is, in effect, one-boxing. In this sense, the Newcomb Problem is a special case of the Prisoner's Dilemma.

#### Exercises

1. Assuming that the probability that Jones and you will make the same decision is 99%, what is the expected value of defecting?
2. On that assumption, what is the expected value of keeping quiet?

## Appendix B: The Tickle Defense

In this section I will tell you about the **Tickle Defense**, which is a way of defending the principle of Expected Value Maximization in the face of cases like the mathematosis example.

Recall our discussion of mathematosis in section 5.3.2 and consider the question of how the mathematosis gene is supposed to cause people to do mathematics. Let us suppose that mathematosis causes a certain urge to do mathematics and that this urge has a distinctive feeling—a *tickle*, if you will. (We'll talk more about whether this is a plausible assumption below.) If that's how the gene works, then the presence or absence of a tickle gives you information about whether you're likely to get the disease.

In particular, if you feel the tickle, that gives you some evidence that you will get the disease. Friends of the Tickle Defense find this interesting because they think that once you've felt the tickle, the decision to do mathematics won't provide you with any *additional* evidence that you'll get the disease.

To put things in probabilistic terms: everyone agrees that the probability of getting the disease ($D$), given that you do mathematics ($M$), is higher than the probability of getting the disease, given that you don't:

$$P(D|M) > P(D|\overline{M})$$

But proponents of the Tickle Defense think for someone who *has felt the tickle* ($T$), the probability of getting the disease, given that one does mathematics, is the *same* as the probability of getting the disease, given that one doesn't do mathematics:

$$P(D|MT) = P(D|\overline{M}T)$$

And from this, it follows that once you've felt the tickle, doing mathematics is probabilistically independent from getting the disease:

$$P(D|MT) = P(D|\overline{M}T) = P(D|T)$$

When $T$ is related to $M$ in the above way, $T$ is said to **screen off** the evidential import of $M$ and $\overline{M}$. That is, any evidence provided by $M$ or $\overline{M}$ on the matter of getting the disease will have already been taken into account once you learn $T$.

Similarly, proponents of the Tickle Defense think that *not having felt the tickle* ($\overline{T}$) screens off the evidential import of $M$ and $\overline{M}$:

$$P(D|M\overline{T}) = P(D|\overline{M}\overline{T}) = P(D|\overline{T})$$

So, says the Tickle Defender, the principle of Expected Value Maximization is right after all. For as long as you make your decision about whether to do mathematics once you've felt the tickle, or once it's clear that you're not feeling it, the expected value of doing mathematics will be *higher* than the expected value of not doing mathematics.

To illustrate how this might come about, let us suppose that having the tickle is very highly correlated with getting the disease:

$$P(D|T)=0.9$$

Since having the tickle screens off the evidential import of $M$ and $\overline{M}$, we have:

$$P(D|MT)=P(D|\overline{M}T)=0.9$$

Let's now suppose that you've felt the tickle. Then your options are:

**$MT$** Doing mathematics in a world in which you've felt the tickle.

**$\overline{M}T$** Not doing mathematics in a world in which you've felt the tickle.

The expected values of these options are as follows, assuming you take the disease to have a value of $-100,000$, doing mathematics to have a value of 90, and the tickle itself to have no intrinsic value.

$$\begin{aligned} EV(MT) &= v(DMT)\cdot p(D|MT)+v(|\overline{D}MT)\cdot p(|\overline{D}|MT) \\ &= (-99910\cdot 0.9)+(90\cdot 0.1) \\ &= -89,910 \end{aligned}$$

$$\begin{aligned} EV(\overline{M}T) &= v(D\overline{M}T)\cdot p(D|\overline{M}T)+v(\overline{D}\overline{M}T)\cdot p(|\overline{D}|\overline{M}T) \\ &= (-100000\cdot 0.9)+(0\cdot 0.1) \\ &= -90,000 \end{aligned}$$

So, once you've felt the tickle, you will see the expected value of doing mathematics as *greater* than the expected value of not doing mathematics, which means that the principle of Expected Value Maximization will tell you to do mathematics.

### B.5.1 The Upshot

Here's what we've concluded so far concerning the Tickle Defense: *if* the mathematosis gene causes people to do mathematics via a tickle, *then* we have a strategy for saving the principle of Expected Value Maximization from the threat of the mathematosis example.

So far, so good. But should we really think that mathematosis works via a distinctive tickle? A friend of the Tickle Defense would say that insofar as you have the intuition that the right thing to do in the mathematosis case is to do mathematics, you should *also* think that mathematosis works via a tickle.

What if we stipulate that the mathematosis gene doesn't generate a tickle of the right kind? (We are, after all, working with made-up examples.) Maybe the gene makes people inclined to do mathematics in ways that are completely indistinguishable from the way in which people who don't have the gene are inclined to do mathematics. A friend of the Tickle Defense would claim that in a world with mathematatosis of this especially stealthy kind, you really shouldn't do mathematics.

What do you think? Is the position plausible?

## Exercises

1. How could the Tickle Defense be used to get the principle of Expected Value Maximization to recommend two-boxing in a Newcomb case?

## Appendix C: Answers to Exercises

### Section 5.1.3

1. Here is one possible method. (I'm sure there are many others.) The predictor asks the subject to participate in a Newcomb case and observes the results. The subject is then asked to take an amnesia-causing drug. After she has done so, the experiment is repeated and the predictor predicts that the subject will make the same decision as before.

### Section 5.2.2

1. There are four possible outcomes, depending on whether you pick $B_1$ or $B_2$ and on whether the coin lands Heads or Tails:

| | Coin lands Heads | Coin lands Tails |
|---|---|---|
| You take bet $B_1$ | $B_1\ H$ | $B_1\ T$ |
| You take bet $B_2$ | $B_2\ H$ | $B_2\ T$ |

And we know that the value of each of these outcomes is as follows:

$v(B_1\ H) = 1,000 \qquad v(B_1\ T) = -200$
$v(B_2\ H) = 100 \qquad v(B_2\ T) = 50$

The expected values of $B_1$ and $B_2$ can be characterized on the basis of these outcomes:

$$EV(B_1) = v(B_1 H) \cdot p(H|B_1) + v(B_1 T) \cdot p(T|B_1)$$

$$EV(B_2) = v(B_2 H) \cdot p(H|B_2) + v(B_2 T) \cdot p(T|B_2)$$

Since the coin is fair, we can fill in numerical values as follows:

$$EV(B_1) = 1,000 \cdot 0.5 + (-200) \cdot 0.5 = 400$$

$$EV(B_2) = (100 \cdot 0.5) + (50 \cdot 0.5) = 75$$

Since $400 > 75$, the expected value of accepting $B_1$ is greater than the expected value of $B_2$. So Expected Value Maximization entails that you should accept $B_1$.

### Section 5.2.3

1. When $p = 80\%$, the expected value of one-boxing is

$$1,000,000 \cdot 0.8 + 0 \cdot 0.2 = 800,000$$

and the expected value of two-boxing is

$$1,001,000 \cdot 0.2 + 1,000 \cdot 0.8 = 200,200 + 800 = 201,000$$

2. If $p$ is the probability that the expert's prediction is accurate, the expected value of one-boxing is

$$1,000,000 \cdot p + 0 \cdot (1-p)$$

and the expected value of two-boxing is

$$1,001,000 \cdot (1-p) + 1,000 \cdot p$$

These two expected values are equal when $p = 0.5005$:

$$\begin{aligned} 1,000,000 \cdot p &= 1,001,000 \cdot (1-p) + 1,000 \cdot p \\ 1,000,000 \cdot p - 1,000 \cdot p &= 1,001,000 \cdot (1-p) \\ 999,000 \cdot p &= 1,001,000 - 1,001,000 \cdot p \\ 2,000,000 \cdot p &= 1,001,000 \\ p &= \frac{1,001,000}{2,000,000} \\ p &= 0.5005 \end{aligned}$$

So as long as $p$ is greater than 50.05%, the expected value of one-boxing will be greater than the expected value of two-boxing.

This means that as long as the predictor is at least 50.05% accurate, Expected Value Maximization entails that one should be a one-boxer.

(That is pretty remarkable. Even if experts that are accurate 99% of the time exist only in science fiction, it is not hard to find a predictor that is accurate 50.05% of the time. I myself have an 80% success rate when performing this experiment on my students!)

3. When the predictor is $0.5 + \epsilon$ accurate, the expected value of one-boxing is

$$l \cdot (0.5 + \epsilon) + 0 \cdot (0.5 - \epsilon)$$

and the expected value of two-boxing is

$$(l + s) \cdot (0.5 - \epsilon) + s \cdot (0.5 + \epsilon)$$

This means that Expected Value Maximization will recommend one-boxing if and only if:

$$\begin{aligned} (l+s) \cdot (0.5-\epsilon) + s \cdot (0.5+\epsilon) &< l \cdot (0.5+\epsilon) \\ (l+s) \cdot (0.5-\epsilon) &< (l-s) \cdot (0.5+\epsilon) \\ l \cdot (0.5-\epsilon) + s \cdot (0.5-\epsilon) &< l \cdot (0.5+\epsilon) - s \cdot (0.5+\epsilon) \\ s \cdot (0.5-\epsilon) + s \cdot (0.5+\epsilon) &< l \cdot (0.5+\epsilon) - l \cdot (0.5-\epsilon) + \\ s \cdot (0.5-\epsilon+0.5+\epsilon) &< l \cdot (0.5+\epsilon-0.5+\epsilon) \\ s &< l \cdot 2\epsilon \\ \frac{s}{2\epsilon} &< l \end{aligned}$$

This means that as long as $\epsilon > 0$, we can guarantee that Expected Value Maximization recommends one-boxing by making $l$ sufficiently big.

(Note that when $s = 1,000$ and $\epsilon = 0.0005$, which are the operative values in the previous exercise, Expected Value Maximization will recommend one-boxing as long as $l > 1,000,000$.)

4. The expected value of choosing the large box is

$$(1000 \cdot (1 - p)) + (0 \cdot p) = 1000 \cdot (1 - p)$$

The expected value of choosing the small box is

$$(100 \cdot (1 - p)) + (0 \cdot p) = 100 \cdot (1 - p)$$

So Expected Value Maximization entails that one should choose the large box (for every case except $p = 1$; in that case, both options have an expected value of 0, so Expected Value Maximization doesn't entail that you ought to choose one option over the other.).

### Section 5.4.1

1. The occurrence of $A$ would make it much more likely that $B$ occurs, since it raises the probability that it rained and, therefore, that they both got soaked. So $A$ and $B$ are not probabilistically independent.

   Since $A$ and $B$ made their decisions independently, we have been given no reason to think that $A$'s occurrence would cause $B$ to occur, or that $B$'s occurrence would cause $A$ to occur. (If either of these events occurs, it will presumably be caused by the relevant subject's decision to go for a walk without an umbrella and by the presence of rain.) So we can expect $A$ and $B$ to be causally independent.

2. Smoking is a cause of lung cancer. So $C$ is not causally independent of $A$.

   Because smoking is a cause of lung cancer, Jones's smoking increases the probability that he will be diagnosed with lung cancer. So $C$ and $A$ are not probabilistically independent.

### Section 5.4.2

1. The indicative conditional is true. For if what Death says is completely reliable, then one way or another, the merchant is going to meet Death tomorrow. So the probability of meeting Death in Damascus conditional of being in Damascus is high.

2. Here is an argument that the counterfactual conditional isn't true. Yes, Death was certainly going to meet the merchant at the appointed date. That's because, as a matter of fact, the merchant was going to go to Aleppo, and Death was going to meet her

there. But given that Death was headed to Aleppo to meet the merchant, the merchant would have missed Death had she stayed in Damascus. (Not all philosophers agree with this argument.)

**Section 5.4.3**

1. No. The two versions of decision theory deliver the same result. When the predictor makes her decision by flipping a coin, the contents of the large box are both probabilistically and causally independent of your decision. So we have $p(F|1B) = p(F) = p(1B \,\Box\!\!\rightarrow F)$ (and similarly for other cases).

**Section A.5.1**

1. We know that Jones will defect. So if you defect, you'll get −9,000, and if you don't, you'll get −10,000. So you'll be better off by 1,000 if you defect.
2. We know that Jones will keep quiet. So if you defect, you'll get 0, and if you don't, you'll get −1,000. So you'll be better off by 1,000 if you defect.

**Section A.5.2**

Let $D_Y$ be the proposition that you defect, $D_J$ the proposition that Jones defects, $Q_Y$ the proposition that you keep quiet, and $Q_J$ the proposition that Jones keeps quiet.

1. Expected value of defecting:

$$EV(D_Y) = v(D_Y D_J) \cdot p(D_J|D_Y) + v(D_Y Q_J) \cdot p(Q_J|D_Y) = -9,000 \cdot 99 + 0 \cdot 0.01 = -8910$$

2. Expected value of keeping quiet:

$$EV(Q_Y) = v(Q_Y D_J) \cdot p(D_J|Q_Y) + v(Q_Y Q_J) \cdot p(Q_J|Q_Y) = -10000 \cdot 0.01 + -1000 \cdot 0.99$$
$$= -1090$$

**Section B.5.1**

1. Let us start by considering a simplified case. Let us assume that you know exactly what method the predictor used to come up with her prediction and that you are in a position to deploy that method yourself. Suppose, for example, that you know the predictor made her decision exclusively on the basis of a report of whether your twin one-boxed or two-boxed in a Newcomb case several years ago, and that you have access to that very same report.

In that case, you can think of looking at the report as the "tickle." For the report constitutes evidence about whether the large box is full, and once you are in possession of such evidence, your actual decision to one-box or two-box will supply no *additional* evidence about whether the large box is full. So—under these very special circumstances—one can construct a version of the Tickle Defense for Newcomb cases.

Another special case in which a version of the Tickle Defense can be constructed is a case in which the subject is able to determine—on the basis of an examination of her mental state *before* making the choice—exactly what her choice will be. In that case, one can think of the result of examining one's mental state as the "tickle." For the result of the examination constitutes evidence about whether the large box is full, and once you are in possession of such evidence, your actual decision to one-box or two-box will supply no *additional* evidence about whether the large box is full.

So, again, we have identified special circumstances in which it is possible to construct a version of the Tickle Defense for the Newcomb case. It is not clear to me, however, that the Tickle Defense is available in general, since it is not clear that one will be able to identify a suitable "tickle" in the general case.

# 6 Probability

In the previous chapter we talked about probabilities informally. In this chapter we'll examine the concept of probability in greater depth, and we'll consider some tricky problems in the foundations of probability theory.

## 6.1 Probability, Subjective and Objective[1]

The concept of probability is ubiquitous. We feel safe traveling by plane because we believe that the probability of an accident is small; we buy fire insurance because we believe that, although the probability of a fire is small, it is not small enough to be ignored. But what *is* probability? What does it mean to say, for example, that the probability the coin will land Heads is 50%?

The answer is that it might mean two different things, depending on whether one understands probability as *subjective probability* or *objective probability*.

**Subjective probability** Claims about subjective probabilities are claims about someone's *beliefs*, and, in particular, about the *degree* to which someone believes something. For example, the claim "Smith's subjective probability that the coin will land Heads is 50%" means something like "Smith believes to degree 0.5 that the coin will land Heads." (I'll have more to say later about what it means to believe something to a degree.)

**Objective probability** Claims about objective probability, in contrast, are *not* claims about what someone believes. When one talks about the objective probability of an event, one is describing a feature of the world that does not depend on the beliefs of any particular subject. Consider, for example, the chemical element Seaborgium. One of its isotopes, $^{256}$Sg, has a half-life of 8.9 seconds. This means that if you take a particle of $^{256}$Sg, the probability that it will decay in the next 8.9 seconds is 50%.

1. An earlier version of this material appeared in Rayo, "¿Qué es la probabilidad?" *Investigación y Ciencia*, June 2011.

When one speaks of probability in this context, one isn't talking about the subjective credences of any particular subject. One is describing an objective feature of the world: the half-life of $^{256}$Sg.

### 6.1.1 How Are Subjective and Objective Probability Related?

There is an important connection between subjective and objective probabilities:

**The Objective-Subjective Connection** The objective probability of $A$ at time $t$ is the subjective probability that a perfectly rational agent would assign to $A$, if she had perfect information about events before or at $t$ and no information about events after $t$.

This principle can be used in two different ways:

**Subjective → Objective** Start with information about what an agent's subjective probabilities ought to be. Then use the Objective-Subjective Connection to get information about what the objective probabilities are.

**Objective → Subjective** Start with information about what the objective probabilities are. Then use the Objective-Subjective Connection to get information about what an agent's subjective probabilities ought to be.

To illustrate the first kind of usage, suppose that you assign a subjective probability of 30% to a particular event. Suppose, moreover, that you have all relevant information about the case at hand and that you've made no mistakes in your reasoning. Then your credences are the credences that a perfectly rational agent would have, if she had perfect information about the past. So the Objective-Subjective Connection entails that the objective probability of the relevant event is 30%.

To illustrate the second kind of usage, suppose you know that $^{256}$Sg has a half-life of 8.9 seconds and therefore that the objective probability that a particle of $^{256}$Sg will decay within the next 8.9 seconds is 50%. Then the Objective-Subjective Connection entails that a perfectly rational agent with access to all available evidence would set her subjective probability that the particle will decay within the next 8.9 seconds at 50%. But such an agent is at least as good at assessing evidence as you are, because she is perfectly rational. And (unless you have access to information about the future) she has at least as much evidence as you do, because she has access to all evidence about the past. So you should defer to her judgment and set your subjective probability that the particle will decay within the next 8.9 seconds at 50%.

#### Exercises

1. We have seen that it follows from the Objective-Subjective Connection, and from the fact that $^{256}$Sg has a half-life of 8.9 seconds, that if you have no information about the future, you should set your subjective probability that a particle of $^{256}$Sg

will decay within the next 8.9 seconds at 50%. Suppose, however, that a trusted time traveler comes back from the future and informs you that she has witnessed the fact that the particle does, in fact, decay within the the next 8.9 seconds. What should your subjective probability that the particle will decay within the next 8.9 seconds be?

## 6.2 Subjective Probability

In this section I'll tell you more about the notion of subjective probability and about what it takes for a subject's subjective probabilities to count as rational.

The central idea behind the notion of subjective probability, recall, is that people's beliefs come in *degrees*. In other words, beliefs aren't just "on" or "off." Instead, you have range of options: for any real number $r$ between 0 and 1, you could, in principle, believe something to degree $r$. For instance, if you are completely confident that it will rain, you believe that it will rain to degree 1; if you are somewhat confident, you might believe it to degree 0.7; if you are fairly doubtful, you might believe it to degree 0.1; and so forth. (The expression "$x$%" is shorthand for "$x/100$"; so believing something to degree 0.5 is the same as believing it to degree 50%.)

A natural way of modeling the credences of a subject $S$ is by using a **credence function**: a function that assigns to each proposition a real number between 0 and 1, representing the degree to which $S$ believes that proposition. In the philosophical literature, a degree of belief is usually called a **credence**. So instead of saying that Smith believes that it will rain to degree 0.6, one says that Smith's *credence* that it will rain is 0.6.

Having the wrong credences can make a subject irrational. Consider, for example, a subject who assigns credence 0.9 both to the proposition that it will rain today and to the proposition that it will not rain today. Such a subject would be confident that it will rain and confident that it will not rain, which is irrational. What sorts of constraints must a subject's credences satisfy in order for the subject to count as perfectly rational? This is an important question because the Objective-Subjective Connection relies on the notion of a perfectly rational subject. We will try to answer it in the remainder of this section.

### 6.2.1 Internal Coherence

A credence function, recall, is a function that assigns to each proposition a real number between 0 and 1, representing the subject's credence in that proposition. As I noted above, not just any assignment of credences to propositions is internally coherent. What does it take for a credence function to count as internally coherent? A standard answer is that it is internally coherent if and only if it is a **probability function**.

A probability function, $p(\ldots)$, is an assignment of real numbers between 0 and 1 to propositions that satisfy the following two coherence conditions:

**Necessity** $p(A) = 1$ whenever $A$ is a necessary truth.

**Additivity** $p(A \text{ or } B) = p(A) + p(B)$ whenever $A$ and $B$ are incompatible propositions.

### Exercises

1. Show that if $p$ is a probability function, $p(\text{Rain})$ and $p(\text{no Rain})$ can't both be 0.9. More generally, show that $p(\overline{A}) = 1 - p(A)$, where $A$ is a proposition and $\overline{A}$ is the negation of $A$.
2. Show that if $p$ is a probability function, then $p(A) \geq p(AB)$, where "$AB$" is short for "$A$ and $B$."

### 6.2.2 Updating by Conditionalization

We have considered a constraint on rational belief: that one's credence function be internally coherent and, more specifically, that it be a probability function. Notice, however, that this doesn't tell us anything about what it takes for one to *update* one's beliefs in a rational way, as one acquires additional information. That will be the topic of this subsection.

Just like one might use an *unconditional* probability function $p(A)$ to talk about the probability of $A$, so one could also use a *conditional* probability function $p(A|H)$ to talk about the probability of $A$ *on the assumption that H obtains*. Suppose, for instance, that $S$ thinks it's unlikely to rain: $p(\text{Rain}) = 0.2$. She thinks its even less likely that there'll be a sudden drop in atmospheric pressure: $p(\text{Drop}) = 0.1$. But $S$ also thinks there's a strong correlation between rain and sudden pressure drops. In particular, she is confident to degree 0.95 in the following conditional statement: it'll rain, *assuming there's a sudden drop in atmospheric pressure*. We can then say that $S$'s *conditional* credence in rain given a pressure drop is 0.95; in symbols: $p(\text{Rain}|\text{Drop}) = 0.95$.

The notion of conditional credence puts us in a position to give an attractive answer to the question of what it takes for a subject to *update* her credences in a rational way, as she acquires additional information:

**Update by Conditionalization** If $S$ is rational, she will update her credences as follows, upon learning that $B$:

$$p^{new}(A) = p^{old}(A|B)$$

where $p^{old}$ is the function describing $S$'s credences before she learned that $B$, and $p^{new}$ is the function describing her credences after she learned that $B$.

Suppose, for example, that $S$ starts out fairly confident that it won't rain: $p^{old}(\text{Rain}) = 0.2$. But she is highly confident that it'll rain, given that there's a sudden pressure drop: $p^{old}(\text{Rain}|\text{Drop}) = 0.95$. Now suppose that $S$ learns that there's been a sudden pressure drop. How confident should she be that it'll rain, in light of the new information? According to Update by Conditionalization, she should be confident to degree 0.95: $p^{new}(\text{Rain}) = 0.95$.

### 6.2.3 Bayes' Law

We have considered two different constraints on rational belief: (i) one's *unconditional* credences can be represented by a probability function, and (ii) one must respond to new information in accordance with one's *conditional* credences. Are there any constraints on how a subject's unconditional credences ought to be related to her conditional credences? Yes! It is natural to think that a subject can only count as fully rational if she satisfies the following substantial principle:

**Bayes' Law** $p(AB) = p(A) \cdot p(B|A)$

In other words: the probability that $A$ and $B$ both occur is the probability that $A$ occurs times the probability that $B$ occurs, given that $A$ occurs.

Notice that whenever the subject assigns non-zero credence to $A$, Bayes' Law entails:

$$p(B|A) = \frac{p(AB)}{p(A)}$$

which allows one to determine the subject's conditional credences on the basis of her unconditional credences. More specifically, one can determine the subject's (conditional) credence in $B$ given $A$ by looking at her (unconditional) credence in $A$ and her (unconditional) credence in $A$ *and* $B$.

#### Exercises

1. Consider a subject $S$ whose unconditional credences regarding rain ($R$) and a sudden drop in atmospheric pressure ($D$) are as follows:

   $$p(R) = 0.2$$
   $$p(D) = 0.1$$
   $$p(RD) = 0.09$$

   Use Bayes' Law to calculate $p(R|D)$ and $p(D|R)$.
2. Now suppose that $p(RD) = 0.02$. What are $p(R|D)$ and $p(D|R)$? How do these conditional probabilities compare with the unconditional probabilities $p(R)$ and $p(D)$?
3. Suppose that $S$ is perfectly rational. (In particular, her credences are described by a probability function, she updates by conditionalizing on her evidence, and she respects Bayes' Law.) Suppose, moreover, that she is certain of $A$ (i.e,. $p(A) = 1$), and

that she assigns a nontrivial probability to $B$ (i.e., $p(B) \neq 0$). Could learning $B$ cause her to be less than certain about $A$?

### 6.2.4 The Principle of Indifference

Although the constraints on rational belief that we have considered so far are nontrivial, they leave a lot of leeway. Notice, in particular, they don't fully constrain a rational subject's *initial credences*: her credences before she has received any information about the way the world is. All our constraints tell us is that her unconditional credences will constitute a probability function and that her conditional and unconditional credences will be related by Bayes' Law. Some people think this is too much leeway and hope to identify further constraints on rational belief—constraints that might help reduce the range of admissible initial credences. It is natural to suppose that the following might be added as a further constraint:

**Principle of Indifference** Consider a set of propositions, and suppose one knows that exactly one of them is true. Suppose, moreover, that one has no more reason to believe any one of them than any other. Then, insofar as one is rational, one should assign equal credence to each proposition in the set.

Suppose, for example, that we have a coin, and that we have no more reason to believe that it will land Heads than that it will land Tails, and vice-versa. The Principle of Indifference tells us that we should divide our credence equally and assign credence 0.5 to both the proposition that the coin will land Heads and the proposition that it will land Tails.

**6.2.4.1 The bad news** The bad news is that it is not clear that there is a coherent way of formulating the Principle of Indifference. It is certainly incoherent in the formulation I used above.

Philosopher Bas van Fraassen came up with a nice way of making the point. Imagine a cube factory. We know that the factory produces cubes with a side-length of less than 1 meter, but we haven't the slightest idea how the cube sizes are chosen. What is the probability that the next cube produced will have a side-length of less than 1/2 meter?

Here is an argument based on the Principle of Indifference. Since the distance between 0 and 1/2 is the same as the distance between 1/2 and 1, our reasons for thinking that the factory will produce a cube with a side-length of less than 1/2 meter are exactly analogous to our reasons for thinking that the factory will produce a cube with a side-length of more than 1/2 meter. So the Principle of Indifference tells us that we should assign credence 0.5 to the proposition that the next cube produced will have a side-length of less than 1/2 meter, and credence 0.5 to the proposition that the next cube will have a side-length of more than 1/2 meter. (To keep things simple, I ignore the possibility that the factory will next produce a cube whose side-length is exactly 1/2 meter.)

Everything sounds plausible so far. But notice that the Principle of Indifference delivers a different conclusion when we focus on *volume* rather than side-length. To see this, we again start with the observation that the distance between 0 and 1/2 is the same as the distance between 1/2 and 1. But this time we we use this observation as grounds for thinking that our reasons for thinking that the factory will next produce a cube with a *volume* of less than 1/2 cubic meter are exactly analogous to our reasons for thinking that the factory will next produce a cube with a volume of more than 1/2 cubic meter. So the Principle of Indifference tells us that we should assign credence 0.5 to the proposition that the next next cube produced will have a volume of less than 1/2 cubic meter, and credence 0.5 to the proposition that the next cube will have a volume greater than 1/2 cubic meter.

The problem, of course, is that cubes with a side-length of 1/2 meter do not have a volume of 1/2 cubic meter, but a volume of 1/8 cubic meter. Accordingly, cubes with a side-length of less than 1/2 meter make up a small fraction of the cubes whose volume is less than 1/2 cubic meter. So if we are 50% confident that the next cube will have a volume of less than 1/2 cubic meter, we should be *less* than 50% confident that the next cube will have a side-length of less than 1/2 meter, which contradicts our earlier result.

Perhaps there is a version of the Principle of Indifference that allows us to avoid this kind of problem. I do not know of any that seem very convincing.

#### Exercises

1. God has picked a real number, and you have no more reason for thinking that She picked one number than you have for thinking She picked any other. Use the Principle of Indifference to show that you ought to assign credence 0 to the proposition that God picked the number $\pi$. (*Hint:* Assume that your credence function is a probability function and don't forget that a credence is always a real number between 0 and 1.)

### 6.2.5 Summing Up

This concludes our discussion of subjective probability. We focused on the question of what a subject's credences must look like in order for the subject to count as fully rational, and we identified three important constraints:

- The subject's credence function must be internally coherent (i.e., it must be a probability function).
- As she acquires additional information, the subject must update her credences by conditionalization.
- The subject's conditional and unconditional credences must be related in accordance with Bayes' Law.

We also considered the hypothesis that the subject's initial credences satisfy a Principle of Indifference, but noted that formulating a consistent version of the principle is not as easy as one might have hoped.

## 6.3 Objective Probability

In this section we will talk about objective probability. I'll start by saying a few words about mathematical constraints on objective probability; I'll then try to address the more philosophical question of what objective probability is.

### 6.3.1 Objective Probability Functions

Because of the Objective-Subjective Connection (section 6.1.1), some of the conclusions we reached in our discussion of subjective probability carry over to the notion of objective probability. Recall, in particular, that in section 6.2.1 we considered the hypothesis that you can't count as a perfectly rational agent unless your credence function is a *probability function*. When the Objective-Subjective Connection is in place, that hypothesis entails that an assignment of objective probabilities must also be a probability function and therefore satisfy the Necessity and Additivity constraints of section 6.2.1.

#### Exercises

1. Suppose we live in a deterministic world. What is the objective probability that a coin toss lands Heads? (Assume that the outcome of the coin toss is determined by the initial conditions of the universe, in conjunction with the laws.)

### 6.3.2 Bayes' Law and Conditionalization

Just as one can speak of conditional *subjective* probabilities, so one can speak of conditional *objective* probabilities. One can speak, for instance, of the objective probability that a pair of coin tosses yields two Heads, given that it yields at least one Head. In symbols: $p(\text{two Heads} \mid \text{at least one Head})$.

How are objective conditional probabilities related to their subjective counterparts? It is natural to answer this question by setting forth a conditional version of the Objective-Subjective Connection:

**The Objective-Subjective Connection (Conditional Version)** The objective probability of $A$ given $H$ at time $t$ is the subjective conditional probability that a perfectly rational agent would assign to $A$ given $H$, if she had perfect information about events before $t$ and no information about events after $t$.

The conditional version of the Objective-Subjective Connection can be used to show that the objective probabilities must evolve in accordance with a version of conditionalization, on the hypothesis that perfectly rational agents update by conditionalization. It can also be used to show that the objective probabilities satisfy Bayes' Law on the hypothesis that you can't count as a perfectly rational agent unless your conditional and unconditional *subjective* probabilities are related by Bayes' Law.

### Exercises

1. Use the conditional version of the Objective-Subjective Connection to show that the objective probabilities must evolve in accordance with a version of conditionalization. More specifically, show that for any proposition $A$ and any times $t_0 < t_1$,

   $$p_{t_1}(A) = p_{t_0}(A|H_{t_1})$$

   where $p_{t_i}$ is the objective probability function at $t_i$ and $H_{t_i}$ consists of perfect information about events before $t_i$. (You may assume that perfectly rational agents update by conditionalization.)

2. Bayes' Law allows us to perform all sorts of interesting computations. Suppose, for example, that you have an urn with two red balls and two black balls, and that you use a random procedure to draw balls from the urn. You draw once and put the ball in your pocket. The probability of getting red on your first draw ($R_1$) is 1/2. But the probability of getting red on the second draw ($R_2$) depends on whether you get red on your first draw. If your first draw is red, the urn will be left with two black balls and one red ball. So the probability of red on your second draw is 1/3: $p(R_2|R_1) = 1/3$. But if your first draw is black ($B_1$), the urn will be left with one black ball and two red balls. So the probability of red on your second draw is 2/3: $p(R_2|B_1) = 2/3$. Accordingly:

   | First draw | Second draw |
   |---|---|
   | $p(R_1) = 1/2$ | $p(R_2\|R_1) = 1/3$ |
   | $p(B_1) = 1/2$ | $p(R_2\|B_1) = 2/3$ |

   Use Bayes' Law to calculate the unconditional probability, $p(R_2)$, of getting red on your second draw.

3. In chapter 4 we talked about probabilistic dependence and independence. These notions can be characterized formally, using the notion of conditional probability. Here is one way of doing so, where $\overline{B}$ is the negation of $B$:

   If $p(A|B) = p(A|\overline{B})$, then $A$ is probabilistically independent of $B$. Otherwise, each of $A$ and $B$ is probabilistically dependent on the other. In fact, there are several equivalent ways of defining probabilistic independence. Show that this is so by using Bayes' Law to verify each of the following:

1. $p(A|B) = p(A|\overline{B})$ if and only if $p(AB) = p(A) \cdot p(B)$.
2. $p(A|B) = p(A|\overline{B})$ if and only if $p(A|B) = p(A)$.

4. Use Bayes' Law to show that $A$ is probabilistically independent of $B$ if and only if $B$ is probabilistically independent of $A$. In other words, verify the following: $p(A|B) = p(A|\overline{B})$ if and only if $p(B|A) = p(B|\overline{A})$.
5. If you toss a fair coin $n$ times, what is the objective probability that it will land Heads on every single toss? (Assume that different tosses are probabilistically independent of one another.)
6. Part of what it is for a coin to have a 50% chance of landing Heads is for it to be *possible* that the coin land Heads. Notice, moreover, that the possibility of a coin's landing Heads does not depend on how many previous coin tosses have landed Heads, since coin tosses are independent of one another. From this, it follows that if you toss a fair coin infinitely many times, it is *possible* (though vanishingly unlikely) that the coin lands Heads every single time. Verify that if the probability of such an outcome is a real number between 0 and 1, then it must be 0. (This is important because it shows that having a probability of 0 is not the same as being impossible; it just means that chances are so unlikely that any positive real number is too big to measure its probability of occurring.)

### 6.3.3 What Is Objective Probability?

Recall that $^{265}Sg$ has a half-life of 8.9 seconds. That means that if you take a particle of $^{265}Sg$ and wait 8.9 seconds, the probability that it will decay is 50%.

When we speak of probability in this context, we are talking about objective probability. To say that $^{265}Sg$ has a half-life of 8.9 seconds is to describe a feature of the external world rather than an aspect of anyone's psychology. But what feature of the world is that? In this section we will try to answer that question.

(A note about terminology: philosophers sometimes refer to objective probability as "objective chance" or simply "chance." I will occasionally follow that practice below.)

#### 6.3.3.1 Frequentism

We will start by considering an answer that is tempting but ultimately unsatisfactory. According to **frequentism**, what it means for a particle of $^{265}Sg$ to have a 50% probability of decaying within the next 8.9 seconds is for the *frequency* of decay of $^{265}Sg$ particles to be 50%. In other words, it means that 50% of the $^{265}Sg$ that exist at a given time decay within 8.9 seconds.

Unfortunately, frequentism cannot be correct. To see this, think of a coin toss as the result of observing a particle of $^{256}Sg$ for 8.9 seconds. If the particle decays within that period, our "coin" is said to have landed Heads; otherwise it is said to have landed Tails.

Now imagine a situation in which only three particles of $^{256}Sg$ ever exist and in which the ensuing "coin tosses" yield Heads, Heads, Tails. Before any of the tosses took place, what was the chance that they would land Heads? The answer ought to be 1/2. But frequentism entails the mistaken conclusion that the answer is 2/3.

A slight improvement on straight frequentism is what is sometimes called hypothetical frequentism. The basic idea is this: if we had performed a sufficient number of coin tosses, we would have gotten Heads 50% of the time. (I will set aside the question of how many tosses would count as "sufficient.") Unfortunately, hypothetical frequentism can't be right either: it is simply not true that a fair coin—a coin with a 50% chance of landing Heads—must land Heads 50% of the time in the long run. It is perfectly possible, for example, for such a coin to land Heads on every single toss. Such an outcome would be extremely unlikely, to be sure. But it is certainly *possible*; that just follows from the fact that the coin has a 50% chance of landing Heads on any given toss.

The fact that a coin is fair does not guarantee that it will land Heads 50% of the time. But a weaker statement is true: if the coin were tossed a sufficiently large number of times, then it would *with very high probability* land Heads approximately 50% of the time.

A rigorous version of this principle is known as the **Law of Large Numbers** and is a theorem of probability theory. It is because of this law that casino owners can feel confident that they will, in the long run, make a profit. A particular table, or a particular slot machine, will have bad nights now and then. But as long as the casino is visited by enough players over a sufficiently lengthy period, it is extremely probable that the casino's gains will outweigh its losses.

The Law of Large Numbers tells us that frequentism has a grain of truth, since it entails that observed frequencies can be a good guide to probabilities. But the Law of Large Numbers also tells us what frequentism got wrong. Frequencies are not necessarily a *perfect* guide to probabilities. The law tells us that if the objective probability of a coin landing Heads is 50%, then it is *very probable* that the coin will, in the long run, land Heads approximately 50% of the time. But that means it is *possible* that it won't. And if frequencies aren't a perfect *guide* to probabilities, they certainly aren't *identical* to probabilities.

### Exercises

Here is a more precise statement of the Law of Large Numbers:

**Law of Large Numbers** Suppose that events of type $T$ have a probability of $p$ of resulting in outcome $O$. Then, for any real numbers $\epsilon$ and $\delta$ larger than zero, there is an $N$ such that the following will be true with a probability of at least $1-\epsilon$:

If $M > N$ events of type $T$ occur, the proportion of them that result in outcome $O$ will be $p \pm \delta$.

1. There is a casino with many slot machines. Each time a customer uses one of the slot machines, the casino has a 49% chance of losing a dollar and a 51% chance of winning a dollar. Use the Law of Large Numbers to show that, if the casino has enough customers at the slot machines, it is at least 99.99% likely to end up with a profit of at least a million dollars.

### 6.3.4 Rationalism

We have seen that frequentism cannot be correct as an account of objective probability. But if frequentism is not an option, what *is* objective probability? One option is to embrace **rationalism**: the view that there is nothing more to objective probability than the Objective-Subjective Connection of sections 6.1.1 and 6.3.2. This means that a claim about objective probability is a claim about what a perfectly rational subject would believe if she had perfect information about the past (but none about the future).

A rationalist account of objective probability is consistent with the idea that objective probabilities are not concerned with the psychology of particular subjects, since the demands of rationality need not depend on what anyone, in fact, believes. It is natural to think that the demands of rationality instead depend on relevant features of the physical world. When we say, for example, that $^{256}$Sg has a half-life of 8.9 seconds, we're describing a feature of the physical world. But we're describing it indirectly by saying that the physical world is such as to demand a certain rational stance. Specifically, the physical world makes it the case that the rational degree of belief to have about whether a particle of $^{256}$Sg will decay in the next 8.9 seconds is 0.5—assuming one has perfect information about the past, but none about the future.

Notice, however, that on a rationalist conception of objective probability, it is not clear that the objective probabilities are always well-defined. To see the worry, note that rationalism entails that the objective probabilities at $t$ can only be well-defined if there is a definite fact of the matter about what the *credences* of a perfectly rational agent should be when she has perfect information about events prior to $t$ (and no information about the future). But recall that we were unable to identify a satisfactory version of the Principle of Indifference (section 6.2.4). In the absence of such a principle, it is not clear that there will always be a definite fact of the matter about what a perfectly rational agent ought to believe.

Notice, in particular, that it is not clear that there is a definite fact of the matter about what a perfectly rational agent's "initial" credences should be—her credences in the absence of any evidence about the way the world is. And if our perfectly rational agent updates by conditionalization, then unclarity about initial credences can translate into unclarity about future credences, since the subject's future credences will only be well-defined if her initial credences, conditional on the relevant evidence, are well-defined.

So the rationalist may need to be open to the possibility that there may not always be a definite fact of the matter about what the objective probabilities are.

I myself am a **localist** about objective probability. I agree with the rationalist that there is no more to the notion of objective probability than the Objective-Subjective Connection and add that the notion of perfect rationality is only well-defined in certain special circumstances; for example, circumstances in which there is an unproblematic way of deploying a Principle of Indifference. I therefore believe that it is only in such circumstances that the objective probabilities are well-defined.

**6.3.4.1 Best system accounts of objective probability** Is there any way of further constraining the notion of objective probability, so as to guarantee that the objective probabilities are well-defined across a large range of cases? In this section I'll tell you about a "best system" account of objective probability, according to which the objective probabilities are the result of striving for an optimal balance of simplicity and strength in our overall theory of the world.

We want our theories to be **strong**; in other words, we want them to provide us with as much information as possible. We also want them to be **simple**; in other words, we want the information to be articulated briefly and systematically. A theory that consisted of a huge list detailing what happens at every corner of the universe at every time would be extraordinarily strong, but it wouldn't be simple. In contrast, a theory whose only principle was "2 + 2 = 4" would be extraordinarily simple, but wouldn't be very strong. Good theories strike a balance between simplicity and strength. They allow us to derive a lot of information from simple principles.

The notion of objective probability might be used to help achieve such a balance. Suppose, for example, that someone flips a coin and that we want to predict the outcome. We could try to come up with a definite prediction by using classical mechanics: we would need precise measurements of all forces acting on the coin, and we would need to perform a complex calculation. Although it is, in principle, possible to do so, it would be exceedingly difficult in practice.

An alternative is to use a probabilistic theory, which states that the objective probability that the coin lands Heads is 50%. Such a theory is not particularly strong. But it is simple enough to be put into practice without specialized measuring equipment or sophisticated computational capabilities. Because a probabilistic theory does conclusively not tell us how the coin will land, it is of limited utility. But it is certainly not useless. It can be used to conclude, for example, that it would be a good idea to accept a bet whereby you win two dollars if the coin lands Heads and lose one dollar if the coin lands Tails. And it might give you enough information to set up a successful casino.

How could one determine if a probabilistic theory is accurate? Suppose someone flips a coin 100,000 times and it lands Heads 31% of the time. This would be extremely unlikely to happen if the probability of Heads was 50%. So there is good reason to reject a probabilistic theory according to which the coin has a 50% chance of landing Heads. If, on the other hand, we observed that 50.023% of the coin tosses landed Heads, we would have little reason to doubt the accuracy of the theory.

Say that a **best system** for a particular phenomenon is a description of the phenomenon that satisfies two conditions: (i) the description is accurate (i.e., its non-probabilistic claims are all true, and its probabilistic claims are reasonably close to the observed frequencies), and (ii) the description delivers an optimal combination of simplicity and strength. (Just what "optimal" means in this context is something that needs to be spelled out.)

Philosopher David Lewis, who we have encountered before, famously argued that the notion of a best system is key to understanding the notion of objective probability. He thought, for example, that all it takes for the objective probability of a quantum event to be $x$ is for our best theory of quantum phenomena to tell us that the objective probability of the event is $x$. In general, he thought that what it is for the objective probability of event $E$ to be $x$ is simply for the best system for the relevant phenomenon to tell us that the objective probability of $E$ is $x$.

Is this story plausible as an account of objective probability? The answer depends, in part, on whether one thinks that there are objective standards for counting a theory as an optimal combination of simplicity and strength. Lewis himself believed that there were objective standards of simplicity. In this view, there is no conflict between a best-system account of objective probability and the claim that objective probabilities are independent of our psychologies. There's also no conflict with rationalism, since one could claim that rationality demands that one's credences be fixed in accordance with the system that is objectively best.

There is, however, room for disagreeing with Lewis about whether there really is such a thing as objective simplicity (or, indeed, an objective fact of the matter about what counts as an "optimal" combination of simplicity and strength). My own view is that simplicity is in the eye of the beholder: what counts as simple for humans may not count as simple for, e.g. let's say, Martians. If this is right, then the best system account of probability fails to deliver a notion of probability that is independent of our psychologies. What it delivers instead is a notion of probability that is partly about objective features of the world and partly about our subjective standards for representing the world in ways that we find simple. (There is a lot more to be said about the best-system account of objective probabilities. If you'd like to know more, have a look at the readings I suggest in section 6.6.)

**6.3.4.2 Primitivism** Suppose you're not comfortable with a localist account of objective probability, according to which the objective probabilities are not generally well-defined. Suppose you also think that a best-systems theory is too subjective to help further constrain objective probability. Is there something to be done?

There is always the option of being a **primitivist** about objective probabilities. A primitivist thinks the notion of objective probability resists full elucidation and is best regarded as a theoretical primitive. She concedes that it might be possible to establish interesting links between objective probability and rationality but thinks that objective probability goes beyond such links.

There is certainly nothing incoherent about primitivism. After all, explanations must end somewhere. Why couldn't they end when we got to the objective probabilities? It is worth noting, however, that primitivism would deliver a somewhat untractable picture of the world. To see this, suppose a primitivist tells you that the objective probability of a certain event $E$ is 30%. To what degree should you believe that $E$ will occur? Unless you have information about the future, one would have thought that the answer is clear: you should believe that $E$ will occur to degree 0.3. The problem with primitivism is that it doesn't make clear what it is about the notion of objective probability that would warrant such a conclusion, because it doesn't explain why rational belief ought to track objective probability. (Notice, in contrast, that the best-systems theorist of section 6.3.4.1 is in a position to claim that the conclusion is warranted, because a subject who adjusts her beliefs to the objective probabilities represents the world with an optimal balance of simplicity and strength, and so representing the world is a constraint of rationality.)

### 6.3.5 Progress Report

Our exploration of the notion of probability has delivered good news and bad news.

The good news is that the Objective-Subjective Connection delivers a simple and powerful way of elucidating the connection between objective and subjective probabilities. It is also possible to specify a number of interesting formal constraints on the credences of a rational agent: they must form a probability function, they must satisfy Bayes' Law, and they must be updated by conditionalization. And the Objective-Subjective Connection allows us to transform these constraints on subjective credences into constraints on objective probability.

The bad news is that we failed to identify a satisfying formulation of the Principle of Indifference. In the absence of such a principle, it is unclear whether we have succeeded in fully characterizing the notion of rational credence. We also found that characterizing the notion of objective probability is harder than one might have thought. And, unfortunately, there is more bad news to come. Before bringing this chapter to an end, I'd like to tell you about a tricky new axiom: the Principle of Countable Additivity. We'll see that it's hard to live with Countable Additivity, and it's hard to live without it.

Then, in the optional section 6.5, I'll tell you about a paradox that brings out some puzzling features of Probabilistic Decision Theory.

## 6.4 The Principle of Countable Additivity

Some philosophers and mathematicians think that the (finite) Additivity principle we considered in section 6.2.1 needs to be strengthened to the following:

**Countable Additivity** Let $A_1, A_2, A_3, \ldots$ be a countable list of propositions, and suppose that $A_i$ and $A_j$ are incompatible whenever $i \neq j$. Then:

$$p(A_1 \text{ or } A_2 \text{ or } A_3 \text{ or } \ldots) = p(A_1) + p(A_2) + p(A_3) + \ldots$$

There are two main reasons to hope for the adoption of Countable Additivity. The first is that it is needed to prove a number of important results in probability theory, including certain versions of the Law of Large Numbers. The second is that probability functions that fail to be countably additive can have mathematically awkward properties. We'll consider one of these properties below. We'll also talk about the fact that adopting Countable Additivity is not without costs. My own view—and that of many others—is that *on balance*, the advantages outweigh the costs. But it is important to be clear that the issue is not entirely straightforward.

(Incidentally, why stop with *Countable* Additivity? Why not go all the way, and require probability distributions to be additive with respect to *any* infinite set of mutually exclusive propositions, regardless of cardinality? The answer is that the resulting mathematics is not nice. Consider, for example, a dart on a random trajectory to the unit interval $[0, 1]$. A principle of Uncountable Additivity would entail that, for some number $r \in [0, 1]$, the probability that the dart lands on $r$ is $x$ for some $x > 0$, which would be an awkward result, since $r$ is smaller than a subinterval of $[0, 1]$ of size $x$. We'll return to this sort of issue in chapter 7.)

### 6.4.1 Life with Countable Additivity[2]

In this section I'll try to give you a sense of why the issue is not entirely straightforward. I'll start by mentioning an awkward consequence of *accepting* Countable Additivity. Then I'll point to an awkward consequence of *not accepting* Countable Additivity.

**6.4.1.1 An awkward consequence of accepting Countable Additivity** Imagine that God has selected a positive integer and that you have no idea which one. For $n$, a positive integer, what credence should you assign to the proposition $G_n$ that God selected $n$? Countable Additivity entails that your credences should remain undefined, unless you're prepared to give different answers for different choices of $n$. To see this, suppose otherwise. Suppose that for some real number $r$ between 0 and 1, you assign credence $r$ to each proposition $G_n$. What real number could $r$ be?

2. My understanding of these matters is much indebted to Vann McGee.

It must either be equal to zero or greater than zero. First, suppose that $r=0$. Then Countable Additivity entails the following:

$$p(G_1 \text{ or } G_2 \text{ or } G_3 \text{ or } \ldots) = p(G_1) + p(G_2) + p(G_3) + \ldots = \underbrace{0+0+0+0\ldots}_{\text{once for each integer}} = 0$$

But $G_1$ or $G_2$ or $G_3$ or ... is just the proposition that God selects some positive integer. So we would end up with the unacceptable conclusion that you're certain that God won't select a positive integer after all.

Now suppose that $r>0$. Then Countable Additivity entails the following:

$$p(G_1 \text{ or } G_2 \text{ or } G_3 \text{ or } \ldots) = p(G_1) + p(G_2) + p(G_3) + \ldots = \underbrace{r+r+r+r\ldots}_{\text{once for each integer}} = \infty$$

But since probabilities are always real numbers between 0 and 1, this contradicts the assumption that $p$ is a probability function.

The moral is that when Countable Additivity is in place, there is no way of assigning probabilities to the $G_n$, unless one is prepared to assign different probabilities to different $G_n$. More generally, Countable Additivity entails that there is no way of distributing probability *uniformly* across a countably infinite set of (mutually exclusive and jointly exhaustive) propositions.

### Exercises

1. As before, God has selected a number. But this time your credences are as follows: your credence that God selected the number 1 is $1/2$, your credence that God selected the number 2 is $1/4$, your credence that God selected the number 3 is $1/8$, and so forth. (In general, your credence that God selected positive natural number $n$ is $1/2^n$.) Assuming your credence function satisfies Countable Additivity, what is your credence that God selected a natural number?

**6.4.1.2 Infinitesimals to the Rescue?** One might be tempted to think that there is an easy way out of this difficulty. Perhaps we could reject the assumption that a probability must be a real number between 0 and 1, and instead allow for *infinitesimal* probabilities: probabilities that are greater than zero but smaller than every positive real number.

On the face of it, bringing in infinitesimal values could make all the difference. For suppose we had an infinitesimal value $\iota$ with the following property:

$$\underbrace{\iota+\iota+\iota+\iota+\iota+\ldots}_{\text{once for each natural number}} = 1$$

Now assign probability $\iota$ to each proposition, $G_n$, that God has selected number $n$. Haven't we succeeded in assigning the same probability to each $G_n$ without having to give up on Countable Additivity?

It is not clear that we have. The problem is not to do with talk of infinitesimals, which were shown to allow for rigorous theorizing by mathematician Abraham Robinson. The problem is to do with what we hope infinitesimals might do for us in the present context. For it is hard to see how they could help unless infinite sums of infinitesimals are defined using limits, in the usual way:

$$\underbrace{\iota+\iota+\iota+\iota+\iota+\ldots}_{\text{once for each natural number}} = \lim_{n\to\infty} n\cdot\iota$$

And this leads to the conclusion that $\iota+\iota+\iota+\iota+\iota+\ldots$ cannot equal 1, as required in the present context. For on any standard treatment of infinitesimals we have $n\cdot\iota<\frac{1}{m}$ for any positive integers $n$ and $m$. And from this, it follows that $\lim_{n\to\infty} n\cdot\iota$ cannot converge to a positive real number.

What if we were to sidestep standard treatments of infinitesimals altogether? Suppose we simply introduce a new quantity $\iota$ with the double stipulation that *(i)* $\iota$ is greater than zero but smaller than any positive real number, and *(ii)* the result of adding $\iota$ to itself countably many times is equal to 1. Even if the details of such a theory could be successfully spelled out, it is not clear that it would put us in a position to assign the same probability to each $G_n$ without giving up on Countable Additivity. For suppose we assume both that $p(G_n)=\iota$, for each $n$, and that Countable Additivity is in place. Then we are left with the unacceptable conclusion that God's selecting an even number and God's selecting an odd number are both events of probability 1. Here is a proof of half of that assertion:

$$\begin{aligned} p(G_1 \text{ or } G_3 \text{ or } G_5 \text{ or } \ldots) &= p(G_1)+p(G_3)+p(G_5)\ldots && \text{[Countable Additivity]}\\ &= \underbrace{\iota+\iota+\iota+\iota+\iota+\ldots}_{\text{once for each natural number}} && [p(G_n)=\iota]\\ &= 1 && \text{[Stipulation } (b)\text{]} \end{aligned}$$

I have not shown that it is impossible to use infinitesimals to construct a probability theory that allows for both Countable Additivity and a uniform probability distribution over a countably infinite set of mutually exclusive propositions. But I hope to have made clear why the prospects of doing so are not as rosy as they might first appear.

**6.4.1.3 An awkward consequence of rejecting Countable Additivity** Wouldn't it be better to give up on Countable Additivity altogether? By giving up on Countable Additivity, one would be free to assign credence 0 to each of a countable set of mutually exclusive propositions. For example, one could assign credence 0 to each of our propositions $G_n$. (Keep in mind that assigning an event credence 0 is not the same as taking the event to be impossible. It is just to say that your degree of belief is so low that no positive real number is small enough to measure it.)

Unfortunately, giving up Countable Additivity can lead to a theory with undesirable mathematical properties. Here is an example, which I borrow from Kenny Easwaran.

Let $\mathbb{Z}^+$ be the set of positive integers and let $X$ and $Y$ be subsets of $\mathbb{Z}^+$. Let $p(X)$ be the probability that God selects a positive integer in $X$, and let $p(X|Y)$ be the probability that God selects a number in $X$ given that She selects a number in $Y$. One might think that the following is an attractive way of characterizing $p(X)$ and $p(X|Y)$:

$$p(X|Y) =_{df} \lim_{n\to\infty} \frac{|X \cap Y \cap \{1, 2, \ldots, n\}|}{|Y \cap \{1, 2, \ldots, n\}|}$$

$$p(X) =_{df} p(X|\mathbb{Z}^+)$$

(Notice that $p(X)$ is finitely additive but not countably additive, since $p(\mathbb{Z}^+) = 1$ but $p(\{k\}) = 0$ for each $k \in \mathbb{Z}^+$. Notice also that $p$ is not well-defined for arbitrary sets of integers. For instance, $\lim_{n\to\infty} \frac{|X \cap \{1, 2, \ldots, n\}|}{|\{1, 2, \ldots, n\}|}$ is not well-defined when $X$ consists of the integers $k$ such that $2^m \leq k < 2^{m+1}$ for some even $m$.)

As it turns out, $p(X)$ has the following awkward property: there is a set $S$ and a partition $E_i$ of $\mathbb{Z}^+$ such that $p(S) = 0$ even though $p(S|E_i) \geq 1/2$ for each $E_i$. For example, let $S$ be the set of squares, and for each $i$ that is not a power of any other positive integer, let $E_i$ be the set of powers of $i$. In other words:

$$\begin{array}{lcc}
S & = & \{1, 4, 9, 16, 25, \ldots\} \\
E_1 & = & \{1\} \\
E_2 & = & \{2, 4, 8, 16, 32, \ldots\} \\
E_3 & = & \{3, 9, 27, 81, 243, \ldots\} \\
\text{[No } E_4\text{, since } 4 = 2^2\text{]} & & \\
E_5 & = & \{5, 25, 125, 625, 3125, \ldots\} \\
 & \vdots &
\end{array}$$

It is easy to verify that $p(S|E_1) = 1$ and $p(S|E_n) = 1/2$ for each $n > 1$, even though $p(S) = 0$.

To see why this is awkward, imagine that Susan's credences are given by $p$. Then Susan can be put in the following situation: there is a sequence of bets such that Susan thinks she ought to take each of the bets, but such that she believes to degree 1 that she will lose money if she takes them all.

Here is one way of spelling out the details. For each $E_i$, Susan is offered the following bet:

$\boldsymbol{B_{E_i}}$: Suppose God selects a number in $E_i$. Then you'll receive \$2 if the selected number is in $S$, and you'll be forced to pay \$1 if the selected number is not in $S$. (If the selected number is not in $E_i$, then the bet is called off and no money exchanges hands.)

We know that $p(S|E_i) \geq 1/2$ for each $E_i$. So the expected value of $B_{E_i}$ on the assumption that the selected number is in $E_i$ will always be positive. (It is at least $\$2 \cdot 1/2 - \$1 \cdot 1/2 = \$0.5$.) So Susan should be willing to accept $B_{E_i}$ for each $E_i$. Notice, however, that Susan believes to degree 1 that if she does accept all such bets, she'll lose money. For Susan believes to degree 1 that God will select a number $x$ outside $S$ ($p(S)=0$). And if $x$ is outside $S$, then Susan loses \$1 on bet $B_{E^x}$ (where $E^x$ is the $E_i$ such that $x \in E_i$), with no money exchanging hands on any other bet.

As it turns out, problems of this general form are inescapable: *they will occur whenever a probability function on a countable set of possibilities fails to be countably additive.* (See section 6.6 for further details.) This suggests that rejecting Countable Additivity is less attractive than one might have thought: it is true that doing so allows one to distribute probability uniformly across a countable set of mutually exclusive propositions, but the sorts of distributions one gets are not especially attractive.

My own view is that, on balance, it is best to accept Countable Additivity. But, as I suggested earlier, it is important to be clear that the issues here are far from straightforward.

## 6.5 Optional: The Two-Envelope Paradox[3]

"Free Money!" reads a sign near the town square. Next to the sign is small man in a white suit. Next to the man is a table, with two envelopes on top. "This must be a joke!" you think as you walk by. "Who could possibly be giving out free money?" A few days later, however, a friend informs you that the sign is for real. The small man in the white suit is actually giving out free money. "There must be some kind of catch!" you say. "How does it work?"

"It's very simple," your friend explains. "The man has placed a check in each envelope. We don't know how much money the checks are made out for—it's different each day. But we do know about the method that the man uses to fill out the checks. The first check is made out for $n$ dollars, where $n$ is a natural number chosen at random. The second check is made out for $2n$ dollars. Unfortunately, you won't know which of the two checks has the larger amount and which one has the smaller."

"That's it?" you ask, incredulous.

"That's it!" your friend replies. "All you need to do is pick one of the envelopes and you'll get to keep the check. I played the game last week, and I ended up with a check for \$1,834,288. I took it to the bank and cashed it without a hitch. Now I'm looking to buy a beach house. It's a shame the man won't let you play more than once!"

Unable to help yourself any longer, you decide to pay a visit to the man in the white suit. "Welcome!" he says, with a big smile. "Select one of these envelopes and its contents will be yours." After a moment's hesitation, you reach for the envelope on the left. How wonderful it would be to have your very own beach house!

3. An earlier version of this material appeared in Rayo, "La Paradoja de los dos sobres," *Investigación y Ciencia*, June 2012.

As you are about to open the envelope, the man interjects: "Are you interested in mathematics?"

"Why, yes. I am," you say. "Why do you ask?"

"You've selected the envelope on your left, but I'll give you a chance to change your mind. Think about it for a moment," the man adds with a smile, "it might be in your interests to do so." You consider the issue for a moment, and conclude that the man is right: you should switch! Your reasoning is as follows:

> I know my envolope contains a certain amount of money: $k$ dollars, say. This means that the other envelope must contain either $2k$ dollars or $k/2$ dollars. These outcomes ought to have equal probability, since the initial number $n$ was selected at random. And if the outcomes have equal probability, I should switch. For although it is true that if I switch, I am just as likely to gain money as I am to lose money, what I'll gain if I gain is more than what I'll lose if I lose. More specifically, if I gain, I'll gain $k$ dollars, and if I lose, I'll lose $k/2$ dollars.
>
> The same point can be made formally. The expected value of switching is as follows:
>
> $$EV(\text{switch})k/2 \cdot 0.5 + 2k \cdot 0.5 = 5/4 \cdot k$$
>
> And, of course, the expected value of staying is $k$. Since $5/4 \cdot k$ is always larger than $k$, Expected Value Maximization entails that I should switch!

"All right!" you cry, "I'll switch!" With a trembling hand, you place your envelope back on the table, still unopened, and pick up the other.

The man in the white suit interrupts, "Excuse me, but are you sure you don't want to switch again? It might be in your interests to do so...."

It is at that point that the problem dawns on you. The exact same reasoning that led you to switch the first time could be used to argue that you should switch again and go back to your original selection!

### Exercises

1. When I speak of "probability" above, should I be understood as talking about subjective probability or objective probability?

### 6.5.1 Paradox

We are before a paradox. We have an apparently valid argument telling us that you should switch *regardless of which of the envelopes you've selected.* And that can't be the right conclusion, since it would lead you to switch envelopes indefinitely. Where have we gone wrong?

The Two-Envelope Paradox, as the problem has come to be known, was first introduced by the Belgian mathematician Maurice Kraitchik in 1953. When I first heard it, I was unimpressed. It seemed to me that the setup relied on a problematic assumption: the assumption that the man in the white suit is able to pick a natural number at random and do so in such a way that different numbers always have the same probability of being selected. (The reasoning above relies on this assumption by assigning equal probability to the outcome in which you get $2k$ dollars by switching and the outcome in which you get $k/2$ dollars by switching.) As we saw in section 6.4, however, there can be no such thing as a *uniform* probability distribution over the natural numbers, on pain of violating Countable Additivity—a principle that I very much hoped to keep in place.

No uniform probability distribution, no paradox! Or so I thought.... I have since come to see that I was wrong. It is a mistake to think that the paradox requires a uniform probability distribution over the natural numbers. As it turns out, there are non-uniform probability distributions that yield the result that switching envelopes always has a higher expected value than staying put. Here is an example, credited to Oxford philosopher John Broome. Take a die and toss it until it lands on 1 or 2. If it first lands on 1 or 2 on the $k$th toss, place $2^{k-1}$ in the first envelope and twice that amount in the second. As you'll be asked to prove below, this setup entails that the expected value of switching is always greater than the expected value of staying put.

#### Exercises

Suppose the man in the white suit uses Broome's method to select natural numbers:

1. What is the probability that the man fills the first envelope with $2^n$ and the second envelope with twice that amount?
2. If your envelope contains \$1, you should definitely switch. Suppose your envelope contains $\$2^k$ for some $k > 0$. Then the expected value of not switching is $\$2^k$. What is the expected value of switching?

### 6.5.2 A Way Out?

My favorite answer to the Two-Envelope Paradox is credited to Broome and to New York University philosopher David Chalmers. One starts by asking a simple question: in a two-envelope setup such as Broome's, what is the expected value of a given envelope? As I'll ask you to verify below, the answer is that each envelope has *infinite* expected value.

This is no accident: one can prove that a two-envelope setup can only lead to trouble when the expected value of the envelopes is infinite. But, as Broome and Chalmers point out, we have independent reasons for thinking that decisions involving options of infinite expected value are problematic.

Consider, for example, the **St. Petersburg Paradox**, which was discovered by the Swiss mathematician Nicolaus Bernoulli. A coin is tossed until it lands Heads. If it lands Heads on the $n$th toss you are given $\$2^n$. How much should you pay for the privilege of playing such a game? The expected value of playing is infinite:

$$EV(\text{play}) = \frac{1}{2^1} \cdot 2^1 + \frac{1}{2^2} \cdot 2^2 + \ldots = 1 + 1 + \ldots = \infty$$

So (assuming you value only money and value \$$n$ to degree $n$) the Principle of Value Maximization entails that you should be willing to pay *any finite amount of money* for the privilege of playing. And this is surely wrong.

I think that Broome and Chalmers are right to think that decisions involving options of infinite expected value are problematic and I believe that this helps answer the Two-Envelope Paradox. I must confess, however, that I do not feel ready to let the matter rest. As with the St. Petersburg Paradox, the Two-Envelope Paradox suggests that our decision-theoretic tools fall short when we try to reason with infinite expected value. But I'm not sure I have a satisfactory way of theorizing about such cases. If I were to assign the Two-Envelope Paradox a "paradoxicality grade" of the kind we used in chapter 3, I would choose a number somewhere between 6 and 8. It is clear that the paradox brings out some important limitations of decision theory, but I do not feel able to gauge just how important they are.

### Exercises

1. Suppose that an envelope is filled using Broom's method: a die is tossed until it lands on 1 or 2. If the die first lands on 1 or 2 on the $k$th toss, the envelope is filled with \$$2^{k-1}$. What is the expected value of the envelope?
2. Here is a possible response to the Two-Envelope Paradox, which relies on material from chapter 2: "We know that the expected value of your envelope is $|\mathbb{N}|$ and that the expected value of the other envelope is twice that: $|\mathbb{N}| \otimes 2$. But we saw in section 2.3.4 that $|\mathbb{N}| \otimes 2 = |\mathbb{N}|$. So the two envelopes have the same expected value. So you have no reason to switch." Is this an adequate response to the paradox?

## 6.6 Further Resources

Many of the ideas discussed in this chapter were originally set forth by David Lewis in his classic paper "A Subjectivist Guide to Objective Chance."

### 6.6.1 Additional Texts

- For a good overview of Bayesian theories of probability, I recommend Kenny Easwaran's "Bayesianism I: Introduction and Arguments in Favor" and "Bayesianism II: Criticism and Applications."
- For a mathematically rigorous introduction to probability theory, see Joseph K. Blitzstein's *Introduction to Probability*.

- For comprehensive treatment of the Principle of Value Maximization, and the ensuing decision theory, see Richard Jeffrey's *The Logic of Decision*.
- The cube factory example that I used to illustrate the difficulty of formulating an adequate Indifference Principle was drawn from Bas van Fraassen's *Laws and Symmetry*.
- If you'd like to know more about the best-systems account of objective probability, I recommend David Lewis's "Humean Supervenience Debugged."
- To learn more about the ways in which probabilities might be derived from a physical theory without bringing in subjective considerations, I recommend Tim Maudlin's "Three Roads to Objective Probability 1."
- My own thoughts about localism are drawn from Crispin Wright's "Wittgensteinian Certainties," which is itself inspired in Ludwig Wittgenstein's *On Certainty*.
- For a discussion of infinitesimals in the context of probability, see Timothy Williamson's "How Probable Is an Infinite Sequence of Heads?"
- In section 6.4.1.3, I mention that failures of Countable Additivity inevitably lead to mathematical awkwardness. The result I have in mind is this: if a probability function on a countable space fails to be countably additive, then there is no conditional probability function extending it that satisfies conglomerability (i.e., the condition that whenever $k_1 \leq p(E|H) \leq k_2$ for each $H$ in a set of mutually exclusive and jointly exhaustive propositions, $k_1 \leq p(E) \leq k_2$). The result is proved in Schervish, Seidenfeld, and Kadane's "The Extent of Non-Conglomerability of Finitely Additive Probabilities" and Hill and Lane's "Conglomerability and Countable Additivity." I learned about it in Kenny Easwaran's "Conditional Probability." If you're interested in this topic, you might also like to have a look at Easwaran's "Why Countable Additivity?"
- For Broome's treatment of the Two-Envelope Paradox, see his "The Two-Envelope Paradox." For David Chalmers's treatment, see his "The St. Petersburg Two-Envelope Paradox."

## Appendix: Answers to Exercises

### Section 6.1.1

1. As long as you trust the time traveler, you should be confident that the coin will land Heads even though the objective probability of decay is 50%. The information that the time traveler gave you is information about events that have not yet happened. This means that you have *better* information than a perfectly rational agent who has perfect information about the past. So you have no reason to defer to the judgment of such an agent. So the Objective-Subjective Connection does not entail that your subjective probabilities should agree with the objective probabilities.

### Section 6.2.1

1. Since it is necessarily true that $A$ or not-$A$, Necessity tells us that

   $$p(A \text{ or } \overline{A}) = 1.$$

   Since $A$ and not-$A$ are incompatible with one another, Additivity tells us that

   $$p(A \text{ or } \overline{A}) = p(A) + p(\overline{A}).$$

   Putting the two together we get

   $$p(A) + p(\overline{A}) = 1.$$

   So:

   $$p(\overline{A}) = 1 - p(A).$$

   And, of course, if $p(\overline{A}) = 1 - p(A)$, $p(\text{Rain})$ and $p(\overline{\text{Rain}})$ can't both be 0.9.
2. Since $A$ is equivalent to $AB$ or $A\overline{B}$, and since $AB$ and $A\overline{B}$ are incompatible, Additivity gives us the following:

   $$p(A) = p(AB) + p(A\overline{B})$$

   But since $p(A\overline{B})$ must be a real number between 0 and 1, this means that

   $$p(A) \geq p(AB).$$

### Section 6.2.3

1. $$p(R|D) = \frac{p(RD)}{p(D)} = \frac{0.09}{0.1} = 0.9$$

   $$p(D|R) = \frac{p(DR)}{p(R)} = \frac{p(RD)}{p(R)} = \frac{0.09}{0.2} = 0.45$$

2. $p(R|D) = \frac{p(RD)}{p(D)} = \frac{0.02}{0.1} = 0.2 = p(R)$

$p(D|R) = \frac{p(DR)}{p(R)} = \frac{p(RD)}{p(R)} = \frac{0.02}{0.2} = 0.1 = p(D)$

So $p(R|D) = p(R)$ and $p(D|R) = p(D)$. As we'll see in section 6.3.2, this must always be the case when $p(RD) = p(R) \cdot p(D)$.

3. No. Learning $B$ cannot cause the subject to lose her certainty in $A$.

To see this, let us assume otherwise and use this assumption to prove a contradiction. We will assume, in particular, that:

$p^{old}(A) = 1 \qquad p^{new}(A) < 1$

where $p^{old}$ is $S$'s credence function before she learns $b$ and $p^{new}$ is $S$'s credence function after she learns that $B$. Since $S$ updates by conditionalizing, this gives us:

$p^{old}(A) = 1 \qquad p^{old}(A|B) < 1$

We now verify the following propositions, where $p$ is a probability function and $\overline{X}$ is the negation of $X$:

(*a*) $p(A) = 1$ entails $p(\overline{A}) = 0$

(*b*) $p(A) = 1$ entails $p(\overline{A}B) = 0$

(*c*) $p(A) = 1$ entails $p(AB) = p(B)$

Proposition (*a*) is an immediate consequence of the fact that $p(\overline{A}) = 1 - p(A)$, which we verified in section 6.2.1. To verify proposition (*b*), note that Additivity entails that $p(\overline{A}) = p(\overline{A}B) + p(\overline{A}\overline{B})$. But we know from proposition (*a*) that $p(\overline{A})$ is zero, so $p(\overline{A}B)$ and $p(\overline{A}\overline{B})$ must both be zero as well (since probabilities are always non-negative real numbers). To verify proposition (*c*), note that Additivity entails that $p(B) = p(AB) + p(\overline{A}B)$. But we know from proposition (*b*) that $p(\overline{A}B)$ is zero, so it must be the case that $p(B) = p(AB)$.

Since we are assuming that $p^{old}$ is a probability function, we may conclude that $p^{old}(B) = p^{old}(AB)$. But note that Bayes' Law entails that $p^{old}(AB) = p^{old}(B) \cdot p^{old}(A|B)$. Since $p^{old}(B) = p^{old}(AB)$, this means that $p^{old}(B) = p^{old}(B) \cdot p^{old}(A|B)$. So as long as $p^{old}(B) \neq 0$, it must be the case that $p^{old}(A|B) = 1$, which contradicts the assumption that $p^{old}(A|B) < 1$.

### Section 6.2.4.1

1. Since you have no more reason for thinking that God picked $\pi$ than you have for thinking that God picked any other number, the Principle of Indifference tells you

that you must assign the same credence to the proposition that God picked $\pi$ and the proposition that God picked $x$, for any other real number $x$.

But, since there are infinitely many real numbers, you must assign credence 0 to each proposition of the form "God picked $x$." For suppose otherwise: suppose that to each proposition of the form "God picked $x$," you assign some number $r > 0$.

Since your credence function is a probability function, it satisfies the principle of Additivity. This means that for any real numbers $x_1, x_2, \ldots, x_n$, your credence in "God picked $x_1$ or $x_2$ or ... or $x_n$" equals your credence in "God picked $x_1$" plus your credence in "God picked $x_2$" ... plus your credence in "God picked ... $x_n$." But we are assuming that your credence in "God picked $x_i$" is $r$, for each $i \leq n$. So your credence in "God picked $x_1$ or $x_2$ or ... or $x_n$" must be $\underbrace{r+r+\ldots+r}_{n}$ which will be greater than 1 for big-enough $n$. And if your credence function is a probability function, no credence can be greater than 1.

## Section 6.3.1

1. The objective probability must be 0 or 1. For a perfectly rational agent with perfect information about the past would be in a position to determine the outcome of the coin toss.

## Section 6.3.2

1. Consider a perfectly rational agent who at each $t_i$ $(i \leq 1)$ has perfect information about events before $t_i$ (and no information about events after $t_i$). By the Subjective-Objective Connection, our subject's credences at each $t_i$ must line up with the objective probabilities at each $t_i$, because perfectly rational agents update their credences by conditionalization

   $$p^{t_1}(A) = p^{t_0}(A|H_{t_1})$$

   where $p^{t_i}$ is the subject's credence function at $t_i$ and $H_{t_i}$ consists of perfect information about events before $t_i$. So the Subjective-Objective Connection yields $p_{t_1}(A) = p_{t_0}(A|H_{t_1})$.

2. 
$$\begin{aligned}
p(R_2) &= p(R_1R_2 \text{ or } B_1R_2) && [R_2 \text{ equivalent to } (R_1R_2 \text{ or } B_1R_2)]\\
&= p(R_1R_2) + p(B_1R_2) && [\text{Additivity}]\\
&= p(R_1)\cdot p(R_2|R_1) + R(B_1)\cdot p(R_2|B_1) && [\text{Bayes' Law}]\\
&= 1/2 \cdot 1/3 + 1/2 \cdot 2/3\\
&= 1/2
\end{aligned}$$

3. (a) Bayes' Law gives us each of the following:

$$p(A|B) = \frac{p(AB)}{p(B)} \quad p(A|\overline{B}) = \frac{p(A\overline{B})}{p(\overline{B})}$$

So the following identities are all equivalent:

$$p(A|B) = p(A|\overline{B})$$

$$\frac{p(AB)}{p(B)} = \frac{p(A\overline{B})}{p(\overline{B})}$$

$$p(\overline{B}) \cdot p(AB) = p(B) \cdot p(A\overline{B})$$

$$(1 - p(B)) \cdot p(AB) = p(B) \cdot p(A\overline{B})$$

$$p(AB) - p(B) \cdot p(AB) = p(B) \cdot p(A\overline{B})$$

$$p(AB) = p(B) \cdot p(A\overline{B}) + p(B) \cdot p(AB)$$

$$p(AB) = p(B) \cdot (p(A\overline{B}) + p(AB))$$

$$p(AB) = p(B) \cdot p(A\overline{B} \text{ or } AB)$$

$$p(AB) = p(B) \cdot p(A)$$

$$p(AB) = p(A) \cdot p(B)$$

(b) Bayes' Law gives us this:

$$p(A|B) = \frac{p(AB)}{p(B)}$$

So the following identities are all equivalent:

$$p(AB) = p(A) \cdot p(B)$$

$$\frac{p(AB)}{p(B)} = p(A)$$

$$p(A|B) = p(A)$$

But, by the previous exercise, we have

$p(A|B) = p(A|\overline{B})$ if and only if $p(AB) = p(A) \cdot p(B)$.

So we conclude

$p(A|B) = p(A|\overline{B})$ if and only if $p(A|B) = p(A)$.

4. Exercise 3 gives us each of the following:

$p(A|B) = p(A|\overline{B})$ if and only if $p(AB) = p(A) \cdot p(B)$

$p(B|A) = p(B|\overline{A})$ if and only if $p(BA) = p(B) \cdot p(A)$

But since $AB$ is equivalent to $BA$, we have $p(AB) = p(BA)$. So

$$p(A|B) = p(A|\overline{B}) \text{ if and only if } p(B|A) = p(B|\overline{A}).$$

5. Answer: $1/2^n$

   Let $H_k$ be the proposition that the coin lands Heads on the $k$th toss. Since the coin is fair, we know that $p(H_k) = 1/2$. And since different coin tosses are independent of one another, exercise 3 from section 6.3.2 entails that the following is true whenever $k \neq l$:

   $$p(H_k H_l) = p(H_k) \cdot p(H_l)$$

   Putting all of this together:

   $$p(H_1 H_2 \ldots H_n) = p(H_1) \cdot p(H_2) \cdot \ldots \cdot (H_n) = \underbrace{1/2 \cdot 1/2 \cdot \ldots \cdot 1/2}_{n} = 1/2^n$$

6. Let $\overset{k\to\infty}{H}$ be the proposition that the coin lands Heads on the $k$th toss and on every toss thereafter. Let us assume that the probability of $\overset{1\to\infty}{H}$ is a real number between 0 and 1; based on that assumption, let us show that the probability must be 0.

   The first thing to note is that, for any $k$, $\overset{1\to\infty}{H}$ is equivalent to $H_1 H_2 \ldots H_{k-1} \overset{k\to\infty}{H}$. But, since different coin tosses are independent of one another, exercise 3 from section 6.3.2 entails the following:

   $$p\left(\overset{1\to\infty}{H}\right) = p\left(H_1 H_2 \ldots H_k \overset{k+1\to\infty}{H}\right) = p\,(H_1 H_2 \ldots H_k) \cdot p\left(\overset{k+1\to\infty}{H}\right)$$

   But, by exercise 5 from section 6.3.2, $p\,(H_1 H_2 \ldots H_k) = 1/2^k$. Since $p\left(\overset{k+1\to\infty}{H}\right)$ cannot be greater than 1, this means that

   $$p\left(\overset{1\to\infty}{H}\right) \leq 1/2^k$$

   for every $k$. But 0 is the only real number between 0 and 1 that is smaller than or equal to $1/2^k$ for every $k$.

### Section 6.3.3.1

1. Set $\epsilon = 0.0001$. Set $\delta$ to be some number smaller than 0.01; say $\delta = 0.005$. It follows from the Law of Large Numbers that there is an $N$ such that the following will be true with a probability of at least $1 - \epsilon = 0.9999$: If the slot machines are used $M > N$ times, then the proportion of them that result in a win for the casino will be $0.51 \pm 0.005$. So, if the machine is played $M$ times, there is a probability of 0.9999 that the casino wins at least 50.5% of the time and therefore ends up with a profit of at

least $M \cdot 0.505 \cdot \$1 - M \cdot 0.495 \cdot \$1 = \$(M \cdot 0.01)$ (which will be more than a million dollars, for large enough $M$).

### Section 6.4.1.1

1. As before, let $G_n$ be the proposition that God selected the number $n$. Then the proposition that God selected some positive integer or other can be expressed as

   $$G_1 \text{ or } G_2 \text{ or } G_3 \text{ or } \ldots.$$

   But, by Countable Additivity,

   $$c_s(G_1 \text{ or } G_2 \text{ or } \ldots) = c_S(G_1) + c_S(G_2) + \ldots = 1/2 + 1/4 + \ldots + 1/2^n + \ldots = 1$$

   (If you'd like to know more about why $1/2 + 1/4 + \ldots + 1/2^n + \ldots = 1$ is true, have a look at the Wikipedia entry on convergent series.)

### Section 6.5

1. What I have in mind is subjective probability and, more specifically, the credences that you ought to have, given the information at your disposal.

   Since there is a fact of the matter about how much money each of the envelopes contains, the objective probability of getting $k/2$ dollars if you switch must be either 0 or 1, rather than 50%, as I suggest in the text. So the difference between objective probability and subjective probability definitely matters.

### Section 6.5.1

1. For each $k$, let $S_k$ be the proposition that the number $2^k$ is selected. Then:

   $$p(S_0) = 1/3$$

   $$p(S_1) = 2/3 \cdot 1/3 = 2/9$$

   $$p(S_2) = 2/3 \cdot 2/3 \cdot 1/3 = 4/27$$

   $$\vdots$$

   $$p(S_n) = (2/3)^n \cdot 1/3 = 2^n/3^{n+1}$$

2. Since your envelope contains $2^k$ dollars ($k > 0$), the number originally selected by the man must have been either $2^{k-1}$ or $2^k$. So there are two relevant states of the world. In the first scenario, the man selected $2^{k-1}$, and the envelope you did not pick contains $2^{k-1}$ dollars. (I call this scenario $O_{2^{k-1}}$.) In the second scenario, the

man selected $2^k$, and the envelope you did not pick contains $2^{k+1}$ dollars. (I call this scenario $O_{2k+1}$.)

Using Broome's method, the probability of selecting $2^{k-1}$ is $2^{k-1}/3^k$, and the probability of selecting $2^k$ is $2^k/3^{k+1}$. So, by Additivity, the probability of selecting one or the other is as follows:

$$\frac{2^{k-1}}{3^k} + \frac{2^k}{3^{k+1}}$$

This allows us to calculate $p(O_{2k-1})$, which is the probability that the man selects $2^{k-1}$, given that he selects $2^{k-1}$ or $2^k$:

$$p(O_{2k-1}) = \frac{\frac{2^{k-1}}{3^k}}{\frac{2^{k-1}}{3^k} + \frac{2^k}{3^{k+1}}} = \frac{2^{k-1}}{2^{k-1} + \frac{2^k}{3}} = \frac{1}{1 + \frac{2}{3}} = \frac{3}{5}$$

We can also calculate $p(O_{2k+1})$, which is the probability that the man selects $2^k$, given that he selects $2^{k-1}$ or $2^k$:

$$p(O_{2k+1}) = 1 - \frac{3}{5} = \frac{2}{5}$$

Accordingly, the expected value of switching ($S$) can be computed as follows:

$$\begin{aligned} EV(S) &= 2^{k-1} \cdot p(O_{2k-1}) + 2^{k+1} \cdot p(O_{2k+1}) \\ &= 2^{k-1} \cdot \frac{3}{5} + 2^{k+1} \cdot \frac{2}{5} \\ &= 2^k \cdot \frac{11}{10} \end{aligned}$$

So the expected value of switching ($2^k \cdot 11/10$ dollars) is larger than the expected utility of not switching ($2^k$ dollars).

### Section 6.5.2

1. We know from exercise 6.5.1 that the probability that the die first lands on one or two on the $k$th toss (and that the envelope is therefore filled with $2^{k-1}$) is $2^{k-1}/3^k$. So the expected value of the envelope is:

$$\begin{aligned} EV(\text{envelope}) &= \frac{2^0}{3^1} \cdot 2^0 + \frac{2^1}{3^2} \cdot 2^1 + \frac{2^2}{3^3} \cdot 2^2 + \ldots \\ &= \frac{(2^0)^2}{3^1} + \frac{(2^1)^2}{3^2} + \frac{(2^2)^2}{3^3} + \ldots \end{aligned}$$

$$= \frac{(2^2)^0}{3^1} + \frac{(2^2)^1}{3^2} + \frac{(2^2)^2}{3^3} + \ldots$$
$$= \frac{4^0}{3^1} + \frac{4^1}{3^2} + \frac{4^2}{3^3} + \ldots$$
$$= \infty$$

To verify that that the final identity holds, note that $\frac{4^k}{3^{k+1}} < \frac{4^{k+1}}{3^{k+2}}$ for any $k$, since $\frac{4^{k+1}}{3^{k+2}} = \frac{4^k}{3^{k+1}} \cdot \frac{4}{3}$.

2. One reason to think that the response is inadequate is that it presupposes that you should always be indifferent between options with infinite expected value. But consider a variant of the St. Petersburg Paradox in which you are given $\$2^{n+1}$ (rather than $\$2^n$) if the coin first lands Heads on the $n$th toss. Playing either version of the game has infinite expected value, but you should obviously prefer to play the new version.

# 7 Non-Measurable Sets

Here are some examples of **additive** notions of size:

- The **probability** that either of two (incompatible) events occurs is the probability that the first occurs plus the probability that the second occurs;
- The **length** of two (non-overlapping) line segments placed side by side is the length of the first plus the length of the second;
- The **mass** of two (non-overlapping) objects taken together is the mass of the first plus the mass of the second.

Notice, in contrast, that the notion of temperature is not additive in this sense: it is not generally the case that the temperature of two (non-overlapping) objects taken together is the result of adding the temperatures of the individual objects.

In this chapter we'll talk about the notion of **measure**, which is a very abstract way of thinking about additive notions of size. We'll also talk about **non-measurable** sets: sets so bizarre that they cannot be assigned a (mathematically respectable) measure. Non-measurable sets have very unusual properties and can be used to prove some pretty remarkable results. One of these results is the Banach-Tarski Theorem, which shows that it is possible to divide a ball into a finite number of pieces and reassemble the pieces (without changing their size or shape) so as to get two balls, each of them of the same size as the original ball. Sounds impossible, right? And yet it's true! We'll prove it in the next chapter.

## 7.1 Measure

One way to get clear about the notion of measure is to start by thinking about the notion of *length*. We will consider the notion of length as it applies to line segments $[a, b]$, where $a$ and $b$ are real numbers such that $a \leq b$. Although it is natural to think of line segments in spatial terms, here we will model them as sets of real numbers. More specifically, we will think of the line segment $[a, b]$ as the set of real numbers $x$ such that

$a \le x \le b$. So, for instance, we take $[1/4, 1/2]$ to be the set $\{x : 1/4 \le x \le 1/2\}$. (*Note:* We will treat $[a, a] = \{a\}$ as a line segment and therefore regard points as special cases of line segments.)

A nice feature of line segments is that there is a simple recipe for characterizing their length:

$$\text{Length}([a, b]) = b - a.$$

For instance, [1/2, 3/4] has length 1/4, because $3/4 - 1/2 = 1/4$; and [2, 2] has length 0, because $2 - 2 = 0$. Now suppose that we wish to generalize the notion of length by applying it to sets of real numbers other than line segments. Suppose, for example, that we wish to assign a "length" to the set

$$[0, \tfrac{1}{4}] \cup [\tfrac{1}{2}, 1].$$

Since $[0, \frac{1}{4}]$ and $[\frac{1}{2}, 1]$ have no points in common, it is natural to take the length of their union to be the sum of their individual lengths, as follows:

$$\text{Length}([0, \tfrac{1}{4}] \cup [\tfrac{1}{2}, 1]) = \text{Length}([0, \tfrac{1}{4}]) + \text{Length}([\tfrac{1}{2}, 1]) = \tfrac{1}{4} + \tfrac{1}{2} = \tfrac{3}{4}$$

As it turns out, there is a systematic way of extending this general idea to a rich and interesting family of sets of real numbers. They are known as the *Borel Sets*, in honor of the great French mathematician Émile Borel. I'll tell you more about them in the next subsection.

### 7.1.1 The Borel Sets

A **Borel Set** is a set that you can get to by performing finitely many applications of the operations of *complementation* and *countable union* on a family of line segments:

- The **complementation operation** takes each set $A$ to its complement, $\overline{A}$ (which is $\mathbb{R} - A$, where $\mathbb{R}$ is the set of real numbers). For instance, the result of applying the complement operation to [0, 1] is the set $\overline{[0,1]} = \mathbb{R} - [0, 1]$, which consists of every real number except for the members of [0, 1].
- The **countable union operation** takes each countable family of sets $A_1, A_2, A_3, \ldots$ to their union, $\bigcup\{A_1, A_2, A_3, \ldots\}$. For instance, the result of applying the countable union operation to the sets $[0, 1/2], [1/2, 3/4], [3/4, 7/8], \ldots$ is the set $[0, 1)$, which consists of the real numbers $x$ such that $0 \le x < 1$. (Note that the round bracket on the right-hand side of "[0, 1)" is used to indicate that the end-point 1 is not included in the set.)

(Formally, the Borel Sets are the members of the smallest set $\mathcal{B}$ such that: (*i*) every line segment is in $\mathcal{B}$, (*ii*) if a set is in $\mathcal{B}$, then so is its complement, and (*iii*) if a countable family of sets is in $\mathcal{B}$, then so is its union.)

To get some practice working with Borel Sets, let us verify that the set of *irrational* real numbers $\overline{\mathbb{Q}}$ is a Borel Set. (Note that $\overline{\mathbb{Q}} = \mathbb{R} - \mathbb{Q}$, where $\mathbb{Q}$ is the set of rational numbers.) What we need to show is that set $\overline{\mathbb{Q}}$ can be generated by applying finitely many operations of complementation and countable union to a family of line segments. This can be done in four steps:

**Step 1:** One gets a line segment by performing *zero* operations of complementation and countable union to that line segment. So every line segment is a Borel Set.

**Step 2:** The singleton set $\{r\}$ $(r \in \mathbb{R})$ is identical to the point-sized line segment $[r, r]$. So, by step 1, $\{r\}$ is a Borel Set.

**Step 3:** Since the set of rational numbers, $\mathbb{Q}$, is countable, it is the countable union of $\{q\}$ $(q \in \mathbb{Q})$. So, by step 2, $\mathbb{Q}$ is a Borel Set.

**Step 4:** $\overline{\mathbb{Q}}$ is the complement of $\mathbb{Q}$. So, by step 3, $\overline{\mathbb{Q}}$ is a Borel Set.

(Note that a procedure of this kind can be used to show that any countable set and its complement is a Borel Set.)

#### Exercises

1. Where $A_1, A_2, A_3, \ldots$ is a countable family of Borel Sets, show that $\bigcap\{A_1, A_2, A_3, \ldots\}$ is a Borel Set. (In general, the **intersection** of a set $\{A_1, A_2, A_3, \ldots\}$ [in symbols: $\bigcap\{A_1, A_2, A_3, \ldots\}$] is the set of individuals $x$ such that $x$ is a member of each set $A_1, A_2, A_3, \ldots$.)
2. Assume that $A$ and $B$ are Borel Sets and show that $A - B$ (i.e., the set of members in $A$ that are not in $B$) is a Borel Set.

### 7.1.2 Lebesgue Measure

As it turns out, there is *exactly one* natural way of extending the ordinary notion of length so that it applies to all Borel Sets. More precisely, there is exactly one function $\lambda$ on the Borel Sets that satisfies these three conditions:

**Length on Segments** $\lambda([a, b]) = b - a$ for every $a, b \in \mathbb{R}$.

(This condition is meant to ensure that $\lambda$ counts as an extension, rather than a modification, of the notion of length.)

**Countable Additivity**

$$\lambda\left(\bigcup\{A_1, A_2, A_3, \ldots\}\right) = \lambda(A_1) + \lambda(A_2) + \lambda(A_3) + \ldots$$

whenever $A_1, A_2, \ldots$ is a countable family of disjoint sets, for each of which $\lambda$ is defined.

(For $A_1, A_2, \ldots$ to be **disjoint** is for $A_i$ and $A_j$ to have no members in common whenever $i \neq j$.)

**Non-Negativity** $\lambda(A)$ is either a non-negative real number or the infinite value $\infty$, for any set $A$ in the domain of $\lambda$.

(Note that when we transition from measuring line segments to measuring Borel Sets, we allow for sets of "infinite length," such as the set of non-negative real numbers, which might be abbreviated $[0, \infty)$, or $\mathbb{R}$, which might be abbreviated $(-\infty, \infty)$.)

The unique function $\lambda$ on the Borel Sets that satisfies these three conditions is called the **Lebesgue Measure**, in honor of another great French mathematician: Henry Lebesgue.

I will not prove that a unique function $\lambda$ satisfying these three conditions exists: I will simply assume that it does and that it is well-defined for every Borel Set. (If you'd like to learn how to prove this result, have a look at the recommended readings in section 7.4.)

#### Exercises

1. Identify the value of $\lambda(\emptyset)$, where $\emptyset$ is the empty set.
2. Identify the value of $\lambda([a, b))$, where $[a, b) = [a, b] - \{b\}$.
3. Identify the value of $\lambda(\mathbb{R})$.
4. Verify that every countable (i.e., finite or countably infinite) set has Lebesgue Measure 0.
5. Show that if $A$ and $B$ are both Borel Sets and $B \subset A$, then $\lambda(B) \leq \lambda(A)$.
6. Countable Additivity gives us an Additivity condition for finite or countably infinite families of disjoint sets. Would it be a good idea to insist on an Additivity condition for *uncountable* families of disjoints sets?

### 7.1.3 Extending Lebesgue Measure beyond the Borel Sets [Optional]

As it turns out, there is a natural way of extending the notion of Lebesgue Measure beyond the Borel Sets while preserving the crucial properties of Countable Additivity and Non-Negativity. (The property of Length on Segments is preserved automatically by any extension of $\lambda$.)

The key idea is that any subset of a Borel Set of measure 0 should be treated as having measure of 0. Formally, one extends the notion of **Lebesgue Measure** in two steps. The first step is to introduce a notion of *Lebesgue measurability*. We say that $A$ is Lebesgue measurable if and only if $A = A^B \cup A^0$, for $A^B$, a Borel Set and $A^0$, a subset of some Borel Set of Lebesgue Measure 0. The second step is to extend the function $\lambda$ by stipulating that $\lambda(A^B \cup A^0) = \lambda(A^B)$.

Note that every Borel Set will count as Lebesgue measurable, since the empty set is a subset of a Borel Set of Lebesgue Measure 0. And note that in extending $\lambda$, we haven't changed its value on Borel Sets. In the exercises below, I'll ask you to verify that the

Lebesgue measurable sets are closed under complements and countable unions, and that the extended version of $\lambda$ satisfies Countable Additivity and Non-Negativity.

It is worth noting that our definition of Lebesgue measurability wouldn't get us beyond the Borel Sets if the only Borel Sets with Lebesgue Measure 0 were countable sets, since every subset of a countable set is countable (and therefore a Borel Set) and since the union of two Borel Sets is always a Borel Set. As it turns out, however, there are *uncountable* Borel Sets of Lebesgue Measure 0 which have subsets that are not Borel Sets. The most famous example of an uncountable set of Lebesgue Measure 0 is the *Cantor Set*, which is named in honor of our old friend from chapter 1, Georg Cantor. We won't get into it here, but you can find plenty of information about it on the web.

The discussion that follows does not turn on whether $\lambda$ is defined only on the Borel Sets, or whether it is defined on the full space of Lebesgue-measurable sets.

### Exercises

In answering the questions below, assume the extended version of $\lambda$:

1. Show that if $A^0$ is a subset of a Borel Set with Lebesgue Measure 0, then $\lambda(A^0) = 0$.
2. Show that Lebesgue measurability is closed under complements.
3. Show that Lebesgue measurability is closed under countable unions.
4. Show that whenever $A$ is Lebesgue measurable, $\lambda(A)$ is either a non-negative real number or $\infty$, and therefore that Non-Negativity is preserved when $\lambda$ is extended to all Lebesgue-measurable sets.
5. Show that our extension of $\lambda$ to the Lebesgue-measurable sets preserves Countable Additivity. In other words, assume that $A_1, A_2, \ldots$ is a finite or countably infinite family of disjoint sets, and show that whenever $A_i$ is Lebesgue measurable for each $i$, we have the following:

$$\lambda\,(A_1 \cup A_2 \cup A_3 \cup \ldots) = \lambda(A_1) + \lambda(A_2) + \lambda(A_3) + \ldots$$

### 7.1.4 Uniformity

A function on the Borel Sets is said to be a **measure** if and only if it satisfies Countable Additivity and Non-Negativity (and assigns the value 0 to the empty set).

The Lebesgue Measure is a measure, but it is not the only one. It is special because it satisfies Length on Segments. This is important because it guarantees that the Lebesgue Measure is **uniform**. Intuitively speaking, what it means for a measure to be uniform is for it to assign the same value to sets that have the same "shape" but occupy different positions on the real line. The formal definition is as follows:

**Uniformity** $\mu(A^c) = \mu(A)$, whenever $\mu(A)$ is well-defined and $A^c$ is the result of adding $c \in \mathbb{R}$ to each member of $A$.

Here is an example. The Lebesgue Measure of $[0, \frac{1}{4}]$ is $\frac{1}{4}$. So the Uniformity of Lebesgue Measure entails that the Lebesgue Measure of $[0, \frac{1}{4}]^{\frac{1}{2}} = [1/2, 3/4]$ must also equal $\frac{1}{4}$.

Uniformity is such a natural property that it is tempting to take it for granted. But there are all sorts of measures that do not satisfy Uniformity. We'll talk about one of them in the next subsection.

**7.1.4.1 Probability measures** So far we have been thinking of the notion of measure as a generalization of the notion of *length*. But, as I suggested at the beginning of this chapter, the notion of measure can also be thought of as a generalization of the notion of *probability*.

Suppose, for example, that we have a random procedure for selecting individual points from the line segment $[0, 1]$. For $A$, a subset of $[0, 1]$, let $p(A)$ be the probability that the next point selected from $[0, 1]$ by our random procedure is a member of $A$. As long as $p(\ldots)$ satisfies Countable Additivity, it will count as a measure on the Borel Sets in [0,1]. But, interestingly, whether $p(\ldots)$ turns out to be the Lebesgue Measure depends on the details of our random selection procedure. The best way to see this is to consider two kinds of selection procedures:

**Standard Coin-Toss Procedure** You toss a fair coin once for each natural number. Each time the coin lands Heads you write down a 0, and each time it lands Tails you write down a 1. This gives you an infinite sequence $\langle d_1, d_2, d_3, \ldots \rangle$. The selection procedure is then as follows: Pick whichever number in $[0, 1]$ has $0.d_1d_2d_3\ldots$ as its binary expansion.

(If you'd like a refresher on binary expansions, have a look at section 1.7.1. As noted in that discussion, rational numbers have two different binary expansions: one ending in 0s and the other ending in 1s. To simplify the present discussion, I assume that the Coin-Toss Procedure is carried out anew if the output corresponds to a binary expansion ending in 1s.)

**Square Root Coin-Toss Procedure** Conduct the Coin-Toss Procedure as before, but this time, pick the *square root* of the number represented by $0.d_1d_2d_3\ldots$. (Suppose, for instance, that a sequence of coin tosses yields 0.0100(0), which is the binary expansion of 1/4. Then you should select 1/2 because $1/2 = \sqrt{1/4}$.)

The Standard Coin-Toss Procedure delivers a uniform probability distribution; and given certain assumptions about the probabilities of sequences of coin tosses, it delivers the Lebesgue Measure. The Square Root Coin-Toss Procedure, in contrast, does not satisfy Uniformity. It can be used to define a measure over $[0, 1]$, but not the Lebesgue Measure.

Here is a nice way of visualizing the difference between the two measures. Suppose we have 1kg of mud. We are asked to pour it on the line segment $[0, 1]$ in such a way that the amount of mud above an interval within $[0, 1]$ is proportional to the probability that the next point selected by one of our coin-toss procedures will fall within that interval. In the case of the Standard Coin-Toss Procedure, our mud distribution will look like this:

In other words, the probability of getting a number within a given interval does not depend on where the interval is located within [0, 1]; it depends only on the size of the interval. When it comes to the Square Root Coin-Toss Procedure, in contrast, our mud distribution will look like this:

In other words: the probability of getting a number within a given interval depends not just on the size of the interval, but also on the *location* of the interval within [0, 1]; the closer to 1 the interval is, the higher the probability of getting a number within that interval.

### Exercises

Answer the following questions with respect to each of our two coin-toss procedures:

1. What is the probability that the next selected point will be in [1/2, 1]?
2. What is the probability that the next selected point will be precisely 1/2? (Assume that the relevant probability is defined and is a real number between 0 and 1.)
3. What is the probability that the next selected point will be in [0, 1/2]? (*Hint:* Use your answer to the previous question and assume that the relevant probabilities satisfy Additivity.)
4. Suppose $\{x_0, x_1, x_2, \ldots\}$ is a countable set of numbers in [0, 1]. What is the probability that the next selected point will be in $\{x_0, x_1, x_2, \ldots\}$? (Assume that the relevant probabilities satisfy Countable Additivity.) *Hint:* Use your answer to question 2.
5. What is the probability that the next point selected will be a member of $[0, 1] - \{1/2\}$? What is the probability that it will be a member of $[0, 1] - \{x_0, x_1, x_2, \ldots\}$?

(Assume that the relevant probabilities are all well defined and that they satisfy Countable Additivity.)

## 7.2 Non-Measurable Sets

In this section we will see that there are subsets of $\mathbb{R}$ that are **non-measurable** in an especially strong sense. Not only do they fall outside the class of Lebesgue-measurable sets, but they also cannot be assigned a measure by any *extension* of the Lebesgue Measure function $\lambda$, unless one gives up on one or more of the principles of Non-Negativity, Countable Additivity, and Uniformity.

(What does it mean for a function $\lambda^+$ to extend $\lambda$? It means that whenever $\lambda(A)$ is defined, $\lambda^+(A)$ is also defined and equal to $\lambda(A)$. For technical reasons, we will impose the additional constraint that $\lambda^+$ only counts as an extension of $\lambda$ if it is closed under intersections; in other words, $\lambda^+(A \cap B)$ must be defined whenever $\lambda^+(A)$ and $\lambda^+(B)$ are both defined.)

To get an idea about why the existence of non-measurable sets in this strong sense is such a striking result, imagine that the Standard Coin-Toss Procedure of section 7.1.4.1 is used to randomly select a point from $[0, 1]$. What is the probability that the selected point will fall within some non-measurable set $A$? Answer: *There is no such thing as the probability that the selected point will turn out to be in A.*

### 7.2.1 The Axiom of Choice

In proving that there are non-measurable sets, we will rely on the Axiom of Choice. (In fact, it is impossible to prove that there are non-measurable sets without some version of the Axiom of Choice, as the American set theorist Robert Solovey proved in 1970.)

The Axiom of Choice is based on a very simple idea. Let $A$ be a set of sets. Then a **choice set** for $A$ is a set that contains exactly one member from each member of $A$. Suppose, for example, that $A$ is the following set:

$$\{\{0, 1\}, \{a, b, c\}, \{3\}, \{e, \pi, \iota\}\}$$

Then $\{1, a, 3, \pi\}$ is a choice set for $A$. (And, of course, there are many others.) The **Axiom of Choice** states that every set has a choice set, as long as each of its members is a non-empty set and as long as no two of its members have any members in common. Here is a more concise definition:

**Axiom of Choice** Every set of non-empty, non-overlapping sets has a choice set.

Upon first inspection, the Axiom of Choice can sound trivial. But it is most certainly not. The great British philosopher Bertrand Russell once suggested a nice way of bringing out the reason that the Axiom of Choice is so powerful. Let $S$ be a set consisting of infinitely many pairs of shoes, and suppose that you'd like to specify a choice set for $S$. Since each pair of shoes consists of a left shoe and a right shoe, one way of doing so is to use "left shoe" as a selection criterion and have one's choice set consist of all and only left shoes in members of $S$.

Now consider a variation of the case. Let $S'$ be a set consisting of infinitely many pairs of *socks*, and consider the question of how one might go about specifying a choice set for $S'$. Assuming there is no difference between left socks and right socks, it is not clear that we are in a position to specify a criterion that would apply to exactly one member of each of our infinitely many pairs of socks: a criterion that might play the same role that "left shoe" played in specifying a choice set for $S$.

The Axiom of Choice assures us that a choice set for $S'$ will exist regardless of our ability to specify such a criterion. That is the reason that the Axiom of Choice is so interesting: it allows us to work with sets whose membership we are not in a position to specify. We'll return to the Axiom of Choice in section 7.2.3.2.

### 7.2.2 The Vitali Sets[1]

The most famous examples of non-measurable sets are the Vitali Sets, named in honor of the Italian mathematician Giuseppe Vitali. In a paper published in 1905, Vitali showed that there is no way of extending the Lebesgue Measure $\lambda$ to apply to a Vitali Set while preserving all three of Non-Negativity, Countable Additivity, and Uniformity. I'll start by giving you an intuitive sense of why the Vitali Sets are not measurable, and then go through the proof.

**7.2.2.1 The intuitive picture** Recall the thought experiment of section 6.4.1.1. God has selected a positive integer and you have no idea which. What should be your credence that God selected the number seventeen? More generally, what should be your credence that She selected a given number $k$?

Our discussion revealed that as long as you think that your credence distribution ought to be *uniform*—as long as you think that the same answer should be given for each value of $k$—it is a consequence of Countable Additivity that your credences must remain undefined. For if $p(\text{God selects } k) = 0$ for each $k$, Countable Additivity entails that $p(\text{God selects a positive integer}) = 0$, which is incorrect. And if $p(\text{God selects } k) = r$ for each $k$, where $r$ is a positive real number, Countable Additivity entails that $p(\text{God selects a positive integer}) = \infty$, which is also incorrect. (We also saw that there are reasons of principle for doubting that infinitesimals can save the day, but we won't have to worry about infinitesimals here, since we're presupposing Non-Negativity, which entails that finite measures must be real numbers.)

The moral of our thought experiment is that, in the presence of Countable Additivity, there is no such thing as a uniform probability distribution over a countably

1. An earlier version of this material appeared in Rayo, "Colecciones no medibles," *Investigación y Ciencia*, October 2012.

infinite set of (mutually exclusive and jointly exhaustive) possibilities. The proof of Vitali's theorem is a version of this same idea. We'll partition $[0, 1)$ into a countable infinity of Vitali Sets and use Uniformity to show that these sets must all have the same measure, if they have a measure at all. We'll then use Countable Additivity and Non-Negativity to show that the Vitali Sets cannot have a measure in the same way we showed that your credences must remain undefined in the thought experiment.

*A nerdy aside:* When I first described the thought experiment, in chapter 6, I didn't say anything about the selection procedure God uses to pick a positive integer. In particular, I didn't clarify whether there could be a selection procedure that doesn't have the feature that some numbers are more likely to be selected than others. You can think of Vitali's theorem as delivering one such procedure. Here's how it works: God starts by partitioning $[0, 1)$ into a countable infinity of Vitali Sets and assigns each member of the partition a distinct positive integer. (Proving that a partition of the right kind exists requires the Axiom of Choice, so God would have to rely on Her superhuman capabilities to identify a suitable partition.) God then uses the Standard Coin-Toss Procedure of section 7.1.4.1 to select a real number in $[0, 1)$, and outputs the integer that corresponds to the member of the partition to which that real number belongs.

**7.2.2.2 Proving the theorem** In this section I'll sketch the proof of the non-measurability of Vitali Sets. (Some of the technical details are assigned as exercises.)

We'll start by **partitioning** $[0, 1)$. In other words, we'll divide $[0, 1)$ into a family of non-overlapping "cells," whose union is $[0, 1)$. The cells are characterized as follows: for $a, b \in [0, 1)$, $a$ and $b$ are in the same cell if and only if $a - b$ is a rational number. For instance, $1/2$ and $1/6$ are in the same cell because $1/2 - 1/6 = 2/3$, which is a rational number. Similarly, $\pi - 3$ and $\pi - 25/8$ are in the same cell because $(\pi - 3) - (\pi - 25/8) = 1/8$, which is a rational number. But $\pi - 3$ and $1/2$ are not in the same cell because $\pi - 7/2$ is not a rational number.

We now have our partition of $[0, 1)$. Let us call it $\mathcal{U}$, because (as I'll ask you to verify below) it has uncountably many cells. The next step of our proof will be to use $\mathcal{U}$ to characterize a second partition of $[0, 1)$. We'll call this second partition $\mathcal{C}$ because it has countably many cells. Each cell of $\mathcal{C}$ will be a set $V_q$ for $q \in \mathbb{Q}^{[0,1)}$ (where $\mathbb{Q}^{[0,1)}$ is the set of rational numbers in $[0, 1)$).

I will now explain which members of $[0, 1)$ to include in a given cell $V_q$ of $\mathcal{C}$. The first part of the process is to pick a *representative* from each cell in $\mathcal{U}$. We therefore need a choice set for $\mathcal{U}$. It is a consequence of Solovey's result, mentioned above, that it is impossible to *define* a choice set for $\mathcal{U}$. In other words, it is impossible to specify a criterion that could be used to single out exactly one member from each cell in $\mathcal{U}$. But it follows from the Axiom of Choice that a choice set for $\mathcal{U}$ must nonetheless exist. And its existence is all we need here, because all we're aiming to show is that non-measurable sets exist.

So we use the Axiom of Choice to show that a choice set for $\mathcal{U}$ exists and therefore that there exists a choice of representatives from each cell in $\mathcal{U}$. We will now use our representatives to populate the cells of $\mathcal{C}$ with members of $[0, 1)$. The first step is to think of $[0, 1)$ as a line segment of length 1 (which is missing one of its endpoints) and bend it into a circle:

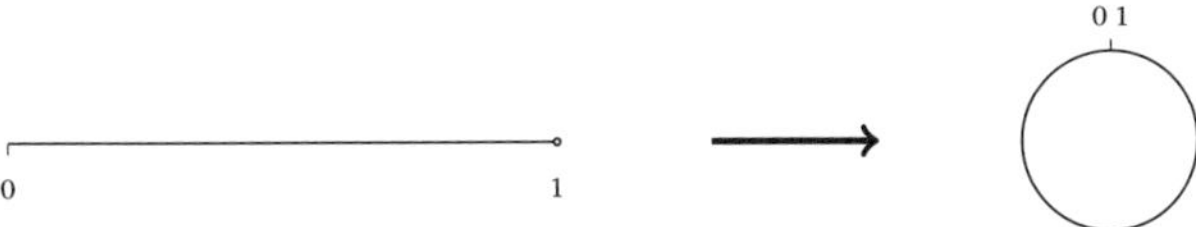


Recall that the difference between any two members in a cell $\mathcal{U}$ is always a rational number. From this, it follows that each member of $[0, 1)$ can be reached by starting at the representative of its cell in $\mathcal{U}$ and traveling some rational distance around the circle, going counterclockwise. Suppose, for example, that 3/4 is in our choice set for $\mathcal{U}$ and has therefore been selected as the representative of its cell in $\mathcal{U}$. (Call this cell $C_{3/4}$.) Now consider a second point in $C_{3/4}$: let's say 1/4. Since 3/4 and 1/4 are in the same cell of $\mathcal{U}$, one can reach 1/4 by starting at 3/4 and traveling a rational distance around the circle, going counterclockwise—in this case a distance of 1:

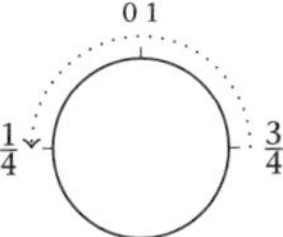


If $a$ is a point in $[0, 1)$, let us say that $\delta(a)$ is the distance one would have to travel on the circle, going counterclockwise, to get to $a$ from the representative for $a$'s cell in $\mathcal{U}$. In our example, $\delta(1/4) = 1/2$.

It is now straightforward to explain how to populate the cells of our countable partition $\mathcal{C}$ with members of $[0, 1)$: each cell $V_q$ $(q \in \mathbb{Q}^{[0,1)})$ of $\mathcal{C}$ is populated with those $a \in [0, 1)$ such that $\delta(a) = q$. As you'll be asked to verify below, this definition guarantees that the $V_q$ $(q \in \mathbb{Q}^{[0,1)})$ cells form a countable partition of $[0, 1)$.

Let a **Vitali Set** be a cell $V_q$ $(q \in \mathbb{Q}^{[0,1)})$ of $\mathcal{C}$. To complete our proof, we need to verify that the Vitali Sets must all have the same measure, if they have a measure at all. The basic idea is straightforward. Recall that $V_q$ is the set of points at a distance of $q$ from their cell's representative, going counterclockwise. From this, it follows that $V_q$ can be obtained by *rotating* $V_0$ on the circle counterclockwise, by a distance of $q$. So one can use Uniformity to show that $V_0$ and $V_q$ have the same measure, if they have a measure at all (and therefore that all Vitali Sets have the same measure, if they have a measure at all).

We are now in a position to wrap up our proof. We have seen that $[0, 1)$ can be partitioned into countably many Vitali Sets and that these sets must all have the same measure, if they have a measure at all. But, for reasons rehearsed in section 7.2.2.1, we know that in the presence of Non-Negativity and Countable Additivity, there can be no such thing as a uniform measure over a countable family of (mutually exclusive and jointly exhaustive) subsets of a set of measure 1. So there can be no way of expanding the notion of Lebesgue Measure to Vitali Sets, without giving up on Non-Negativity, Countable Additivity, or Uniformity.

### Exercises

1. Verify that $\mathcal{U}$ is a partition of $[0, 1)$ by showing that the relation $R$, which holds between $a$ and $b$ if and only if $a-b$ is a rational number, satisfies these three properties:
   a) **Reflexivity** For every $x$ in $[0, 1)$, $xRx$.
   b) **Symmetry** For every $x$ and $y$ in $[0, 1)$, if $xRy$, then $yRx$.
   c) **Transitivity** For every $x$, $y$, and $z$ in $[0, 1)$, if $xRy$ and $yRz$, then $xRz$.
2. Show each cell of $\mathcal{U}$ has only countably many members.
3. Show that there are uncountably many cells in $\mathcal{U}$.
4. To verify that $\mathcal{C}$ is a partition of $[0, 1)$, show that each of the following is the case:
   a) Every real number in $[0, 1)$ belongs to some $V_q$ $(q \in \mathbb{Q}^{[0,1)})$.
   b) No real number in $[0, 1)$ belongs to more than one $V_q$ $(q \in \mathbb{Q}^{[0,1)})$.
5. In showing that the Vitali Sets all have the same measure, if they have a measure at all, we proceeded somewhat informally by thinking of $[0, 1)$ as a circle. When $[0, 1)$ is instead thought of as a line segment, one can get from $V_0$ to $V_q$ by adding $q$ to each member of $V_0$ and then subtracting 1 from any points that end up outside $[0, 1)$. More precisely, $V_0$ can be transformed into $V_q$ in three steps. One first divides the points in $V_0$ into two subsets, depending on whether they are smaller than $1-q$:

   - $V_{\overset{\leftarrow}{0}} = V_0 \cap [0, 1-q)$
   - $V_{\overset{\rightarrow}{0}} = V_0 \cap [1-q, 1)$

   Next, one adds $q$ to each member of $V_{\overset{\leftarrow}{0}}$ and $q-1$ to each member of $V_{\overset{\leftarrow}{0}}$, yielding $(V_{\overset{\leftarrow}{0}})^q$ and $(V_{\overset{\rightarrow}{0}})^{q-1}$ respectively. Finally, one takes the union of the resulting sets, as follows: $(V_{\overset{\leftarrow}{0}})^q \cup (V_{\overset{\rightarrow}{0}})^{q-1}$. Verify that $V_q = (V_{\overset{\leftarrow}{0}})^q \cup (V_{\overset{\rightarrow}{0}})^{q-1}$ $(q \in \mathbb{Q}^{[0,1)})$.

(Note that from this it follows that $V_0$ and $V_q$ have the same measure, if they have a measure at all, and therefore that all Vitali Sets have the same measure, if they have a measure at all. For it follows from Uniformity that $V_{\overleftarrow{0}}$ and $V_{\overrightarrow{0}}$ must have the same measures as $(V_{\overleftarrow{0}})^q$ and $(V_{\overrightarrow{0}})^{q-1}$ respectively, if they have measures at all. And it follows from Countable Additivity that the measure of $V_0$ must be the sum of the measures of $V_{\overleftarrow{0}}$ and $V_{\overrightarrow{0}}$, if the latter have measures, and that the measure of $V_q$ must be the sum of the measures of $(V_{\overleftarrow{0}})^q$ and $(V_{\overrightarrow{0}})^{1-q}$, if the latter have measures.)

### 7.2.3 What to Do About Non-Measurability?

In proving Vitali's non-measurability theorem, we have learned a horrifying truth. We have learned that there is a precise sense in which it is impossible to assign a respectable measure to certain subsets of [0, 1]. How should one make sense of such a result? In this section we will explore some possible answers.

**7.2.3.1 An experiment to the rescue?** The intimate connection between measurability and probability raises an intriguing possibility. What if there was an "experiment" that allowed us to assign a measure to a Vitali Set $V$? Imagine a situation in which we are able to apply the Coin-Toss Procedure again and again, to select points from [0, 1]. Couldn't we count the number of times we get a point in $V$ and use this information to assign a measure to $V$? Unfortunately, it is not clear that such an experiment would be helpful. In this section I'll try to explain why.

Let me begin with a preliminary observation. It goes without saying that it is impossible to carry out such an experiment in practice. Each application of the Coin-Toss Procedure requires an infinite sequence of coin tosses. But let us idealize and suppose that we are able to apply the Coin-Toss Procedure (and indeed apply it infinitely many times). Notice, moreover, that in proving the theorem, we didn't actually lay down a definite criterion for membership in $V$. All we did was show that some suitable set $V$ must exist. But let us idealize once more and imagine that we are able to fix upon a specific Vitali Set $V$ by bypassing the Axiom of Choice and selecting a particular member from each cell in $\mathcal{U}$.

Would such an idealized experiment help us assign a measure to $V$? It seems to me that it would not. The first thing to note is that it is not enough to *count* the number of times the Coin-Toss Procedure yields points in $V$ and compare it with the number of times it yields points outside $V$. To see this, imagine applying the Coin-Toss Procedure $\aleph_0$-many times and getting the following results:

- Number of times the procedure yielded a number in $V$: $\aleph_0$.
- Number of times the procedure yielded a number outside $V$: $\aleph_0$.

Should you conclude that $P(V) = 50\%$? No! Notice, for example, that in such an experiment, one would expect to get $\aleph_0$-many numbers in $[0, 1/4]$ and $\aleph_0$-many numbers outside $[0, 1/4]$. But that is no reason to think that

$$p\left(\left[0,\frac{1}{4}\right]\right)=p\left(\overline{\left[0,\frac{1}{4}\right]}\right)=50\%$$

If the experiment is to be helpful, we need more than just information about the *cardinality* of outputs within $V$ and outside $V$. We need a method for determining the *proportion* of total outputs that fall within $V$. And, as the case of $[0, 1/4]$ illustrates, cardinalities need not determine proportions.

One might wonder whether there is a strategy that one could use to try to identify the needed proportions "experimentally." Suppose one carries out the Coin-Toss Procedure infinitely many times, once for each natural number. One then determines, for each finite $n$, how many of the first $n$ tosses fell within a given subset $A$ of $[0, 1]$. (Call this number $|A_n|$.) Finally, one lets the measure of $A$ be calculated experimentally, as follows:

$$\mu(A)=\lim_{n\to\infty}\frac{|A_n|}{n}$$

The hope is then that if one actually ran the experiment, one would get the result that $\mu([0, 1/4]) = 1/4$. More generally, one might hope to get the result that $\mu(A)$ turns out to be the Lebesgue Measure of $A$ whenever $A$ is a Borel Set (or, indeed, a Lebesgue-measurable set).

Let us suppose that we run such an experiment in an effort to identify a measure for our Vitali Set $V$. What sort of results one might expect? One possibility is that $\mu(V)$ fails to converge. In that case the experiment would give us no reason to think that $V$ has a measure. But suppose we run the experiment again and again, and that $\mu(V)$ always converges to the same number $m \in [0, 1]$. Would this give us any grounds for thinking that $m$ is the measure of $V$? It is not clear that it would. For consider the following question: What values might our experiment assign to $\mu(V_q)$ $(q \in \mathbb{Q}^{[0,1)})$?

There is no comfortable answer to this question. If we get the result that $\mu(V_q) = m$ for each $q \in \mathbb{Q}^{[0,1)}$, it will thereby be the case that $\mu$ fails to be countably additive, since there is no real number $m$ such that $m + m + \ldots = 1$. So, to the extent that we treat Countable Additivity as an axiom, we will have grounds for thinking that something has gone wrong with our experiment. If, on the other hand, we get the result that $\mu(V_q)$ is not uniformly $m$ for each $q \in \mathbb{Q}^{[0,1)}$, it will thereby be the case that $\mu$ fails to be uniform, since the $V_q$ $(q \in \mathbb{Q}^{[0,1)})$ are essentially copies of $V$, just positioned elsewhere in $[0, 1)$. So, again, to the extent that we treat Uniformity as an axiom, we will have grounds for thinking that something has gone wrong with our experiment.

The moral is that it is not clear that we could use an experiment to help find a measure for $V$, even when we take certain idealizations for granted. For should we find that $\mu(V)$ converges to some value, it is not clear that we would be left with grounds for assigning a measure to $V$ rather than grounds for thinking that something went wrong with the experiment. In retrospect, this should come as no surprise. The non-measurability of $V$ follows from three very basic features of Lebesgue Measure: Non-Negativity, Countable Additivity, and Uniformity. So we know that no version of our experiment could deliver a value for $V$ without violating at least one of these assumptions. And the price of violating the assumptions is that it's no longer clear that the experiment delivers something worth calling a Lebesgue Measure.

**7.2.3.2 Is there room for revising our assumptions?** When we proved that the Vitali Sets are non-measurable, what we proved is that there is no way of extending the Lebesgue Measure function $\lambda$ to a function that is defined for the Vitali Sets while satisfying all three principles of Non-Negativity, Countable Additivity, and Uniformity. Why is it so important that these assumptions be satisfied? What would be so wrong about leaving Lebesgue Measure behind and instead focusing on a notion of measure that gives up on one or more of the principles of Non-Negativity, Countable Additivity, and Uniformity?

When it comes to Uniformity, I think there is a simple answer: giving up on Uniformity means *changing the subject*. The whole point of our enterprise is to find a way of extending the notion of Lebesgue Measure without giving up on Uniformity.

What about Non-Negativity and Countable Additivity? As it turns out, abandoning these constraints won't help us avoid the phenomenon of non-measurability. When we discuss the Banach-Tarski Theorem in the next chapter, we will see that there are sets that can be shown to be non-measurable without presupposing Non-Negativity or Countable Additivity.

Could we give up on the Axiom of Choice? In order for a Vitali Set to count as an example of a non-measurable set, it has to exist. And, as I mentioned earlier, the Vitali Sets cannot be shown to exist without the Axiom of Choice:

**Axiom of Choice** Every set of non-empty, non-overlapping sets has a choice set.

Because of this, one might be tempted to sidestep the phenomenon of non-measurability altogether by giving up on the Axiom of Choice. Such a temptation might seem especially strong in light of the fact that the Axiom of Choice can be used to prove all sorts of bizarre results. We used it in chapter 3 to find an incredible solution to Bacon's Puzzle, and we'll soon use it to prove one of the most perplexing, and famous, of all its consequences: the Banach-Tarski Theorem.

Unfortunately, there are regions of mathematics that would be seriously weakened without the Axiom of Choice. For example, there would be no way of proving that, for any two sets, either they have the same size or one is bigger than the other:

$|A| = |B|$ or $|A| < |B|$ or $|B| < |A|$

It is hard to know what to do about the Axiom of Choice. When you think about some of its more bizarre consequences, it's tempting to give it up. But it certainly seems plausible on its face. And when you're trying to work within certain areas of mathematics, you feel like you can't live without it. My own view is that we should learn to live with the Axiom of Choice. But there is no denying that it delivers some strange results.

## 7.3 Conclusion

We have proved Vitali's non-measurability theorem, which shows that there are non-measurable sets. More precisely, it shows that there is a family of sets one cannot assign a measure to while preserving all three principles of Non-Negativity, Countable Additivity, and Uniformity.

This is, to my mind, a horrifying result. Is there anything that can be done about it? We saw that giving up Uniformity isn't an option, on pain of changing the subject, and that Non-Negativity and Countable Additivity aren't really required to prove non-measurability results.

I noted that the Axiom of Choice is required to prove non-measurability results. So there's always the option of giving up the Axiom of Choice. But we saw that doing so would come with definite costs. My own view is that we should learn to live with the Axiom of Choice, but I won't pretend that it's an easy choice.

## 7.4 Further Resources

- My discussion of measure theory is drawn from Robert Ash's *Probability and Measure Theory*. I recommend it wholeheartedly, especially if you're interested in learning about probability from a measure-theoretic perspective.
- For a weird consequence of the Axiom of Choice not discussed here, see Frank Arntzenius, Adam Elga, and John Hawthorne's "Bayesianism, Infinite Decisions, and Binding."

## Appendix: Answers to Exercises

### Section 7.1.1

1. The key observation is that

$$\bigcap\{A_1, A_2, A_3, \ldots\} = \overline{\bigcup\{\overline{A_1}, \overline{A_2}, \overline{A_3}, \ldots\}}$$

where $\overline{A} = \mathbb{R} - A$. One can therefore verify that $\bigcap\{A_1, A_2, A_3, \ldots\}$ is a Borel Set as follows: Since each of $A_1, A_2, A_3, \ldots$ is a Borel Set, we can use complementation to show that each of $\overline{A_1}, \overline{A_2}, \overline{A_3}, \ldots$ is a Borel Set. But by countable union, $\bigcup\{\overline{A_1}, \overline{A_2}, \overline{A_3}, \ldots\}$ must also be a Borel Set. So, by complementation, $\bigcap\{A_1, A_2, A_3, \ldots\}$ must also be a Borel Set.

2. The key observation is that $A - B = A \cap \overline{B}$. Since the complement of a Borel Set is a Borel Set and since (as verified in the previous exercise) the intersection of Borel Sets is a Borel Set, $A - B$ is a Borel Set.

### Section 7.1.2

1. Since $[0, 1]$ and the empty set are both Borel Sets, $\lambda([0, 1])$ and $\lambda(\emptyset)$ are both well-defined. So we can use Countable Additivity to get

$$\lambda([0, 1]) = \lambda([0, 1]) + \lambda(\emptyset).$$

But it follows from Length on Segments that $\lambda([0, 1]) = 1$. So $\lambda(\emptyset) = 1 - 1 = 0$.

2. Since $[a, b]$ and $\{b\}$ are Borel Sets, it follows from exercise 2 in section 7.1.1 that $[a, b)$ is a Borel Set. So we know that both $\lambda([a, b))$ and $\lambda(\{b\})$ are defined. We can therefore use Countable Additivity to get the following:

$$\lambda([a, b]) = \lambda([a, b)) + \lambda(\{b\})$$

But since $\{b\} = [b, b]$ and $\lambda([b, b]) = b - b = 0$ (by Length on Segments), $\lambda(\{b\}) = 0$. So $\lambda([a, b)) = \lambda([a, b]) = b - a$.

3. Since each set $[n, n+1)$ is a Borel Set for each integer $n$, we know that $\lambda([n, n+1))$ is defined for each $n$. We can therefore use Countable Additivity to get the following:

$$\lambda(\mathbb{R}) = \ldots \lambda([-2, -1)) + \lambda([-1, 0)) + \lambda([0, 1)) + \lambda([1, 2)) + \ldots$$

But the previous answer entails that $\lambda([a, a+1)) = 1$ for each $a$. So we have:

$$\lambda(\mathbb{R}) = \ldots 1 + 1 + 1 + 1 \ldots$$

Since no real number is equal to an infinite sum of ones, Non-Negativity entails that $\lambda(\mathbb{R}) = \infty$.

4. Let $A$ be the countable set $\{a_0, a_1, a_2, \ldots\}$. Since each $\{a_i\}$ is a Borel Set, we know that each $\lambda(\{a_i\})$ is defined. We can therefore use Countable Additivity to get:

$$\lambda(\{a_0, a_1, a_2, \ldots\}) = \lambda(\{a_0\}) + \lambda(\{a_1\}) + \lambda(\{a_2\}) + \ldots$$

But since $\{a_i\} = [a_i, a_i]$, it follows from [Length on Segments] that $\lambda(\{a_i\}) = 0$ for each $i$. So we have the following:

$$\lambda(\{a_0, a_1, a_2, \ldots\}) = 0 + 0 + 0 + \ldots = 0$$

5. Since $A$ and $B$ are both Borel Sets, so is $A - B$. So Countable Additivity entails

$$\lambda(A) = \lambda(A - B) + \lambda(B).$$

But Non-Negativity entails that $\lambda(A - B)$ is either a non-negative real number or $\infty$. In either case, it follows that $\lambda(B) \leq \lambda(A)$.

6. Recall that for any $x \in \mathbb{R}$, $\lambda(\{x\}) = 0$. So an Uncountable Additivity principle would entail that $\lambda([0, 1]) = 0$, as follows:

$$\lambda\left([0, 1]\right) = \lambda\left(\bigcup_{x\in[0,1]} (\{x\})\right) = \sum_{x\in[0,1]} \left(\lambda\left(\{x\}\right)\right) = \sum_{x\in[0,1]} (0) = 0$$

## Section 7.1.3

1. We are assuming that $A^0$ is a subset of a Borel Set of measure 0, since $\emptyset$ is a Borel Set, and since $A^0 = \emptyset \cup A^0$, and since the stipulation we used to extend $\lambda$ yields $\lambda(A^0) = \lambda(\emptyset)$. But we showed in exercise 1 of section 7.1.2 that $\lambda(\emptyset) = 0$.

2. Since $A$ is Lebesgue measurable for each $i$, we know that $A = A^B \cup A^0$ for each $i$, where $A^B$ is a Borel Set and $A^0$ is a subset of a Borel Set $B$ of measure 0.

   Let us now show that $\overline{A}$ is Lebesgue measurable. Notice, first, that $\overline{A^B \cup B}$ is a Borel Set (since $A^B$ and $B$ are Borel Sets, and Borel Sets are closed under complements and countable unions). Notice, also, that $(B \cap \overline{A^B} \cap \overline{A^0})$ is a subset of a Borel Set of measure 0 (since $B$ is a Borel Set of measure 0). This means that one can show that $\overline{A}$ is Lebesgue measurable by verifying the following equation:

$$\overline{A} = \overline{A^B \cup B} \cup (B \cap \overline{A^B} \cap \overline{A^0})$$

   which can be done in the following steps:

$$A = A^B \cup A^0$$

$$A = ((A^B \cup B) \cap \overline{B}) \cup A^B \cup A^0$$

$$A = ((A^B \cup B) \cap \overline{B}) \cup ((A^B \cup B) \cap A^B) \cup A^0$$

$$A = ((A^B \cup B) \cap \overline{B}) \cup ((A^B \cup B) \cap A^B) \cup ((A^B \cup B) \cap A^0)$$

$$A = (A^B \cup B) \cap (\overline{B} \cup A^B \cup A^0)$$

$$A = (A^B \cup B) \cap \overline{(B \cap \overline{A^B} \cap \overline{A^0})}$$

$$\overline{A} = \overline{A^B \cup B} \cup (B \cap \overline{A^B} \cap \overline{A^0})$$

3. Suppose that $A_1, A_2, A_3, \ldots$ is a countable family of Lebesgue-measurable sets. We will verify that $A_1 \cup A_2 \cup A_3 \cup \ldots$ is Lebesgue measurable.

   Since $A_i$ is Lebesgue measurable for each $i$, we know that $A_i = A_i^B \cup A_i^0$ for each $i$, where $A_i^B$ is a Borel Set and $A_i^0$ is a subset of a Borel Set $B_i$ of measure 0.

   We also know that

   $$A_1 \cup A_2 \cup A_3 \cup \ldots = (A_1^B \cup A_2^B \cup A_3^B \cup \ldots) \cup (A_1^0 \cup A_2^0 \cup A_3^0 \cup \ldots).$$

   This means that all we need to do is to check that $A_1^B \cup A_2^B \cup A_3^B \cup \ldots$ is a Borel Set and that $A_1^0 \cup A_2^0 \cup A_3^0 \cup \ldots$ is a subset of a Borel Set of measure 0. The former follows from the fact that each of $A_1^B, A_2^B, A_3^B, \ldots$ is a Borel Set and from the fact that Borel Sets are closed under countable unions, and the latter follows from these five observations: *(i)* $A_1^0 \cup A_2^0 \cup A_3^0 \cup \ldots \subset B_1 \cup B_2 \cup B_3 \cup \ldots$; *(ii)* $B_1 \cup B_2 \cup B_3 \cup \ldots$ is a Borel Set, since each of the $B_1, B_2, B_3, \ldots$ is a Borel Set and Borel Sets are closed under countable unions; *(iii)* $B_1 \cup B_2 \cup B_3 \cup \ldots = B_1 \cup (B_2 - B_1) \cup (B_3 - (B_1 \cup B_2)) \cup \ldots$, *(iv)* since each of the $B_1, B_2, B_3, \ldots$ has measure zero, each $B_k - (B_1 \cup \ldots \cup B_{k-1}) (k > 1)$ has measure zero (by exercise 5 of section 7.1.2), and *(v)* $B_1 \cup (B_2 - B_1) \cup (B_3 - (B_1 \cup B_2)) \cup \ldots$ has measure zero, since Countable Additivity holds for disjoint Borel Sets.

4. Since $A$ is Lebesgue measurable, $A = A^B \cup A^0$, where $A^B$ is a Borel Set and $A^0$ is a subset of a Borel Set of measure 0. We know, moreover, that $\lambda(A) = \lambda(A^B)$. Since Non-Negativity applies to Borel Sets, $\lambda(A^B)$ is either a non-negative real number or $\infty$, so the same must be true of $\lambda(A)$.

5. Since each of the $A_i$ is Lebesgue measurable, we know that $A_i = A_i^B \cup A_i^0$, where $A_i^B$ is a Borel Set and $A_i^0$ is a subset of a Borel Set $B_i$ of measure 0.

   We verified in exercise 3 that

   $$A_1 \cup A_2 \cup A_3 \cup \ldots = (A_1^B \cup A_2^B \cup A_3^B \cup \ldots) \cup (A_1^0 \cup A_2^0 \cup A_3^0 \cup \ldots)$$

   where $A_1^B \cup A_2^B \cup A_3^B \cup \ldots$ is a Borel Set and $A_1^0 \cup A_2^0 \cup A_3^0 \cup \ldots$ is a subset of a Borel Set of measure 0.

   So the stipulation we used to extend $\lambda$ entails that

   $$\lambda(A_1 \cup A_2 \cup A_3 \cup \ldots) = \lambda(A_1^B \cup A_2^B \cup A_3^B \cup \ldots).$$

   Since $A_1 \cup A_2 \cup A_3 \cup \ldots$ are pairwise disjoint, $A_1^B \cup A_2^B \cup A_3^B \cup \ldots$ are pairwise disjoint. So, by Countable Additivity on Borel Sets, we have:

   $$\lambda(A_1^B \cup A_2^B \cup A_3^B \cup \ldots) = \lambda(A_1^B) + \lambda(A_2^B) + \lambda(A_3^B) + \ldots.$$

Putting all of this together gives us

$$\lambda(A_1 \cup A_2 \cup A_3 \cup \ldots) = \lambda(A_1^B) + \lambda(A_2^B) + \lambda(A_3^B) + \ldots.$$

But the stipulation we used to extend $\lambda$ entails that $\lambda(A_i) = \lambda(A_i^B)$ for each $i$. So we have

$$\lambda(A_1 \cup A_2 \cup A_3 \cup \ldots) = \lambda(A_1) + \lambda(A_2) + \lambda(A_3) + \ldots$$

which is what we wanted.

## Section 7.1.4.1

1. Let us start with the Standard Coin-Toss Procedure. The first thing to note is that the Standard Coin-Toss Procedures yield a number in $[1/2, 1]$ if and only if the first toss lands Tails.

   To verify this, note that the real number $1/2$ is represented by the binary expansion $0.1(0)$. (It is also represented by the binary expansion $0.0(1)$, but this latter expansion is stipulated not to count as a valid output of our Coin-Toss Procedures.) Accordingly, each valid binary sequence not of the form $0.1d_2d_3\ldots$ represents a real number smaller than $1/2$, and each valid binary sequence of the form $0.1d_2d_3\ldots$ represents a real number equal to or larger than $1/2$.

   With this observation in place, the answer is straightforward. Since (on standard assumptions about the probabilities of coin tosses) the probability that the first toss lands Tails is $1/2$, then the probability that one gets a number in $[1/2, 1]$ as output is $1/2$.

   On the Square Root Coin-Toss Procedure, in contrast, one gets a number in $[1/2, 1]$ if and only if at least one of the first two tosses lands tails.

   To verify this, consider two different cases: (i) the first toss can land Tails, and (ii) the first toss lands Heads and the second toss lands Tails. If the first toss lands Tails, then the sequence of coin tosses will yield a number of the form $0.1d_2d_3\ldots$, which represents a real number greater than or equal to $1/2$. Since $(1/\sqrt{2})^2 = 1/2$, this means that the Square Root Coin-Toss Procedure will yield a number greater than or equal to $1/\sqrt{2}$. If, on the other hand, the first toss lands on heads and the second toss lands on tails, then the sequence of coin-tosses will yield a number of the form $0.01d_2d_3\ldots$, which represents a real number greater than or equal to $1/4$ but smaller than $1/2$. Since $(1/2)^2 = 1/4$ and $(1/\sqrt{2})^2 = 1/2$, this means that the Square Root Coin-Toss Procedure will yield a number greater than or equal to $1/2$ but smaller than $1/\sqrt{2}$. Putting these two cases together yields the desired result: one gets a number in $[1/2, 1]$ if and only if at least one of the first two tosses lands Tails.

   With this observation in place, the answer is straightforward. Since (on standard assumptions about the probabilities of coin tosses) the probability of getting Tails in at least one of the first two tosses is 3/4, the probability that one gets a number in [1/2, 1] as output is 3/4.

2. Start with the Standard Coin-Toss Procedure. The unique binary expansion of 1/2 that is a valid output of our coin-toss procedures is 0.1(0). So the only way for 1/2 to be selected is for our sequence of coin tosses to result in exactly the following sequence:

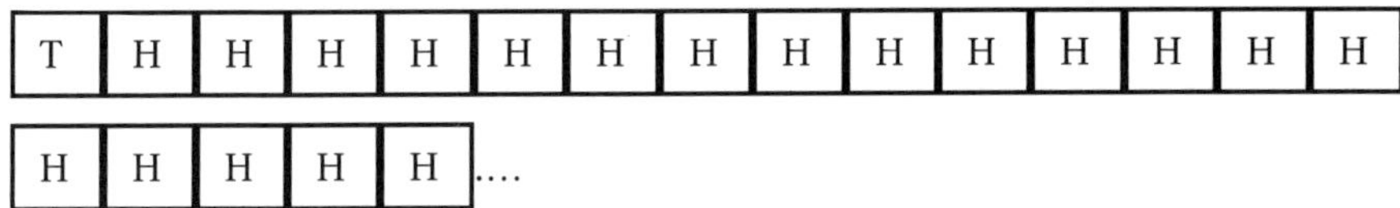


And what is the probability that this will happen? Zero! One can verify this by noting that each of the following must be true:

- The probability must be smaller than 1, since it requires our first toss to land Tails, which is an event of probability 1/2.
- The probability must be smaller than 1/2, since it requires our first two tosses to land Tails and then Heads, which is an event of probability 1/4 .
- The probability must be smaller than 1/4, since it requires our first three tosses to land Tails-Heads-Heads, which is an event of probability 1/8.
- And so forth: the probability must be smaller than $1/2^n$ for each $n \in \mathbb{N}$.

Since we are assuming that the relevant probability exists, since probabilities are real numbers between 0 and 1, and since the only such number smaller than $1/2^n$ for each $n \in \mathbb{N}$ 0, the probability of selecting 1/2 must be 0.

Notice, moreover, that 1/2 is not a special case. For any $x$ in [0, 1], the probability that our Coin-Toss Procedures yield precisely $x$ as output is always zero. (It is worth keeping in mind, however, that, as we saw in chapter 4, saying that an event has a probability of zero is *not* the same as saying that the event can't happen. To see this, note our procedure always yields *some* number in [0, 1] as output and that *whatever* number we get, the probability of getting that number is zero. But it happened anyway. For an event with well-defined probability to have a probability of zero is for the event to be so vanishingly unlikely that the probability of its occurrence is too small to be measured by a positive real number, no matter how small.)

Let us now consider the Square Root Coin-Toss Procedure. Since $(1/2)^2 = 1/4$, the only way for the Square Root Coin-Toss Procedure to yield 1/2 is for our sequence of coin tosses to yield a sequence that represents $1/\sqrt{4}$ in binary notation. But an analogue of the argument above shows that the probability of getting such a sequence is zero.

3. We first consider the Standard Coin-Toss Procedure. There are only two ways for a number in $[0, 1/2]$ to be selected. The first is for the first toss to land Heads, in which case the procedure yields a number in $[0, 1/2)$ (i.e., a number $x$ such that $0 \leq x < 1/2$). The second is to get exactly the following sequence:

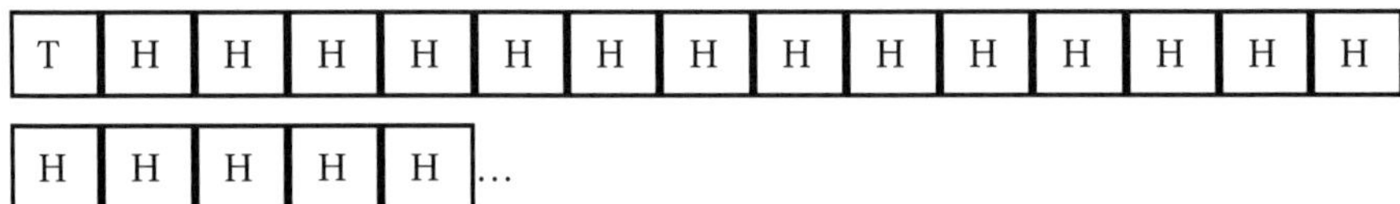

in which case the procedure yields the number 1/2.

The probability of the first outcome is 1/2, and we know from the previous exercise that the probability of the second is 0. Since $[0, 1/2] = [0, 1/2) \cup \{1/2\}$, and since $[0, 1/2)$ and $\{1/2\}$ have no elements in common, Additivity yields

$$p\left(\left[0, \frac{1}{2}\right]\right) = p\left(\left[0, \frac{1}{2}\right)\right) + p\left(\left\{\frac{1}{2}\right\}\right) = \frac{1}{2} + 0 = \frac{1}{2}$$

Let us now consider the Square Root Coin-Toss Procedure. Since $(1/2)^2 = 1/4$, there are two ways for a number in $[0, 1/2]$ to be selected. The first is for the first two tosses to land Heads, in which case the sequence of coin tosses corresponds to a number in $[0, 1/4)$ and the Square Root Coin-Toss Procedure yields a number in $[0, 1/2)$. The second is to get exactly the following sequence:

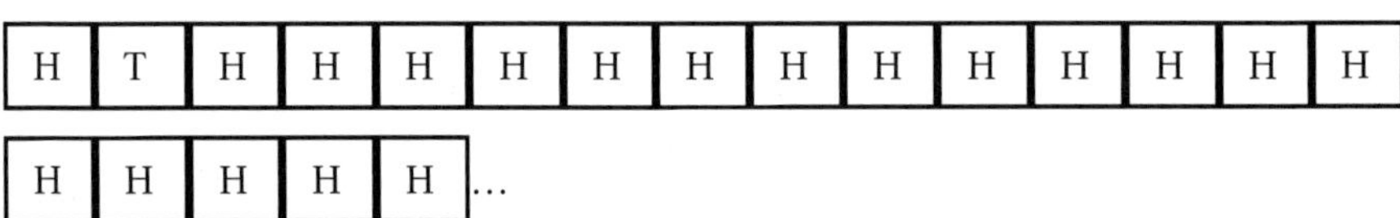

in which case our coin-toss sequence represents the number 1/4 and the Square Root Coin-Toss Procedure yields the number 1/2.

The probability of the first outcome is 1/4, and we know from the previous exercise that the probability of the second is 0. Since $[0, 1/2] = [0, 1/2) \cup \{1/2\}$, and since $[0, 1/2)$ and $\{1/2\}$ have no elements in common, Additivity yields

$$p\left(\left[0, \frac{1}{2}\right]\right) = p\left(\left[0, \frac{1}{2}\right)\right) + p\left(\left\{\frac{1}{2}\right\}\right) = \frac{1}{4} + 0 = \frac{1}{4}.$$

4. The following argument works for either version of the Coin-Toss Procedure. We know from question 2 that $p(\{x_i\}) = 0$ for each individual point $x_i$. But by Countable Additivity,

$$p(\{x_0, x_1, x_2, \ldots\}) = p(\{x_0\}) + p(\{x_1\}) + p(\{x_2\}) + \ldots.$$

So $p(\{x_0, x_1, x_2, \ldots\})$ must be 0.

5. The following arguments work for either version of the Coin-Toss Procedure.

Notice, first, that $p([0, 1]) = 1$, since the next valid output of either procedure is guaranteed to be a number in $[0, 1]$. We know, moreover, from question 2 that $p(\{1/2\}) = 0$. Since we are assuming that $p([0, 1] - \{1/2\})$ is well defined, Additivity yields

$$p([0, 1]) = p\left([0, 1] - \left\{\frac{1}{2}\right\}\right) + p\left(\left\{\frac{1}{2}\right\}\right).$$

So we know that

$$1 = p\left([0, 1] - \left\{\frac{1}{2}\right\}\right) + 0$$

and therefore that

$$p\left([0, 1] - \left\{\frac{1}{2}\right\}\right) = 1.$$

A similar argument yields the result that $p([0, 1] - \{x_0, x_1, x_2, \ldots\}) = 1$, since we know from question 4 that $p(\{x_0, x_1, x_2, \ldots\}) = 0$.

### Section 7.2.2.2

1. First, reflexivity. We want to show that every number in $[0, 1]$ differs from itself by a rational number. Easy: every real number differs from itself by 0, and 0 is a rational number.

Next, symmetry. For $x, y \in [0, 1]$ we want to show that if $x - y \in \mathbb{Q}$, then $y - x \in \mathbb{Q}$. If $x - y = r$, then $y - x = -r$. But if $r \in \mathbb{Q}$, then $-r \in \mathbb{Q}$.

Finally, transitivity. For $x, y, z \in [0, 1]$ we want to show that if $x - y \in \mathbb{Q}$ and $y - z \in \mathbb{Q}$, then $x - z \in \mathbb{Q}$. Suppose $x - y = r$ and $y - z = s$. Then $y = x - r$ and $y = s + z$. So $x - r = s + z$, and therefore $x - z = s + r$. But if $r, s \in \mathbb{Q}$, then $s + r \in \mathbb{Q}$.

2. Let $C$ be a cell in $\mathcal{U}$, and let $a \in C$. Every number in $C$ differs from $a$ by some rational number. Since there are only countably many real numbers, this means that there must be at most countably many numbers in $C$.

3. We know from the previous exercise that each cell of $\mathcal{U}$ has only countably many objects, and we know from chapter 1 that the union of countably many countable sets is countable. So if there were only countably many cells in $\mathcal{U}$, there would have to be countably many objects in the union of all families. In other words, there would have to be countably many objects in $[0, 1]$, which we know to be false.

4. (a) We verify that every real number in $[0, 1)$ belongs to some $V_q$ ($q \in \mathbb{Q}^{[0,1)}$).

Let $a \in [0, 1)$. Since $\mathcal{U}$ is a partition, $a$ must be in some cell $C$ of $\mathcal{U}$. Let $r$ be $C$'s representative. By the definition of $\mathcal{U}$, $a - r$ is a rational number. So $\delta(a)$, which is the distance one would have to travel on the circle to get from $r$ to $a$ going counterclockwise, must be a rational number in $[0, 1)$. So by the definition of $\mathcal{C}$, $a$ must be in $V_{\delta(a)}$.

(b) We verify that no real number in $[0, 1)$ belongs to more than one $V_q$ $(q \in \mathbb{Q}^{[0,1)})$. Suppose otherwise. Then some $a \in [0, 1)$ belongs to both $V_q$ and $V_p$ $(p \neq q,\ q, p \in \mathbb{Q}^{[0,1)})$. By the definition of $\mathcal{C}$ this means that $\delta(a)$ must be equal to both $p$ and $q$, which is impossible.

5. We verify that $V_q = (V_{\overleftarrow{0}})^q \cup (V_{\overrightarrow{0}})^{q-1}$ $(q \in \mathbb{Q}^{[0,1)})$.

Let $a \in V_q$, and let $r$ be the representative of $a$'s cell in $\mathcal{U}$. $V_q$, recall, is the set of $x \in [0, 1)$ such that $\delta(x) = q$. There are two cases, $r + q < 1$ and $r + q \geq 1$. Let us consider each of them in turn, and show that $a \in V_q$ if and only if $a \in (V_{\overleftarrow{0}})^q \cup (V_{\overrightarrow{0}})^{q-1}$.

- Assume $r + q < 1$. Since $\delta(a) = q$, one would have to travel a distance of $q$ on the circle, going counterclockwise, to get to $a$ from $r$. Since we have $r + q < 1$, this means that $a = r + q$.

  Now suppose that $r \in V_0$. We have $r \in V_{\overleftarrow{0}}$, since $V_{\overleftarrow{0}} = V_0 \cap [0, 1 - q)$ and $r + q < 1$. So $a = r + q$ entails $a \in (V_{\overleftarrow{0}})^q$. Suppose, conversely, that $a \in (V_{\overleftarrow{0}})^q$. Then $a = r + q$ entails that $r \in V_{\overleftarrow{0}}$ and therefore that $r \in V_0$.

- Assume $r + q \geq 1$. Since $\delta(a) = q$, one would have to travel a distance of $q$ on the circle, going counterclockwise, to get to $a$ from $r$. But $r + q \geq 1$, so one must cross the 0 point on the circle when traveling from $r$ to $a$. Since since $q < 1$ (and since the circle has a circumference of 1), this means that $a = r + q - 1$.

  Now suppose that $r \in V_0$. We have $r \in V_{\overrightarrow{0}}$, since $V_{\overrightarrow{0}} = V_0 \cap [1 - q, 1)$ and $r + q \geq 1$. So $a = r + q - 1$ entails $a \in (V_{\overrightarrow{0}})^{q-1}$. Suppose, conversely, that $a \in (V_{\overrightarrow{0}})^{q-1}$. Then $a = r + q - 1$ entails that $r \in V_{\overleftarrow{0}}$ and therefore that $r \in V_0$.

# 8 The Banach-Tarski Theorem

In the last chapter we saw that the Axiom of Choice can be used to show that non-measurable sets exist. In this chapter we'll use the Axiom of Choice to prove the following:

**The Banach-Tarski Theorem** It is possible to decompose a ball into a finite number of pieces and reassemble the pieces (without changing their size or shape) so as to get two balls, each of the same size as the original.

It is easy to see that this theorem must involve non-measurable sets. For suppose otherwise. Suppose that each of the finitely many pieces mentioned in the theorem has a definite volume. Is the sum of the volumes of the pieces equal to the volume of one ball or two? However we answer this question, we'll end up contradicting (finite) Additivity. So at least one of the pieces must have no definite volume. But volume is just a special case of Lebesgue Measure (or its three-dimensional variant).

The original version of the Banach-Tarski Theorem was proved in 1924 by Stefan Banach and Alfred Tarski, but it has since been refined by others. Notably, Raphael M. Robinson showed that the procedure can be carried out using only five pieces, and that it is impossible to carry it out with fewer.

## 8.1 Three Warm-Up Cases

Before tackling the theorem itself, it'll be useful to consider some warm-up cases. They are all constructions in which one decomposes an object into pieces and reassembles the pieces so as to end up with more than we started with. But they are much simpler than the Banach-Tarski Theorem and help illustrate some of the key ideas in the proof.

### 8.1.1 Warm-Up Case 1: A Line

$[0, \infty)$ is the set of non-negative real numbers. Let us consider the result of removing the number 1 from that set. Is it possible to decompose $[0, \infty) - \{1\}$ into two distinct parts and then to reassemble the parts (without changing their size or shape) so as to get back $[0, \infty)$, including the number 1?

Yes! In fact, it is very easy to do so. We'll use a trick that is not too different from the trick we used to find room for Oscar in Hilbert's Hotel, back in chapter 1. Let us start by drawing a picture of $[0,\infty) - \{1\}$:

Now focus on a particular "piece" of $[0,\infty) - \{1\}$; namely, the set $\{2, 3, 4, \ldots\}$. In the diagram below I have separated that piece from the rest of $[0,\infty) - \{1\}$, to make it easier to see:

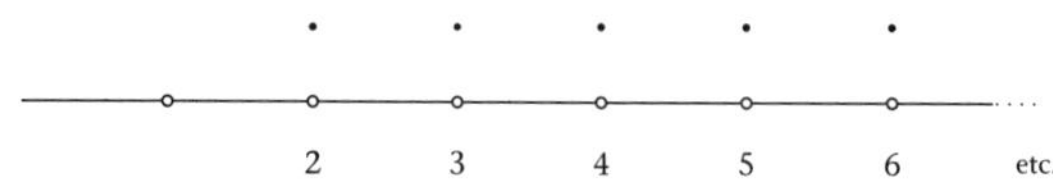


Now translate the piece one unit to the left:

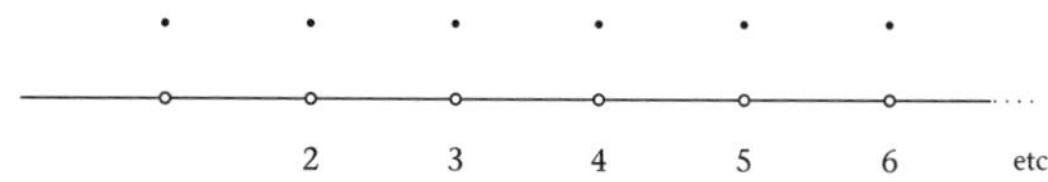


Combine the translated piece with the rest of $[0,\infty) - \{1\}$:

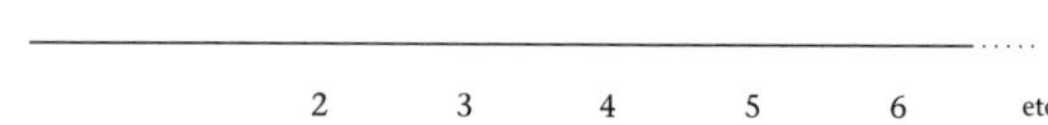


Success! We have decomposed the set $[0,\infty) - \{1\}$ into two pieces, reassembled the pieces (without changing their size or shape), and ended up with the full set $[0,\infty)$.

### 8.1.2 Warm-Up Case 2: A Circle

In warm-up case 1 we made essential use of the fact that the set $[0,\infty) - \{1\}$ extends infinitely in one direction. We will now perform a similar decomposition involving an object of finite length. Let $S^1$ be the circle of radius 1, and consider the result of removing a single point $p$ from $S^1$:

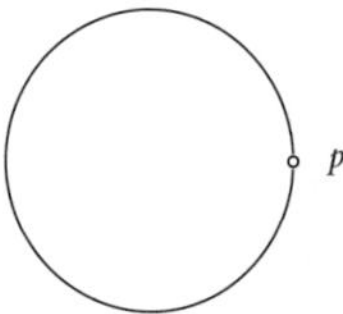

Is it possible to decompose $S^1 - \{p\}$ into two distinct parts and then reassemble the parts (without changing their size or shape) so as to get $S^1$?

Yes! But this time we'll have to be clever about how we decompose $S^1 - \{p\}$. Let us characterize a subset $B$ of $S^1 - \{p\}$ by going clockwise around the circle. The first member of $B$ is the point $p_1$ which is one unit clockwise from $p$. The second member of $B$ is the point $p_2$ which is one units clockwise from $p$. And so forth. Here are the locations of the first six members of $B$:

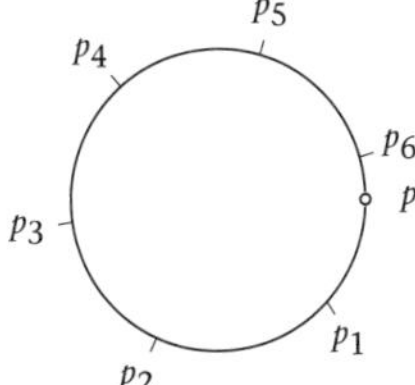


Since the radius of $S^1$ is one unit, its circumference is $2\pi$ units. And, as I'll ask you to prove below, the fact that $2\pi$ is an irrational number can be used to show that the points $p_1, p_2, p_3, \ldots$ are all distinct. In other words, if one starts at $p$, moves one unit clockwise, moves another unit clockwise, and then moves yet another and another, and so forth, one will never return to a location one had already visited.

Let us now *rotate* each point in $B$ one unit counterclockwise around the center of the circle. The result of such a rotation is the *unbroken* circle $S^1$: $p_1$ ends up at $p$, $p_2$ ends up at $p_1$, $p_3$ ends up at $p_2$, and so forth.

Success! We have decomposed our broken circle into two pieces, reassembled them (without changing their size or shape), and ended up with an unbroken circle!

**Exercises**

1. Show that the points $p_1, p_2, p_3, \ldots$ are all distinct.

### 8.1.3 Warm-Up Case 3: The Cayley Graph

The point of the last two warm-up cases was to give you a general flavor of how it is possible to decompose an object into pieces and then to reassemble the pieces so as to end up with more than you started with. Now I'd like to introduce you to a construction that is closely analogous to the construction we'll use in proving the Banach-Tarski Theorem. It is also an example of dividing an object into finitely many pieces, reassembling the pieces, and getting two copies of the original object. There is, however, an important respect in which our example will be different than the Banach-Tarski Theorem. In our example, but not in the theorem, we'll be allowed to change the *size* of the pieces as we move them around.

Consider the **Cayley Graph**, which is named after its creator, British mathematician Arthur Cayley:

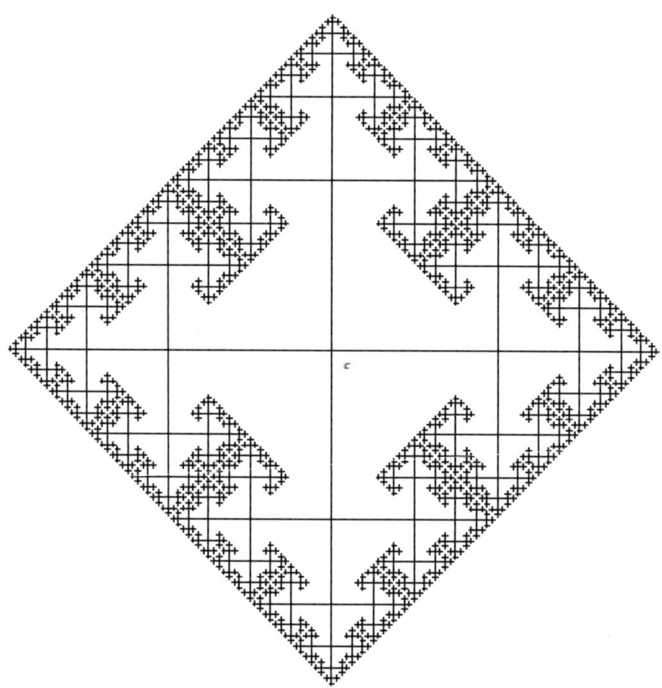


The best way to think of a Cayley Graph is as an infinite collection of paths. A **Cayley Path** is a finite sequence of steps, starting from a central point, $c$. Each step is taken in one of four directions: up, down, right, or left, with the important restriction that one is forbidden from following opposite directions in adjacent steps. (One cannot, for example, follow an "up" step with a "down" step, or a "left" step with a "right" step.) The steps in a Cayley Path span smaller and smaller distances: in the first step of a Cayley Path you advance one unit; in the second step you advance half a unit; in the third step you advance a quarter of a unit; and so forth. An **endpoint** (or vertex) of the Cayley Graph is the endpoint of some Cayley Path (i.e., the result of starting from point $c$ and taking each step in that path). We count the "empty" path as a Cayley Path, and therefore count the starting point $c$ as an endpoint of the empty path.

Let $C$ be the set of Cayley Paths, and let $C^e$ be the set of endpoints of paths in $C$. (In general, if $X$ is a set of Cayley Paths, we will let $X^e$ be the set of endpoints of paths in $X$.) We will see that it is possible to divide $C^e$ into a finite number of components, rearrange the components, and end up with two copies of $C^e$. (As advertised previously, we will allow some of our pieces to change size as we move them around.)

The first step of our construction is to partition the set of Cayley Paths into four disjoint cells:

- $U$ is the set of Cayley Paths that have "up" as their first step;
- $D$ is the set of Cayley Paths that have "down" as their first step;
- $L$ is the set of Cayley Paths that have "left" as their first step;
- $R$ is the set of Cayley Paths that have "right" as their first step.

(I am simplifying things slightly by ignoring the empty path and its endpoint, but you'll be asked to find a version of the construction that does not require this simplification in an exercise below.)

Think of each of our four sets as a different "quadrant" of the Cayley Graph. In the diagram below, I've used a different shade of gray for each.

We will now move from talking about Cayley Paths to talking about their *endpoints*. Focus your attention on $R^e$, the set of endpoints of paths in $R$. You'll see that it is an exact copy of $C^e - L^e$, except that it is a bit shrunken in size and translated slightly to the right. This means that we can use $R^e$ and $L^e$ to create a perfect copy of $C^e$: we start by expanding $R^e$ and then translate it slightly to the left to meet $L^e$, as in the following diagram. (I've included Cayley Paths along with their endpoints to make the diagram easier to read.)

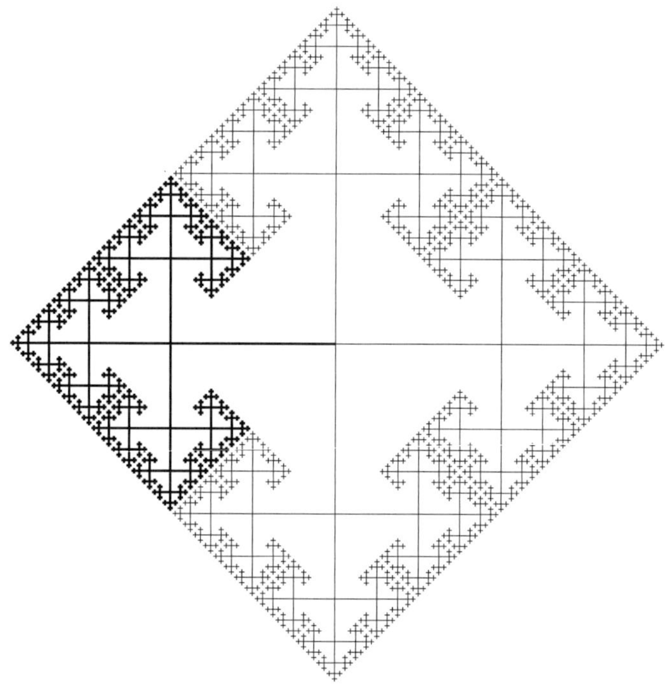

Similarly, we can use $U^e$ and $D^e$ to create a perfect copy of $C^e$: we start by expanding $D^e$ and then translate it slightly upward to meet $U^e$.

Success! We have decomposed $C^e$ into finitely many pieces, reassembled them (changing sizes but not shapes), and ended up with *two* copies of $C^e$!

## Exercises

1. I simplified the construction above by ignoring the empty path $\langle\rangle$, whose endpoint is the graph's central vertex. Find a version of the construction that does not require this simplification. *Hint:* Proceed by dividing $C^e$ into the following components: $U^e, D^e$, $(L \cup S)^e$, and $(R - S)^e$, where $S$ is the set of Cayley Paths that contain only "right" steps:

   $S = \{\langle\rangle, \langle\text{right}\rangle, \langle\text{right, right}\rangle, \langle\text{right, right, right}\rangle, \ldots\}$

### 8.1.4 A More Abstract Description of the Procedure

In this section we will replicate the procedure of the preceding section in a more abstract setting. It is worth doing so because the more abstract version of the procedure is exactly analogous to the procedure we'll use to prove the Banach-Tarski Theorem.

First, some notation: if $X$ is a set of Cayley Paths, let $\overleftarrow{X}$ be the set that results from eliminating the first step from each of the Cayley Paths in $X$. Now recall that $R$ is the set of Cayley Paths that start with a "right" step. Since one can get a valid Cayley Path by appending a "right" step at the beginning of any Cayley Path that does not start with a "left" step, $\overleftarrow{R} = C - L$. (For analogous reasons, $\overleftarrow{D} = C - U$.) So we have the following:

($\alpha$) $C = \overleftarrow{R} \cup L$

($\beta$) $C = \overleftarrow{D} \cup U$

But since every Cayley Path has a unique endpoint, equations ($\alpha$) and ($\beta$) immediately entail the following:

($\alpha'$) $C^e = \left(\overleftarrow{R}\right)^e \cup L^e$

($\beta'$) $C^e = \left(\overleftarrow{D}\right)^e \cup U^e$

Notice, however, that equations ($\alpha'$) and ($\beta'$) deliver the result of the previous section. This is because (as you'll be asked to verify below) we have each of the following:

1. $C^e$ is decomposed into $U^e$, $D^e$, $L^e$, and $R^e$ (ignoring the central vertex).
2. One can get from $R^e$ to $\left(\overleftarrow{R}\right)^e$, and from $D^e$ to $\left(\overleftarrow{D}\right)^e$, by performing a translation together with an expansion.

So it follows from equations ($\alpha'$) and ($\beta'$) that $C^e$ can be decomposed into $U^e$, $D^e$, $L^e$, and $R^e$ (ignoring the central vertex) and recombined into two copies of $C^e$ by carrying out suitable translations and expansions.

This procedure is exactly analogous to the procedure we'll use to prove the Banach-Tarski Theorem. The only difference is that we'll work with a variant of the Cayley Graph in which "following a Cayley Path" has a different geometrical interpretation than it does here. We'll think of our Cayley Graph as wrapped around the surface of a ball, and we'll think of each step of a Cayley Path as corresponding to a certain *rotation* of the ball.

The key reason we'll be able to prove our result in the new setting is that the truth of equations ($\alpha'$) and ($\beta'$) is independent of how we interpret "following a Cayley Path." We'll need to proceed carefully, though, because the truth of claims 1 and 2 above does depend on how we interpret "following a Cayley Path." Fortunately, this won't be a problem, because of the following:

- We'll be able to recover claim 1 because our interpretation of "following a Cayley Path" will be such that different Cayley Paths have different endpoints.
- We won't be able to recover claim 2 as it is stated above, but we'll get something even better: our new interpretation of "following a Cayley Path" will yield the result that one can get from $R^e$ to $\left(\overleftarrow{R}\right)^e$, and from $D^e$ to $\left(\overleftarrow{D}\right)^e$, by performing a suitable rotation—no expansions required. This means that we'll be able to reassemble $U^e$, $D^e$, $L^e$, and $R^e$ into two copies of $C^e$ *without changing their size.*

### Exercises

1. Verify that $C^e$ is decomposed into $U^e$, $D^e$, $L^e$, and $R^e$ (ignoring the central vertex).
2. Verify that one can get from $R^e$ to $\left(\overleftarrow{R}\right)^e$, and from $D^e$ to $\left(\overleftarrow{D}\right)^e$, by performing a translation together with an expansion.
3. As in the case of section 8.1.3, I have simplified the construction by ignoring the empty path, whose endpoint is the center of the Cayley Graph. The exercise at the end of section 8.1.3 shows that it is possible to revise the construction of section 8.1.3 in such a way that this simplification is not required. Show that the same trick can be used in the more abstract setting of section 8.1.4. *Hint:* As before, proceed by dividing $C^e$ into the following components: $U^e, D^e, (L \cup S)^e$, and $(R - S)^e$, where $S$ is the set of Cayley Paths that contain only "right" steps:

   $S = \{\langle\rangle, \langle\text{right}\rangle, \langle\text{right, right}\rangle, \langle\text{right, right, right}\rangle, \ldots\}$

   As usual $X^e$ is the set of endpoints of Cayley Paths in $X$.

## 8.2 The Theorem

With three warm-up cases under our belts, we're ready to tackle the Banach-Tarski Theorem. (I'll skip some of the technical details. If you'd like a more explicit version of the proof, have a look at the resources listed in section 8.4.)

### 8.2.1 Cayley on a Ball

The first step of the proof is to construct a *modified* version of the Cayley Graph—a version of the graph that is wrapped around the surface of a ball.

The "center" of our graph is an arbitrary point $c$ on the surface of our ball. As in the case of the original graph, our modified graph consists of an infinite number of Cayley Paths. As before, each Cayley Path starts from $c$ and is the result of taking a finite sequence of steps. As before, there are four different kinds of steps: up, down, right, and left. And, as before, one is not allowed to follow a step with its inverse. But there is an important difference. In the original Cayley Graph, a step was the result of traversing a certain distance on the plane. In our modified Cayley Graph, in contrast, a step is the result of performing a *rotation* on the sphere.

To see what the relevant rotations consist in, imagine that you're holding the ball in your hand and consider the following axes:

- The $x$-axis runs from your right to your left through the center of the ball.
- The $y$-axis is orthogonal to the $x$-axis. It runs from the wall in front of you to the wall behind of you, through the center of the ball.
- The $z$-axis is orthogonal to the other two. It runs from the ground to the sky, through the center of the ball.

For a given angle $\theta$ (which will be chosen carefully—more about this later), we will be concerned with the following rotations:

- An "up" rotation is a counterclockwise rotation of $\theta$ degrees about the $x$-axis. (When you're holding the ball in front of you, you perform this rotation by rotating the ball from bottom to top.)
- A "down" rotation is a clockwise rotation of $\theta$ degrees about the $x$ axis. (When you're holding the ball in front of you, you perform this rotation by rotating the ball from top to bottom.)
- A "right" rotation is a counterclockwise rotation of $\theta$ degrees about the $z$-axis. (When you're holding the ball in front of you, you perform this rotation by rotating the ball from left to right.)
- A "left" rotation is a clockwise rotation of $\theta$ degrees about the $z$-axis. (When you're holding the ball in front of you, you perform this rotation by rotating the ball from right to left.)

To each such rotation $\rho$ corresponds a function $f_\rho$, which takes each point $p$ on the surface of the ball to the point on the surface of the ball whose current location (relative to an external reference frame) would come to be occupied by $p$ were rotation $\rho$ to be performed. For instance, $f_{\text{right}}$ is the function that takes each point $p$ to the point that is currently $\theta$ degrees east of $p$.

Now return to our modified version of the Cayley Graph. I said earlier that each path of the graph is the result of taking a sequence of up, down, right, and left steps (without ever following a step with its inverse). I am now in a position to explain what it means to "follow" one of these paths on the surface of the ball. Consider the following example of a Cayley Path:

$\langle$up, right, up, up, left, down$\rangle$

In the original version of the Cayley Graph, one follows this path by starting at the graph's center and moving through the plane: first upward, then rightward, then upward again, and so forth. In our modified version of the Cayley Graph, one "follows" the path by starting at the graph's "central" point $c$ and applying each of the rotations in the path in reverse order. Accordingly, the endpoint of the path above is:

$$f_{\text{up}}(f_{\text{right}}(f_{\text{up}}(f_{\text{up}}(f_{\text{left}}(f_{\text{down}}(c))))))$$

In general, the endpoint of Cayley Path $\langle \rho_1, \rho_2, \ldots, \rho_n \rangle$ is the point $f_{\rho_1}(f_{\rho_2}(\ldots f_{\rho_n}(c)\ldots))$.

### Exercises

1. By representing rotations as matrices, give an analytic characterization of each of our four rotations (up, down, right, left) and the corresponding functions ($f_{\text{up}}, f_{\text{down}}, f_{\text{right}}, f_{\text{left}}$). *Note:* This is a difficult exercise unless you're comfortable with trigonometry and matrix multiplication. If you're not, you might like to use the answer at the end of this chapter to learn more about how to work out the details of our rotations.

### 8.2.2 The Key Property of Rotations

When the rotation $\theta$ is chosen properly, we get an important result. (Let us assume, for concreteness, that $\theta = \arccos(1/3) \approx 70.53°$, since that turns out to be one of the choices of a rotation angle that works.) The result is that *different Cayley Paths always have different endpoints*; In other words:

**Key Property** If $\langle \rho_1, \rho_2, \ldots, \rho_n \rangle$ and $\langle \sigma_1, \sigma_2, \ldots, \sigma_m \rangle$ are distinct Cayley Paths, then

$$f_{\rho_1}(f_{\rho_2}(\ldots f_{\rho_n}(c)\ldots)) \neq f_{\sigma_1}(f_{\sigma_2}(\ldots f_{\sigma_m}(c)\ldots)).$$

(Annoyingly, the Key Property holds for *almost* every choice of our central point $c$ but not quite every choice. I'll come back to this in section 8.2.4. Until then, I'll simply assume that $c$ has been chosen properly.)

The Key Property is, in effect, a sophisticated version of a point we came across in warm-up case 2, when we saw that one can never return to one's starting point by traveling around the unit circle in steps of one unit. The proof is long and tedious, so I won't go through it here, but it is spelled out in the second of the two readings I list in section 8.4.

I will, however, illustrate the Key Property with a couple of examples. We'll need to start with a system of geographical coordinates: a system of lines of latitude and longitude. We will let the "North Pole" and the "South Pole" of our ball be the top-most and bottom-most points of the ball respectively. A line of latitude (or a "parallel") is a circle on the ball's surface, whose points are at a uniform distance from the ball's North Pole. The ball's "equator" is its largest parallel. A line of longitude (or a "meridian") is a great circle that crosses the poles. Finally, the ball's "prime meridian" is an arbitrarily selected meridian; for the sake of concreteness, I'll count the meridian that crosses the point on the ball closest to you as the prime meridian.

We will think of this as an "external" coordinate system. In other words, our coordinates do not change their locations with respect to the room as you rotate the ball. You can think of the coordinates as being projected onto the ball by lasers, rather than painted on the ball. So, for instance, the North and South Poles of our ball will always be the top-most and bottom-most points of the ball, regardless of how the ball moves about as we perform particular rotations.

Let us think of the point that is currently at the intersection of the ball's equator and prime meridian as our "central" point, $c$. We will mark its location on the ball with a pencil and see what happens to our mark as we perform particular rotations. For the sake of concreteness, let us follow the Cayley Path $\langle$right, up$\rangle$.

One follows a path by applying each of the rotations in the path in reverse order. So we start by performing an up rotation. In other words, we rotate the ball $\theta$ degrees about the $x$-axis (which runs from your left to your right, through the center of the ball). The effect of this rotation is that the mark goes from being located at $(0°, 0°)$ (which is the intersection of the prime meridian and the equator), to being located at ($\theta°$ North, $0°$) (which is a latitude of $\theta°$ North on the prime meridian). On Earth, this corresponds to the location you would get to if you traveled north on the prime meridian until you reached the latitude of northern Norway. Next, we perform a right rotation. In other words, we rotate the ball $\theta$ degrees about the $z$-axis (which runs from the ground to the sky, through the center of the ball). This causes the pencil mark to travel east on the $\theta$th parallel until it reaches ($\theta°$ North, $\theta°$ East). On Earth, this corresponds to the location of the Yamal peninsula in northern Siberia.

Note that even though our up and right rotations were both rotations of $\theta$ degrees, they did not cause our mark to travel the same distance across the surface of the ball. In general, the distance traveled by our mark during a particular rotation depends on its distance from the poles of rotation (i.e., the points at which the axes of rotation intersect the surface of the ball). If our mark is located close to a pole of rotation, it will travel around a small circle (specifically, the unique circle on the surface of the ball that contains our mark and is centered on the pole of rotation), and it will cover a distance corresponding to $\theta/360$ of the circumference of that circle. If, on the other hand, our mark is equidistant between the two poles of rotation, it will travel around the great circle of points equidistant from those poles, and it will cover a distance corresponding to $\theta/360$ of the circumference of that larger circle.

Now go back to our example. When you performed the initial up rotation, the mark was exactly equidistant between the poles of rotation: it was at $(0°, 0°)$ and the poles of an up rotation (which is a rotation about the $x$-axis) are at ($0°$, $90°$ East) and ($0°$, $90°$ West). So the mark traveled a distance of $\theta/360$ of the circumference of a great circle. In contrast, when you performed the right rotation, the mark was close to one of the axes of rotation: it was at ($\theta°$ North, $0°$), and the poles of a right rotation (which is a rotation about the $z$-axis) are at the ball's North and South Poles. So the mark traveled a relatively short distance: namely, $\theta/360$ of the circumference of the ball's $\theta$th parallel (recall that $\theta \approx 70.53°$). The radius of this parallel is a third of the radius of the ball (because $\theta = \arccos(1/3)$), so the distance traveled by the mark on the second rotation was a third of the distance traveled in the first rotation.

We have considered the result of following the Cayley Path ⟨right, up⟩. Let us compare this with the result of following the Cayley Path ⟨up, right⟩. As usual, we proceed by applying each of the rotations in the path in reverse order. Our pencil mark starts out at $(0°, 0°)$, which is equidistant between the ball's North and South Poles. So a right rotation will cause our mark to travel a distance of $\theta/360$ of the circumference of a great circle, and end up at ($0°$, $\theta°$ East), which, on Earth, corresponds to a location in the middle of the Indian Ocean, west of the Maldives. This point is close to ($0°$, $90°$ East), which is one of the poles of an up rotation. So applying such a rotation will cause our mark to travel a relatively short distance and end up close to ($18.32°$ North, $83.28°$ East), which, on Earth, corresponds to a location near the village of Garbham on the east coast of India.

The lesson of our examples is that the Cayley Paths ⟨right, up⟩ and ⟨up, right⟩ have different endpoints. In the first case, our mark starts out traveling a long distance "up" and then travels a short distance "right"; in the second case, our mark starts out traveling a long distance "right" and then travels a short distance "up." This illustrates the key property of our rotations: as long as $c$ is chosen properly, different Cayley Paths will always have different endpoints.

### Exercises

1. Use the answer to exercise 1 from section 8.2.1 to calculate the values of $f_{\text{right}}(f_{\text{up}}((0,1,0)))$ and $f_{\text{up}}(f_{\text{right}}((0,1,0)))$. (Assume that $\theta = \arccos(1/3) \approx 70.53°$.) *Note:* This is a difficult exercise unless you're comfortable with trigonometry and matrix multiplication. If you're not, you might find it helpful to look at the answer at the end of this chapter.

### 8.2.3 Duplicating the Graph

Let $C$ be the set of Cayley Paths, and let $C^e$ be the set of endpoints of paths in $C$. Since we are considering a version of the Cayley Graph that is wrapped around the surface of the ball, $C^e$ consists of every point $f_{\rho_1}(f_{\rho_2}(\ldots f_{\rho_n}(c)\ldots))$ for $\langle \rho_1, \rho_2, \ldots, \rho_n \rangle$, a Cayley Path.

In this section we will prove a preliminary version of our main result. We will show that there is a decomposition of $C^e$ into finitely many pieces which is such that the pieces can be rearranged—without modifying their sizes or shapes—so as to get two copies of $C^e$. (The reason this is only a preliminary version of the main result is that $C^e$ contains only countably many points from the ball's surface, and we're ultimately looking for a decomposition of the entire ball.)

Just as we did in section 8.1.4, we begin by considering the following four subsets of $C$:

- $U$ is the set of Cayley Paths that have "up" as their first step;
- $D$ is the set of Cayley Paths that have "down" as their first step;
- $L$ is the set of Cayley Paths that have "left" as their first step;
- $R$ is the set of Cayley Paths that have "right" as their first step.

These sets constitute a partition of $C$ (ignoring the empty path $\langle\rangle$). As before, we define $\overleftarrow{X}$ as the result of deleting the first step from each Cayley Path in $X$. And, as before, we get the following:

($\alpha$) $C = \overleftarrow{R} \cup L$

($\beta$) $C = \overleftarrow{D} \cup U$

which entail:

($\alpha'$) $C^e = \left(\overleftarrow{R}\right)^e \cup L^e$

($\beta'$) $C^e = \left(\overleftarrow{D}\right)^e \cup U^e$

Notice, moreover, that equations ($\alpha'$) and ($\beta'$) deliver our preliminary result, as long as we have the following:

1. $C^e$ is decomposed into $U^e$, $D^e$, $L^e$, and $R^e$ (ignoring the central vertex).
2. One can get from $R^e$ to $\left(\overleftarrow{R}\right)^e$, and from $D^e$ to $\left(\overleftarrow{D}\right)^e$, by performing a rotation.

For when claims 1 and 2 are in place, it follows from equations ($\alpha'$) and ($\beta'$) that $C^e$ can be decomposed into $U^e$, $D^e$, $L^e$, and $R^e$ (ignoring the central point $c$) and recombined into two copies of $C^e$ by performing suitable rotations.

In order to verify claims 1 and 2, however, we'll need to proceed differently than in warm-up case 3, because the details depend essentially on our interpretation of "following a Cayley Path." Fortunately, it is all very straightforward:

- If you worked out exercise 1 of section 8.1.4, you know that all it takes for claim 1 to be true is for different Cayley Paths to have different endpoints. And this is indeed the case: it is the Key Property of section 8.2.2.
- The proof of claim 2 is the nicest feature of the entire construction. Let $\langle \rho_{\text{right}}, \rho_2, \rho_3, \ldots, \rho_n \rangle$ be a path in $R$. Its endpoint is

  $$f_{\text{right}}(f_{\rho_2}(f_{\rho_3}(\ldots f_{\rho_n}(c)\ldots))).$$

  The corresponding path in $\overleftarrow{R}$ is $\langle \rho_2, \rho_3, \ldots, \rho_n \rangle$, and its endpoint is:

  $$f_{\rho_2}(f_{\rho_3}(\ldots f_{\rho_n}(c)\ldots)).$$

So one can get from the endpoint of $\langle \rho_2, \rho_3, \ldots, \rho_n \rangle$ to the endpoint of $\langle \rho_{\text{right}}, \rho_2, \rho_3, \ldots, \rho_n \rangle$ by applying $f_{\text{right}}$, which is a right rotation! And, of course, the same is true for every other path in $R$. So we get the result that one can move from the entire set $\left(\overleftarrow{R}\right)^e$ to the entire set $R^e$ by performing a right rotation (and, therefore, that one can move from $R^e$ to $\left(\overleftarrow{R}\right)^e$ by performing a left rotation).

For analogous reasons, we get the result that one can move from $\left(\overleftarrow{D}\right)^e$ to $D^e$ by performing a down rotation (and, therefore, that one can move from $D^e$ to $\left(\overleftarrow{D}\right)^e$ by performing an up rotation).

This concludes our preliminary result. We have verified that it is possible to decompose $C^e$ into four separate components $(U^e, D^e, L^e, R^e)$ and then reassemble them to create two copies of $C^e$, *without modifying sizes or shapes*. All we need to do is rotate $R^e$ and $D^e$.

**Exercises**

1. Recall that we have simplified things slightly by ignoring the "center" of the graph. Use the fact that $U, D, L, R$ constitute a partition of $C$ (minus the empty path $\langle\rangle$) to show that $U^e, D^e, L^e, R^e$ constitute a partition of $C^e$ (minus the "center" point $c$). *Hint:* Use the technique described in the answer to exercise 8.1.4, and avail yourself of the fact (mentioned in section 8.2.2) that different Cayley Paths always have different endpoints.

### 8.2.4 Many Cayleys on a Ball

The next step is to duplicate the entire surface of our ball. The basic idea is straightforward. We have seen that there is a procedure that can be used to duplicate the endpoints of a modified Cayley Graph that is wrapped around the surface of the ball. (Section 8.2.3.) What we'll do now is work with an infinite family of modified Cayley Graphs, whose endpoints *jointly* cover the surface of the ball without overlapping. We'll then duplicate the surface of the ball by applying our duplication procedure simultaneously to each of our infinitely many graphs.

We begin by partitioning the surface of the sphere into cells. This is done by stipulating that two points are in the same cell if and only if one can get from one point to the other by applying a finite sequence of left, right, up, and down rotations.

Note that a cell is just the set of endpoints of some modified Cayley Graph, except that no endpoint has been designated as a center. So all one needs to do to characterize a modified Cayley Graph on the basis of a cell is to designate one of the members of the cell as a center. Unfortunately, there turns out to be no way of specifying a condition that would succeed in singling out a unique member from an arbitrarily given cell. So our proof must make use of the Axiom of Choice. Rather than specifying a center for each cell, we use the Axiom of Choice to prove that a set of cell-centers exists, and therefore that there exists a family, $\mathcal{G}$, of modified Cayley Graphs, whose endpoints cover the surface of the sphere without overlapping.

Showing that $\mathcal{G}$ must exist is the hard part. The remainder is straightforward. For each graph $g$ in $\mathcal{G}$, let $C_g$ be the set of paths in $g$. As usual, we decompose $C_g$ (minus the empty path) into four non-overlapping pieces:

- $U_g$ is the set of paths in $g$ that have "up" as their first step;
- $D_g$ is the set of paths in $g$ that have "down" as their first step;
- $L_g$ is the set of paths in $g$ that have "left" as their first step;
- $R_g$ is the set of paths in $g$ that have "right" as their first step.

Now, the upshot of section 8.2.3 is that $C_g^e$ (minus its center) can be decomposed into pieces and that the pieces can be rearranged so as to get two copies of $C_g^e$. We can get a first copy of $C_g^e$ using $L_g^e$ and $R_g^e$ because

$$C_g^e = \left(\overleftarrow{R_g}\right)^e \cup L_g^e$$

and because one can move from $R_g^e$ to $\left(\overleftarrow{R}\right)^e$ by performing a left rotation. And we can get a second copy of $C_g^e$ using $D_g^e$ and $U_g^e$ because

$$C_g^e = \left(\overleftarrow{D_g}\right)^e \cup U_g^e$$

and because one can move from $D_g^e$ to $\left(\overleftarrow{D}\right)^e$ by performing an up rotation.

The fact that this is true for each modified Cayley Graph $g$ in $\mathcal{G}$ can be used to generate a version of the construction that applies *simultaneously* to every graph $\mathcal{G}$. Let $U_{\mathcal{G}}^e$ be the union of $U_g^e$ for every $g$ in $\mathcal{G}$ (and similarly for $D_{\mathcal{G}}^e$, $L_{\mathcal{G}}^e$, and $R_{\mathcal{G}}^e$). Then one can generate a first copy of the surface of the sphere (minus the centers of graphs in $\mathcal{G}$) by combining $L_{\mathcal{G}}^e$ plus a rotated version of $R_{\mathcal{G}}^e$. And one can generate a second copy of the surface of the sphere (minus the centers of graphs in $\mathcal{G}$) by combining $U_{\mathcal{G}}^e$ plus a rotated version of $D_{\mathcal{G}}^e$. As usual, one can deal with the centers by using the technique described in the answer to the exercise at the end of section 8.1.4.

(I have been ignoring an important complication. As noted in section 8.2.3, it is essential to our procedure that different Cayley Paths have different endpoints. But, as noted in section 8.2.2, we only get this Key Property for certain choices of $c$. The reason is that, for each Cayley Path $\langle \rho_1, \rho_2, \rho_3, \ldots \rangle$, the rotation $f_{\rho_1}(f_{\rho_2}(f_{\rho_r}(\ldots)))$ does not change the location of the points intersecting its axis of rotation. So whenever $c$ is such that some such point is the endpoint of some Cayley Path, we will have different Cayley Paths sharing an endpoint. Fortunately there are only countably many such points (because there are only countably many Cayley Paths), and therefore there are only countably many graphs in $\mathcal{G}$ containing such points. One can deal with the vertices of these graphs separately by applying a sophisticated version of the trick we used in warm-up case 2 to turn a circle missing a point into a complete circle. For details, have a look at the resources in section 8.4).

Success! We have managed to decompose the surface of our ball into finitely many pieces and then reassemble the pieces—without changing their sizes or shapes—so as to get two copies of the original surface.

And now that we know how to duplicate the *surface* of our ball, it is easy to duplicate the ball itself. We can use the same procedure as before, but rather than working with points on the surface of the ball, we work with the lines that connect the center of the ball with each point.

That's the Banach-Tarski Theorem!

**Reward** As a reward for your efforts, let me tell you a joke. (I don't know who came up with it originally; I heard it from one of my students, Anthony Liu.)

*Question:* What's an anagram of "Banach-Tarski"?
*Answer:* Banach-Tarski Banach-Tarski.

### Exercises

1. The procedure I describe above to duplicate the entire ball leaves out the center of the ball. Address this point by showing that one can start with a ball that is missing its central point, decompose it into two pieces, and then reassemble the pieces (without changing their size or shape) so as to get a full ball.

## 8.3 Paradox?

The Banach-Tarski Theorem is an astonishing result. We have decomposed a ball into finitely many pieces, moved around the pieces without changing their size or shape, and then reassembled them into two balls of the same size as the original.

I think the theorem teaches us something important about the notion of volume. As noted earlier, it is an immediate consequence of the theorem that some of the Banach-Tarski pieces must lack definite volumes and, therefore, that not every subset of the unit ball can have a well-defined volume. A little more precisely, the theorem teaches us that there is no way of assigning volumes to the Banach-Tarski pieces while preserving three-dimensional versions of the principles we called Uniformity and (finite) Additivity in chapter 7.

(*Proof:* Suppose that each of the (finitely many) Banach-Tarski pieces has a definite finite volume. Since the pieces are disjoint, and since their union is the original ball, Additivity entails that the sum of the volumes of the pieces must equal the volume of the original ball. But Uniformity ensures that the volume of each piece is unchanged as we move it around. Since the reassembled pieces are disjoint, and since their union is two balls, Additivity entails that the sum of their volumes must be twice the volume of the original ball. But since the volume of the original ball is finite and greater than zero, it is impossible for the sum of the pieces to equal both the volume of the original ball and twice the volume of the original ball.)

If I were to assign the Banach-Tarski Theorem a paradoxicality grade of the kind we used in chapter 3, I would assign it an 8. The theorem teaches us that although the notion of volume is well-behaved when we focus on ordinary objects, there are limits to how far it can be extended when we consider certain extraordinary objects—objects that can only be shown to exist by assuming the Axiom of Choice.

## 8.4 Further Resources

- My discussion of the Banach-Tarski Paradox is drawn from two excellent papers, one by Avery Robinson and the other by Tom Weston. They both have the same title: "The Banach-Tarski Paradox."
- My student Sydney Gibson created a wonderful applet for visualizing Banach-Tarski rotations. It is available at http://web.mit.edu/arayo/www/Cayley/viz.html.

## Appendix: Answers to Exercises

### Section 8.1.2

1. Suppose that $p_i = p_j$ for some $i$ and $j$ such that $i < j$. Then it must be the case that you can start at $p_i$ and return to $p_i$, advancing $j - i$ units clockwise. Since the circumference of $S^1$ is $2\pi$, this means that $j - i = n \cdot 2\pi$, where $n$ is some positive integer. But this is impossible, since it entails that $2\pi = \frac{j-i}{n}$, and we know that $2\pi$ is an irrational number.

### Section 8.1.3

1. $C^e$ can be partitioned into the following four components

   $U^e, D^e, (L \cup S)^e, (R - S)^e$

   We produce our first copy of $C^e$ from $(L \cup S)^e$ and $(R - S)^e$. The key insight is that $(R - S)^e$ is an exact copy of $C^e - (L \cup S)^e$, except that it is a bit shrunken in size and translated slightly to the right. So we can create a perfect copy of $C^e$ by expanding $(R - S)^e$ and then translating it slightly to the left to meet $(L \cup S)^e$.

   We produce the second copy of $C^e$ from $U^e$ and $D^e$, as in the main text.

### Section 8.1.4

1. We need to do two things. First, we need to verify that $U^e \cup D^e \cup L^e \cup R^e = C^e$. Second, we need to verify that there is no overlap between $U^e$, $D^e$, $L^e$, and $R^e$. The first claim follows from $U \cup D \cup L \cup R = C$, since every Cayley Path has a unique endpoint. (As usual, we ignore the empty path.) The second claim follows from the fact that there is no overlap between $U$, $D$, $L$, and $R$, since different Cayley Paths always have different endpoints.
2. Consider the transition from $R^e$ to $\left(\overleftarrow{R}\right)^e$. (The transition from $D^e$ to $\left(\overleftarrow{D}\right)^e$ is analogous.) When one removes an initial right step from a given path, the resulting path is translated one unit to the left and doubled in size. But since $\overleftarrow{R}$ is the result of eliminating the initial "right" step from each path in $R$, this means that $\left(\overleftarrow{R}\right)^e$ is none other than the translated and expanded version of $R^e$ that figured in section 8.1.3.
3. The following identities hold:
   - $C = \overleftarrow{D} \cup U$
   - $C = \overleftarrow{(R - S)} \cup L \cup S$

So we have the following:

- $C^e = \left(\overleftarrow{D}\right)^e \cup U^e$
- $C^e = \left(\overleftarrow{R-S}\right)^e \cup (L \cup S)^e$

Notice, moreover, that claims 1 and 2 from the main text continue to hold: $U^e, D^e, (L \cup S)^e, (R-S)^e$ is a decomposition of $C^e$, and $\left(\overleftarrow{R-S}\right)^e$ is a translated and expanded version of $(R-S)^e$.

### Section 8.2.1

1. Following Weston's "The Banach-Tarski Paradox," we represent a rotation $\rho$ as a matrix and let $f_\rho((a,b,c))$ be:

$$\rho \begin{pmatrix} a \\ b \\ c \end{pmatrix}$$

An up rotation (i.e., a counterclockwise rotation of $\theta$ degrees about the $x$-axis) is represented by the matrix

$$\begin{pmatrix} 1 & 0 & 0 \\ 0 & \cos(\theta) & -\sin(\theta) \\ 0 & \sin(\theta) & \cos(\theta) \end{pmatrix}$$

and a down rotation (i.e., a clockwise rotation of $\theta$ degrees about the $x$-axis) is represented by the following matrix:

$$\begin{pmatrix} 1 & 0 & 0 \\ 0 & \cos(\theta) & \sin(\theta) \\ 0 & -\sin(\theta) & \cos(\theta) \end{pmatrix}$$

A right rotation (i.e., a counterclockwise rotation of $\theta$ degrees about the $z$-axis) is represented by the matrix

$$\begin{pmatrix} \cos(\theta) & -\sin(\theta) & 0 \\ \sin(\theta) & \cos(\theta) & 0 \\ 0 & 0 & 1 \end{pmatrix}$$

and a left rotation (i.e., a clockwise rotation of $\theta$ degrees about the $z$-axis) is represented by the following matrix:

$$\begin{pmatrix} \cos(\theta) & \sin(\theta) & 0 \\ -\sin(\theta) & \cos(\theta) & 0 \\ 0 & 0 & 1 \end{pmatrix}$$

Accordingly,

$$f_{\text{up}}((a,b,c)) = \begin{pmatrix} 1 & 0 & 0 \\ 0 & \cos(\theta) & -\sin(\theta) \\ 0 & \sin(\theta) & \cos(\theta) \end{pmatrix} \begin{pmatrix} a \\ b \\ c \end{pmatrix} = \begin{pmatrix} a \\ b\cdot\cos(\theta) - c\cdot\sin(\theta) \\ b\cdot\sin(\theta) + c\cdot\cos(\theta) \end{pmatrix}$$

$$f_{\text{down}}((a,b,c)) = \begin{pmatrix} 1 & 0 & 0 \\ 0 & \cos(\theta) & \sin(\theta) \\ 0 & -\sin(\theta) & \cos(\theta) \end{pmatrix} \begin{pmatrix} a \\ b \\ c \end{pmatrix} = \begin{pmatrix} a \\ b\cdot\cos(\theta) + c\cdot\sin(\theta) \\ -b\cdot\sin(\theta) + c\cdot\cos(\theta) \end{pmatrix}$$

$$f_{\text{right}}((a,b,c)) = \begin{pmatrix} \cos(\theta) & -\sin(\theta) & 0 \\ \sin(\theta) & \cos(\theta) & 0 \\ 0 & 0 & 1 \end{pmatrix} \begin{pmatrix} a \\ b \\ c \end{pmatrix} = \begin{pmatrix} a\cdot\cos(\theta) - b\cdot\sin(\theta) \\ a\cdot\sin(\theta) + b\cdot\cos(\theta) \\ c \end{pmatrix}$$

$$f_{\text{left}}((a,b,c)) = \begin{pmatrix} \cos(\theta) & \sin(\theta) & 0 \\ -\sin(\theta) & \cos(\theta) & 0 \\ 0 & 0 & 1 \end{pmatrix} \begin{pmatrix} a \\ b \\ c \end{pmatrix} = \begin{pmatrix} a\cdot\cos(\theta) + b\cdot\sin(\theta) \\ -a\cdot\sin(\theta) + b\cdot\cos(\theta) \\ c \end{pmatrix}$$

## Section 8.2.2

1. Since $\cos(\arccos(1/3)) = 1/3$ and $\sin(\arccos(1/3)) = \frac{2\sqrt{2}}{3}$, we will be focusing on the following rotations:

$$\text{up} = \rho^x_\theta = \frac{1}{3}\begin{pmatrix} 3 & 0 & 0 \\ 0 & 1 & -2\sqrt{2} \\ 0 & 2\sqrt{2} & 1 \end{pmatrix} \qquad \text{down} = \rho^x_{-\theta} = \frac{1}{3}\begin{pmatrix} 3 & 0 & 0 \\ 0 & 1 & 2\sqrt{2} \\ 0 & -2\sqrt{2} & 1 \end{pmatrix}$$

$$\text{right} = \rho^z_\theta = \frac{1}{3}\begin{pmatrix} 1 & -2\sqrt{2} & 0 \\ 2\sqrt{2} & 1 & 0 \\ 0 & 0 & 3 \end{pmatrix} \qquad \text{left} = \rho^z_{-\theta} = \frac{1}{3}\begin{pmatrix} 1 & 2\sqrt{2} & 0 \\ -2\sqrt{2} & 1 & 0 \\ 0 & 0 & 3 \end{pmatrix}$$

Let us use these matrices to calculate $f_{\text{right}}(f_{\text{up}}((0,1,0)))$. We first calculate $f_{\text{up}}((0,1,0))$:

$$f_{\text{up}}((0,1,0)) = \frac{1}{3}\begin{pmatrix} 3 & 0 & 0 \\ 0 & 1 & -2\sqrt{2} \\ 0 & 2\sqrt{2} & 1 \end{pmatrix} \begin{pmatrix} 0 \\ 1 \\ 0 \end{pmatrix} = \left(0, \frac{1}{3}, \frac{2\sqrt{2}}{3}\right)$$

This point lies $\arcsin\left(\frac{2\sqrt{2}}{3}\right) = \theta \approx 70.53$ degrees north of the equator, $\arccos\left(\frac{\frac{1}{3}}{\cos\left(\arcsin\left(\frac{2\sqrt{2}}{3}\right)\right)}\right) = 0$ degrees east of the prime meridian. We then use the resulting values to calculate $f_{\text{right}}(f_{\text{up}}((0,1,0)))$:

$$f_{\text{right}}\left(\left(0, \frac{1}{3}, \frac{2\sqrt{2}}{3}\right)\right) = \frac{1}{3}\begin{pmatrix} 1 & -2\sqrt{2} & 0 \\ 2\sqrt{2} & 1 & 0 \\ 0 & 0 & 3 \end{pmatrix}\begin{pmatrix} 0 \\ \frac{1}{3} \\ \frac{2\sqrt{2}}{3} \end{pmatrix} = \left(-\frac{2\sqrt{2}}{9}, \frac{1}{9}, \frac{2\sqrt{2}}{3}\right)$$

This point lies $\arcsin\left(\frac{2\sqrt{2}}{3}\right) = \theta \approx 70.53$ degrees north of the equator, $\arccos\left(\frac{\frac{1}{9}}{\cos\left(\arcsin\left(\frac{2\sqrt{2}}{3}\right)\right)}\right) = \theta \approx 70.53$ degrees east of the prime meridian.

Let us now calculate $f_{\text{up}}(f_{\text{right}}((0,1,0)))$. We first calculate $f_{\text{right}}((0,1,0))$:

$$f_{\text{right}}((0,1,0)) := \frac{1}{3}\begin{pmatrix} 1 & -2\sqrt{2} & 0 \\ 2\sqrt{2} & 1 & 0 \\ 0 & 0 & 3 \end{pmatrix}\begin{pmatrix} 0 \\ 1 \\ 0 \end{pmatrix} = \left(-\frac{2\sqrt{2}}{3}, \frac{1}{3}, 0\right)$$

This point lies on the equator, $\arccos\left(\frac{1}{3}\right) = \theta \approx 70.53$ degrees east of the prime meridian. We then use the resulting values to calculate $f_{\text{up}}(f_{\text{right}}((0,1,0)))$:

$$f_{\text{up}}\left(\left(-\frac{2\sqrt{2}}{3}, \frac{1}{3}, 0\right)\right) = \frac{1}{3}\begin{pmatrix} 3 & 0 & 0 \\ 0 & 1 & -2\sqrt{2} \\ 0 & 2\sqrt{2} & 1 \end{pmatrix}\begin{pmatrix} -\frac{2\sqrt{2}}{3} \\ \frac{1}{3} \\ 0 \end{pmatrix} = \left(-\frac{2\sqrt{2}}{3}, \frac{1}{9}, \frac{2\sqrt{2}}{9}\right)$$

This point lies $\arcsin\left(\frac{2\sqrt{2}}{9}\right) \approx 18.32$ degrees north of the equator, $\arccos\left(\frac{\frac{1}{9}}{\cos\left(\arcsin\left(\frac{2\sqrt{2}}{9}\right)\right)}\right) \approx 83.28$ degrees east of the prime meridian.

### Section 8.2.3

1. To verify that $C^e$ (minus the "center," $c$) is partitioned into $U^e, D^e, L^e, R^e$, we need to check two things: (*i*) that $U^e, D^e, L^e, R^e$ are all distinct from one another, and (*ii*) that their union is $C^e$ (minus the empty path $\langle\rangle$). The situation in (*i*) is an immediate consequence of the fact that different Cayley Paths always have different endpoints, and the situation in (*ii*) follows from the fact that $C$ (minus the empty path) is the union of $U, D, L, R$.

### Section 8.2.4

1. Consider a circle that lies entirely within the ball. Assume, moreover, that the circle is missing a point and that the missing point is located at the center of the ball. We know from warm-up case 2 that one can start with a circle that is missing a single point, rotate one of its subsets, and get an unbroken circle. The result of performing that same rotation in the present context is that we go from a ball missing its center to a complete ball.

# III Computability and Gödel's Theorem

# 9 Computability

Suppose that you have a function $f$ from natural numbers to natural numbers and that you would like to program a computer to compute it. In other words: you would like to build a computer program that always yields $f(n)$ as output when it is given a natural number $n$ as input. As it turns out, not every function from natural numbers to natural numbers is computable in this sense. In this chapter I will give you examples of functions so complex that no computer could possibly compute them, no matter how powerful.

## 9.1 Turing Machines

Let me start by introducing you to **Turing Machines**, which are computers of a particularly simple sort. They are named after their creator, the great British mathematician and war hero Alan Turing.

A Turing Machine's **hardware** consists of two components, a memory tape and a tape-reader:

**The Memory Tape** You can think of the tape as a long strip of paper, divided into cells:

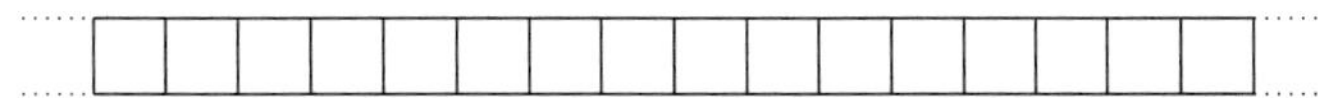

The tape is assumed to be infinite in both directions. (Alternatively, one might assume a finite tape and an assistant who is ready to add paper to either end, as needed.)

**The Tape-Reader** At any given time, the machine's tape-reader is positioned on a cell of the memory tape and is able to perform any of the following functions:

- Read the symbol written on the cell.
- Write a new symbol on the cell.

- Move one cell to the left.
- Move one cell to the right.

We will indicate the position of the reader in our diagrams by placing an arrow beneath the cell at which the reader is positioned, as shown:

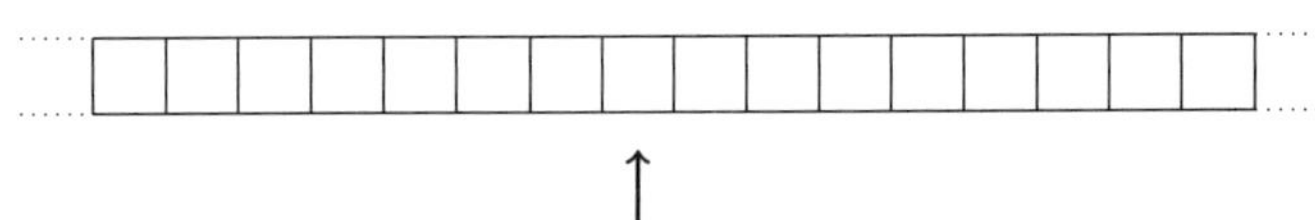

A Turing Machine's **software** (i.e., the computer program it implements) consists of a finite list of **command lines**. Each command line is a sequence of five symbols corresponding to the following parameters:

⟨current state⟩ ⟨current symbol⟩ ⟨new symbol⟩ ⟨direction⟩ ⟨new state⟩

Think of a command line as encoding the following instruction:

> If you are in ⟨current state⟩ and your reader sees ⟨current symbol⟩ written on the memory tape, replace ⟨current symbol⟩ with ⟨new symbol⟩. Then move one step in direction ⟨direction⟩ and go to ⟨new state⟩.

The parameters of a command line are to be filled in as follows:

- The ⟨current state⟩ and ⟨new state⟩ parameters are to be filled with numerals (e.g., 0, 1, 2).
- The ⟨current symbol⟩ and ⟨new symbol⟩ parameter can be filled with letters or numerals (e.g., *a*, 0), or with the special symbol "_" which is used to indicate a blank.
- The ⟨direction⟩ parameter is to be filled with "l" (for "left"), "r" (for "right"), or "*" (for "don't move").

For instance, the command line "7 0 1 *r* 2" is interpreted as the following instruction:

> If you are in state 7 and your reader sees "0" written on the memory tape, replace the "0" with a "1." Then move one step to the right and go to state 2.

And the command line "0 _ *a* * 10" is interpreted as:

> If you are in state 0 and your reader sees a blank, replace the blank with an "*a*." Then stay put and go to state 10.

### 9.1.1 The Turing Machine's Operation

A Turing Machine always starts out in state 0 and runs in stages. At each stage, the machine performs the instruction corresponding to the symbol on which its reader is positioned and to the command line that matches its current state (or the first such command line, if there is more than one). It then goes on to the next stage. If at any stage the machine is unable to find a command line that matches its present state and the symbol on which its reader is positioned, the machine halts. (To make Turing Machine programs easier for humans to understand, programmers sometimes use "halt" as a name for a non-existing state.)

### 9.1.2 An Example

Consider the Turing Machine whose program consists of the following commands:

0 _ *k* ∗ 1
0 *a* *o* *r* 0

As always, our machine starts out in state 0. We'll suppose, moreover, that its tape and reader have been initialized as follows:

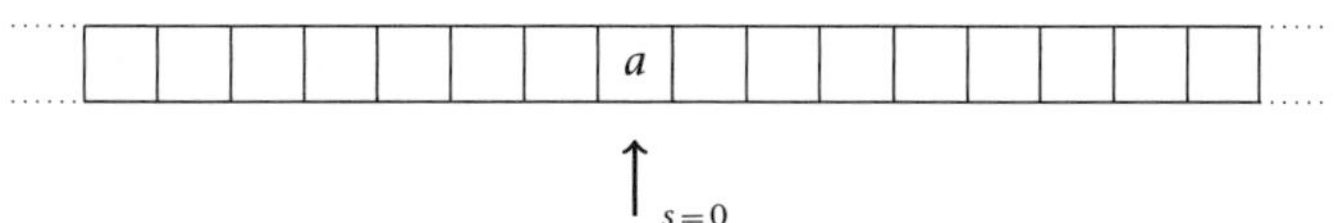


(I subscript the arrow with "$s=0$" to indicate that the Turing Machine is in state 0 at the time depicted by the diagram.) This is what happens when the Turing Machine is set in motion:

**Step 1** The machine starts out in state 0 and its reader sees "*a*." So the machine will look for a command line starting with "0 *a*" (or the first such command line, if there is more than one). It finds "0 *a o r* 0" and follows that instruction by replacing the "*a*" with an "*o*," moving its reader one cell to the right and remaining in state 0:

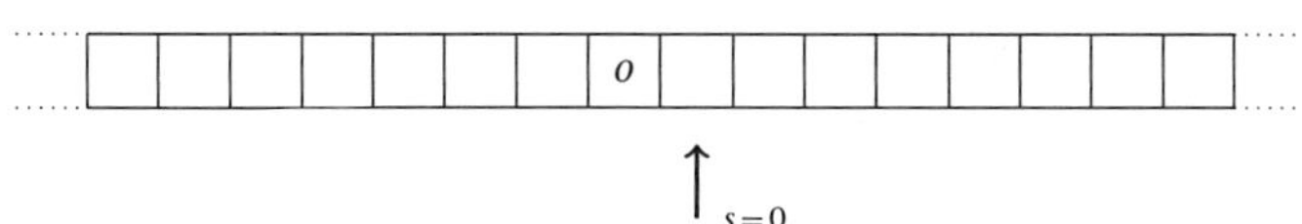


**Step 2** The machine is still in state 0 but is now reading a blank. So it looks for a command line starting with "0 _" (or the first such command line, if there is more than one). It finds "0 _ *k* ∗ 1," and follows that instruction by replacing the blank with a "*k*," remaining in its current position, and switching to state 1:

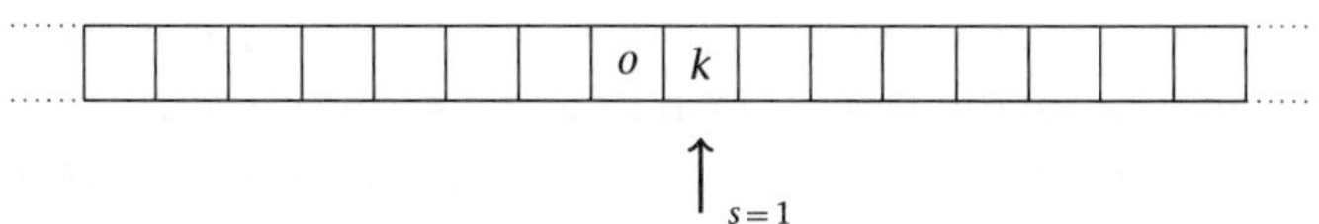

**Step 3** The machine is now in state 1, reading a "*k*." It has no command line that starts with "1 *k*," so it halts.

## Exercises

In working through the exercises below, I recommend using a Turing Machine simulator. There are several good simulators online, but you might find the one at http://morphett.info/turing/turing.html especially helpful, since it uses the same notation as this book.

1. Assume a blank tape, and design a Turing Machine that writes "1" on two consecutive cells and halts.
2. Assume a tape that consists of a finite sequence of zeroes and ones surrounded by blanks, and a reader positioned at the left-most member of the sequence. Design a Turing Machine that replaces the zeroes with ones, replaces the ones with zeroes, and halts.
3. Assume a tape that contains a sequence of $n$ ones ($n > 0$) and is otherwise blank. Assume a reader is positioned at the left-most member of the sequence. Design a Turing Machine that doubles the length of the string of ones, and halts with the reader at the left-most one.
4. Assume a tape that contains a sequence of $n$ ones ($n > 0$) and is otherwise blank. Assume a reader is positioned at the left-most member of the sequence. Design a Turing Machine that replaces the sequence of $n$ ones with a sequence of zeroes and ones that names $n$ in binary notation.
5. Assume the tape contains a sequence of zeroes and ones and is otherwise blank, with the reader positioned at the left-most member of the sequence. Design a Turing Machine that replaces the original sequence with a sequence of $n$ ones, where the original sequence names $n$ in binary notation.

### 9.1.3 Computing Functions on a Turing Machine

I opened this chapter by explaining what it is for a computer to **compute** a function $f$, from natural numbers to natural numbers:

> For a computer to compute $f$ is for it to return $f(n)$ as output when given $n$ as input, for any natural number $n$.

Turing Machines can be used to implement a version of this idea. We can think of a Turing Machine as taking number $n$ ($n \geq 0$) as **input** if it starts out with a tape that contains only a sequence of $n$ ones (with the reader positioned at the left-most one, if $n > 0$). And we can think of the Turing Machine as delivering number $f(n)$ as **output** if it halts with a tape that contains only a sequence of $f(n)$ ones (with the reader positioned at the left-most one, if $n > 0$). Finally, we can say that a Turing Machine **computes** a function $f(x)$ if and only if it delivers $f(n)$ as output whenever it is given $n$ as input. (For a function to be **Turing-computable** is for it to be computed by some Turing Machine.)

Notice, incidentally, that a similar definition could be used to define computability for functions from $n$-tuples of natural numbers to natural numbers. For example, we can think of a Turing Machine as taking a sequence of natural numbers $\langle n_1, \ldots, n_k \rangle$ as **input** if it starts out with a tape that contains only a sequence composed of the following: a sequence of $n_1$ ones (or a blank, if $n_1 = 0$), followed by a blank, followed by a sequence of $n_2$ ones (or a blank, if $n_2 = 0$), followed by a blank, followed by ..., followed by a sequence of $n_k$ ones (or a blank, if $n_k = 0$), with the reader positioned at the left-most one of the left-most sequence of ones (unless $n_1 = 0$, in which case the reader is positioned at the blank corresponding to $n_1$).

### 9.1.4 The Fundamental Theorem

The reason Turing Machines are so valuable is that they make it possible to prove the following theorem:

**Fundamental Theorem of Turing Machines** A function from natural numbers to natural numbers is Turing-computable if and only if it can be computed by an ordinary computer, assuming unlimited memory and running time.

(What I mean by "ordinary computer" is what sometimes gets referred to as a "register machine," but the details won't matter for present purposes.) The proof of the Fundamental Theorem is long and clunky, but the basic idea is straightforward. One shows that each kind of computer can be *simulated* using the other, as follows:

- One shows that every Turing-computable function is computable by an ordinary computer (given unlimited memory and running time) by showing that one can program an ordinary computer to simulate any given Turing Machine.
- One shows that every function computable by an ordinary computer (given unlimited memory and running time) is Turing-computable by showing that one can find a Turing Machine that simulates any given ordinary computer.

In fact, computer scientists tend to think that something stronger than the Fundamental Theorem is true:

**Church-Turing Thesis** A function is Turing-computable if and only if it can be computed algorithmically.

For a problem to be solvable **algorithmically** is for it to be possible to specify a finite list of instructions for solving the problem such that:

1. Following the instructions is guaranteed to yield a solution to the problem in a finite amount of time.
2. The instructions are specified in such a way that carrying them out requires no ingenuity of any kind: they can be followed mechanistically.

3. No special resources are required to carry out the instructions: they could, in principle, be carried out by an ordinary human (albeit a human equipped with unlimited supplies of pencils and paper and patience) or by a machine built from transistors.
4. No special physical conditions are required for the computation to succeed (no need for faster-than-light travel, special solutions to Einstein's equations, etc.).

Intuitively speaking, you can think of the Church-Turing Thesis as stating that a function is Turing-computable if and only if there is a *finite* way of specifying what its values are. So, in particular, a function that fails to be Turing-computable is a function that is too complex for its values to be finitely specified.

I mentioned above that computer scientists tend to think that the Church-Turing Thesis is true. But that is not because they are able to prove it. It's actually not clear what a formal proof of the Church-Turing Thesis would look like. The problem is that such a proof would require a formal characterization of the notion of an algorithmic computability, and it is not clear that one could formalize the notion of algorithmic computability without thereby restricting its scope. Notice, for example, that the notion of Turing-computability is itself one natural way of formalizing the notion of algorithmic computability. On such a formalization, the Church-Turing Thesis is certainly true. But it is also trivial.

The reason the Church-Turing Thesis tends to be regarded as true is that any sensible formalization of the notion of algorithmic computability that anyone has ever been able to come up with is provably equivalent to the notion of Turing-computability.

Programming with Turing Machines can be farily cumbersome. But Turing Machines are so incredibly simple that theorizing *about* Turing Machines tends to be pretty straightforward. In light of the Fundamental Theorem, this means that theorizing about Turing Machines is an extremely useful way of gaining insight about computers more generally. Assuming the Church-Turing Thesis is true, it also means that theorizing about Turing Machines is an extremely useful way of gaining insight about algorithmic methods more generally.

### 9.1.5 Coding Turing Machines

One can use natural numbers as *codes* for Turing Machines. In this section I'll describe one way of doing so. Two preliminary observations:

1. The only differences between Turing Machines that we'll care about here are differences in their programs. Accordingly, we will only count Turing Machines as different if they instantiate different programs.

   (On the other hand, we will treat Turing Machines as different even if there are differences in their programs that make no difference to their behavior. As long as their programs consist of different lists of symbols, the machines will be counted as distinct.)

2. I will simplify the discussion by focusing on **one-symbol** Turing Machines, in which the only symbol allowed on the tape is "1" (plus blanks). This might look like a substantial assumption, but it's actually not. As you'll be asked to verify in one of the exercises below, every function that is computed by a many-symbol Turing Machine is also computed by some one-symbol Turing Machine.

The coding scheme we will discuss in this section proceeds in stages. First, we'll code a Turing machine as a sequence of symbols; then we'll code the sequence of symbols as a sequence of numbers; finally, we'll code the sequence of numbers as a single number. So the process appears as follows:

Turing Machine → Sequence of symbols → Sequence of numbers → Unique number

Let us consider each stage in turn.

**Stage 1 Turing Machine → Sequence of symbols** Since we are counting Turing Machines with the same program as one and the same, we can think of a Turing Machine program as a sequence of command lines. Each command line is a sequence of symbols. So all we need to do to encode a Turing Machine as a sequence of symbols is concatenate its commands into a single string. For instance, the Turing Machine whose program is:

$$\begin{array}{ccccc} 0 & \_ & 1 & * & 0 \\ 0 & 1 & \_ & l & 1 \end{array}$$

gets coded as this sequence of symbols:

$$0 \; \_ \; 1 \; * \; 0 \; 0 \; 1 \; \_ \; l \; 1$$

**Stage 2 Sequence of symbols → Sequence of numbers** We have managed to code each Turing Machine as a sequence of symbols. We will now code each of these symbols as a number. Recall that each command line in a Turing Machine program results from filling out the following parameters:

⟨current state⟩ ⟨current symbol⟩ ⟨new symbol⟩ ⟨direction⟩ ⟨new state⟩

We will employ the following coding scheme:

- The parameters ⟨current state⟩ and ⟨new state⟩ will always be filled in with numerals (which are names of numbers). We'll code each numeral using the number it represents:

$$\begin{array}{ccc} \text{“0”} & \rightarrow & 0 \\ \text{“1”} & \rightarrow & 1 \\ \vdots & & \vdots \end{array}$$

- The parameters ⟨current symbol⟩ and ⟨new symbol⟩ will always be filled in with "1" or "_." (Recall that we are working with one-symbol Turing Machines.) We'll use the following codes:

  "_" → 0
  "1" → 1

- The parameter ⟨direction⟩ will always be filled with "r," "*," or "l." We'll use the following codes:

  "r" → 0
  "*" → 1
  "l" → 2

So, for instance, the sequence of symbols

0 _ 1 * 0 0 1 _ *l* 1

gets transformed into this sequence of numbers:

0 0 1 1 0 0 1 0 2 1

**Stage 3 Sequence of numbers → Unique number** The final step in our coding scheme is a method for coding each sequence of numbers as a single number. There are many ways of doing so. The method we will consider here is credited to the great Austrian-American logician and philosopher Kurt Gödel, about whom you'll hear more in the next chapter. It is extremely simple. One codes the sequence of numerals $\langle n_1, n_2, \ldots, n_k \rangle$ as the number

$$p_1^{n_1+1} \cdot p_2^{n_2+1} \cdot \ldots \cdot p_k^{n_k+1}$$

where $p_i$ is the $i$th prime number. For instance, the sequence of numbers $\langle 0, 0, 1, 2, 0 \rangle$ gets coded as follows:

$$2^{0+1} \cdot 3^{0+1} \cdot 5^{1+1} \cdot 7^{2+1} \cdot 11^{0+1} = 2 \cdot 3 \cdot 25 \cdot 343 \cdot 11 = 565,950$$

(Why add one to exponents? Because otherwise, our coding scheme would be ambiguous. For instance, it would use 1 as a code for each of the following sequences: $\langle 0 \rangle$, $\langle 0, 0 \rangle$, $\langle 0, 0, 0 \rangle$, etc.)

To ensure that this coding scheme gives us what we want, we need to verify that different Turing Machines are always assigned different codes. Fortunately, this is an immediate consequence of the following theorem (which we first encountered in chapter 1):

**Fundamental Theorem of Arithmetic** Every positive integer greater than 1 has a unique decomposition into primes.

We now have identified a coding-scheme that assigns a different number to every (non-empty) one-symbol Turing Machine. Since the discussion below will presuppose that a fixed coding scheme is in place, let us agree that it is to be the coding scheme described in this section, with a small qualification. A feature of the scheme so far is that not every number codes a valid sequence of command lines. That turns out to be a nuisance. So we will stipulate that any number that doesn't code a valid sequence of command lines is to be treated as a code for the "empty" Turing Machine, which has no command lines. The result is a coding scheme in which: (*i*) every number codes some Turing Machine, and (*ii*) every Turing Machine is coded by some number.

### Exercises

1. In the coding system above, what number gets assigned to the Turing Machine whose program consists of the following command line?

   0 1 _ ∗ 1

2. Use the coding system above to spell out the program of Turing Machine number 97,020.
3. Show that every function that can be computed on an $n$-symbol Turing Machine can also be computed on a one-symbol Turing Machine (which is allowed only ones and blanks on the tape).
4. As it turns out, every function that can be computed on a Turing Machine whose tape is infinite in both directions can also be computed on a Turing Machine whose tape is infinite in only one direction. Suppose that $M^{\infty^2}$ is a Turing Machine that computes function $f(n)$. Sketch a procedure for transforming $M^{\infty^2}$ into a Turing Machine $M^{\infty}$ that computes $f(n)$ while using a tape that is infinite in only one direction.

## 9.2 Non-Computable Functions

Can every function from natural numbers to natural numbers be computed by a Turing Machine? Turing proved that the answer to this question is a resounding no. (Assuming the Church-Turing Thesis, you can think of this result as showing that some functions are so complex that their values cannot be finitely specified.)

The easiest way to see this is to note that there are more functions from natural numbers to natural numbers than there are Turing Machines. (*Proof:* In showing that each Turing Machine can be coded by a different natural number, we showed that there are no more Turing Machines than there are natural numbers. But as you were asked to show in chapter 1, there are more functions from natural numbers to natural numbers than there are natural numbers.)

In this section I'll give you a couple examples of functions that are not Turing-computable.

### 9.2.1 The Halting Function

The most famous example of a function that is not Turing-computable is the **Halting Function**. There are two different versions of the Halting Function. The first, $H(n,m)$, is a function from pairs of natural numbers to $\{0,1\}$ and is defined as follows:

$$H(n,m) = \begin{cases} 1, & \text{if the } n\text{th Turing Machine halts when given input } m; \\ 0, & \text{if otherwise.} \end{cases}$$

Consider, for example, the Turing Machine whose program is the following:

$0 \ \_ \ \_ \ r \ 0$

This is the 2,310th Turing Machine according to our scheme, since

$$2^{0+1} \cdot 3^{0+1} \cdot 5^{0+1} \cdot 7^{0+1} \cdot 11^{0+1} = 2 \cdot 3 \cdot 5 \cdot 7 \cdot 11 = 2310.$$

When given input 0 (i.e., the empty string), this machine will start moving to the right and never stop. So the the 2,310th Turing Machine does not halt on input 0, which means that $H(2310, 0) = 0$. In contrast, when given input $n$ ($n > 0$), our Turing Machine halts immediately, since it has no command line telling it what to do when reading a "1" in state 0. This means, in particular, that it halts given input 2,310. So $H(2310, 2310) = 1$.

The second version of the Halting Function, $H(n)$, is a function from natural numbers to $\{0,1\}$. It can be defined using the first version of the Halting Function:

$$H(n) = H(n,n)$$

So, for example, $H(2310) = 1$, since we have seen that $H(2310, 2310) = 1$.

In what follows we'll verify that $H(n)$ is not Turing-computable. This entails that $H(n,m)$ is not Turing-computable, as you'll be asked to verify below.

#### 9.2.1.1 A proof of non-computability

Let us show that the Halting Function is not Turing-computable. (The result is credited to Alan Turing.) We'll proceed by *reductio*. More specifically, we'll assume that there is a Turing Machine $M^H$ that computes $H(n)$ and show that the existence of $M^H$ would allow us to construct an "impossible" Turing Machine $M^I$.

Here is how $M^I$ works when given input $k$:

**Step 1** $M^I$ uses $M^H$'s program as a subroutine to check whether the $k$th Turing Machine halts on input $k$.

**Step 2** If the answer is yes (i.e., if $M^I$'s reader is on a "1" after performing the subroutine), then $M^I$ goes on an infinite loop; otherwise, it halts.

To see why $M^I$ is impossible, let $M^I$ be the $i$th Turing Machine and consider the question of whether $M^I$ halts given input $i$. There are two possibilities:

**$M^I$ halts on input $i$.** But wait! By the definition of the Halting Function, $H(i) = 1$. So it follows from the definition of $M^I$ that $M^I$ will go on an infinite loop, given input $i$. So if we assume that $M^I$ does halt on input $i$, we get the conclusion that it doesn't halt.

**$M^I$ doesn't halt on input $i$.** But wait! By the definition of the Halting Function, $H(i) = 0$. So it follows from the definition of $M^I$ that $M^I$ will halt, given input $i$. So if we assume that $M^I$ doesn't halt on input $i$, we get the conclusion that it does halt.

We have seen that $M^I$ halts if and only if it doesn't halt, which is impossible. This means that $M^I$ can't really exist. So $M^H$ can't really exist. So the Halting Function isn't Turing-computable.

### Exercises

1. Our construction of $M^I$ assumes that Turing Machines are able to carry out certain operations. To verify that these assumptions are correct, design a Turing Machine, $M$, that behaves as follows: given 1 as input, $M$ goes off on an infinite loop; given a blank as input, $M$ halts.
2. Show that $H(n, m)$ is not Turing-computable by showing that if it were Turing-computable, then $H(n)$ would be Turing-computable too.
3. As it turns out, there is a **Universal Turing Machine**, $M^U$, which does the following:
   - If the $m$th Turing Machine halts given input $n$, leaving the tape in configuration $p$, then $M^U$ halts given input $\langle m, n \rangle$, leaving the tape in configuration $p$.
   - If the $m$th Turing Machine never halts given input $n$, then $M^U$ never halts given input $\langle m, n \rangle$.

   Give an informal description of how a Universal Turing Machine might be implemented. (*Hint*: When $M^U$ is given $\langle m, n \rangle$ as input, it uses its tape to create a simulation of the behavior of Turing Machine $m$ on input $n$.)
4. Could a Universal Turing Machine be used to compute the Halting Function? More specifically: Could we construct a Turing Machine that does the following, given input $n$? First, our machine uses the Universal Turing Machine as a subroutine, to figure out what Turing Machine $n$ would do on an empty input. Our machine then prints out a one or a zero, depending on whether the simulation halts.

### 9.2.2 The Busy Beaver Function

My favorite non-computable function is the Busy Beaver Function. It relies on the notion of Turing Machine **productivity**, which is defined as follows:

> If a Turing Machine yields output $k$ on an empty input, its productivity is $k$; otherwise, its productivity is zero.

The **Busy Beaver Function**, $BB(n)$, is then defined as follows:

> $BB(n)$ = the productivity of the most productive (one-symbol) Turing Machine with $n$ states or fewer.

Note that for each $n$, there are only finitely many one-symbol Turing Machines with $n$ states or fewer, so the Busy Beaver Function is guaranteed to be well-defined for each $n$.

One reason I like the Busy Beaver Function is that it is independent of one's scheme for coding Turing Machines as natural numbers. Notice, in contrast, that the Halting Function makes sense only with respect to a particular coding scheme.

**9.2.2.1 A proof of non-computability** Let us verify that the Busy Beaver Function is not Turing-computable. (The result is credited to Hungarian mathematician Tibor Radó.) As in the case of the Halting Function, we'll proceed by *reductio*. More specifically, we'll assume that there is a Turing Machine $M^{BB}$ that computes the Busy Beaver Function, and show that the existence of $M^{BB}$ would allow us to construct an "impossible" Turing Machine $M^I$. Here is how $M^I$ works when given an empty input:

**Step 1** For a particular number $k$ (to be specified below), $M^I$ prints a sequence of $k$ ones on the tape and brings the reader to the beginning of the resulting sequence.

**Step 2** $M^I$ duplicates the initial sequence of $k$ ones and brings the reader to the beginning of the resulting sequence of $2k$ ones.

**Step 3** $M^I$ applies $M^{BB}$'s program as a subroutine, which results in a sequence of $BB(2k)$ ones.

**Step 4** $M^I$ adds an additional one to the sequence of $BB(2k)$ ones, returns to the beginning of the sequence, and halts, yielding $BB(2k)+1$ as output.

We will now prove that our impossible machine $M^I$ is indeed impossible. Here is the intuitive idea behind the proof. At step 3 of its operation, $M^I$ produces as long a sequence of ones as a machine with $2k$ states could possibly produce. But we'll see that $M^I$ is built using no more than $2k$ states. So at step 3 of its operation, $M^I$ produces as long a sequence of ones as it itself could possibly produce. This means that when, at step 4, $M^I$ adds a one to the sequence, it produces a *longer* string of ones than it itself could possibly produce, which is impossible.

Formally speaking, we show that $M^I$ is impossible by verifying that its existence would allow us to prove each of the following inequalities (where $\mathring{M}^I$ is the number of states in $M^I$'s program):

**(E1)** $BB(\mathring{M}^I) \geq BB(2k)+1$

**(E2)** $BB(\mathring{M}^I) < BB(2k)+1$

**Proof of (E1)** The construction of $M^I$ guarantees that $M^I$ delivers an output of $BB(2k)+1$, given an empty input. So we know that the productivity of $M^I$ is $BB(2k)+1$. But, by definition, $BB(\mathring{M}^I)$ is the productivity of the most productive Turing Machine with $\mathring{M}^I$ states or fewer. Since $M^I$ has $\mathring{M}^I$ states, and since its productivity is $BB(2k)+1$, it must therefore be the case that $BB(\mathring{M}^I) \geq BB(2k)+1$. In other words, (E1) must be true.

**Proof of (E2)** We will now show that, when $k$ is chosen appropriately, (E2) must be true. We start with some notation:

- Let $a_k$ be the number of states $M^I$ used to implement step 1 of its procedure. (As you'll be asked to verify below, it is possible to build a Turing Machine that outputs $k$ on an empty input using $k$ states or fewer. So we may assume that $a_k \leq k$.)
- Let $b$ be the minimum number of states $M^I$ uses to implement step 2 of its procedure. (Since a fixed number of states can be used to duplicate sequences of ones of any length, $b$ can be assumed to be a constant, independent of $k$.)
- Let $c$ be the number of states $M^I$ uses to implement step 3 of its procedure. ($c$ is simply the number of states in $M^{BB}$, and is therefore a constant, independent of $k$.)
- Let $d$ be the number of states $M^I$ uses to implement step 4 of its procedure. (Since a fixed number of states can be used to add an extra one at the end of a sequence of any length, $d$ can be assumed to be a constant, independent of $k$.)

This allows us to count the number of states required by $M^I$: $a_k+b+c+d$. Since we know that $a_k \leq k$, the following must be true:

$$\mathring{M}^I \leq k+b+c+d$$

Let us now assume that $k$ is chosen so that it is at least as big as $b+c+d$. We then have $\mathring{M}^I \leq 2k$. From this, it follows that $BB(\mathring{M}^I) \leq BB(2k)$, and therefore that $BB(\mathring{M}^I) < BB(2k)+1$. In other words, (E2) must be true.

**Summary** By looking at $M^I$'s output, we showed that (E1) must be true. But by analyzing $M^I$'s program, we came to the contrary conclusion that (E2) must be true. So $M^I$ cannot exist. So $M^{BB}$ cannot exist. So the Busy Beaver Function isn't Turing-computable.

### Exercises

1. Show that for any natural number $k$, it is possible to build a Turing Machine that outputs $k$ on an empty input, using $k$ states or fewer.
2. Show that for some fixed constant $c$, it is possible to build a one-symbol Turing Machine that uses $|\log_2 k| + c$ states and outputs $k^2$ on an empty input.

3. Determine the productivity of each of the following Turing Machines:

   a) 0 _ 1 r 0        b) 0 _ 1 r 1

4. Show that $BB(n)$ grows faster than any Turing-computable function. More specifically, show that for any Turing-computable function $f(n)$, there is an $m$ such that, for every $n > m$, $BB(n) > f(n)$.

## 9.3 Efficiency[1]

So far, our attention has been focused on the distinction between functions that are Turing-computable (like addition and multiplication) and functions that are not (like the Halting Function and the Busy Beaver Function). If the Church-Turing Thesis is true, this is the same as the distinction between functions that are computable and functions that are not. As it turns out, however, the question of whether a function is computable is not the only interesting question in computer science. In fact, some of the most exciting research in computer science today is concerned with the question of how *efficiently* the values of a computable function can be computed. In this section I'd like to give you a taste of what efficiency is all about by discussing the $P = NP$ problem.

### 9.3.1 Exponential Time and Polynomial Time

Here is the story of the king and the sage. The sage invented chess. The king liked the game so much that he offered the sage a reward. "All I want is wheat," said the sage. "Then I'll give you as much wheat as you want!" said the king. The sage smiled and said, "I would like 1 grain of wheat for the first square of a chessboard, 2 grains for the second square, 4 grains for the third, and so forth, for each of the sixty-four squares." The king agreed, without realizing that he had committed to delivering more wheat than was contained in his entire kingdom.

In order to fulfill his promise with respect to the $n$th square of the chess-board, the king would have to deliver $2^{n-1}$ grains of wheat. So, in order to fulfill his promise with respect to the last square on the board, the king would need $2^{63}$ ($\approx 9.2 \times 10^{18}$) grains of wheat. That is an enormous number, much greater than the yearly production of wheat on Earth. Notice, moreover, that the sage's method will quickly generate even larger numbers. If the board had contained 266 squares instead of 64, the king would have committed to delivering more grains of wheat than there are atoms in the observable universe (about $10^{80}$, according to some estimates).

The moral of our story is that functions of the form $2^x$ can grow extremely fast. This has important consequences related to the efficiency of computer programs. Suppose, for example, that I wish to program a computer to solve the following problem:

1. An earlier version of this material appeared in Rayo, "P = NP," *Investigación y Ciencia*, April 2010.

I pick a finite set of integers, and the computer must determine whether the numbers in some non-empty subset of this set add up to 0. For instance, if I pick the set $\{104, -3, 7, 11, -2, 14, -8\}$, the computer should respond Yes because the members of $\{-3, 11, -8\}$ add up to zero.

This problem is known as the **subset sum problem**. The simplest way to write a computer program to solve it for any initial set is using a "brute force" technique: the computer generates all the subsets of the original set, one by one, and determines for each one the sum of the members. If at any point, it gets 0 as a result, it outputs Yes; otherwise, it answers No.

Unless the initial set is very small, however, this method is not efficient enough to be used in practice. This is because if the initial set has $n$ members, the brute force approach could require more than $2^n$ operations to yield a verdict. (An $n$-membered set has $2^n - 1$ non-empty subsets, and our computer may have to check them all before being in a position to yield an answer.) We therefore face the same problem that was faced by the king of our story. As of October 2014, the world's fastest computer was the Tianhe-2 at Sun Yat-sen University, Guangzhou, China. The Tianhe-2 is able to execute $33.86 \times 10^{15}$ operations per second. But even at that speed, it could take more than seventy years to solve the subset sum problem using the brute force method when the initial set has 86 members.

An algorithm like the brute force method, which requires up to $2^n$ operations to solve the subset sum problem for an input of size $n$, is said to be of **exponential time**. In general, an algorithm is said to be of exponential time if, given an input of size $n$, it must perform up to $2^{p(n)}$ operations to solve the problem. (Here and below, $p(n)$ is a fixed polynomial of the form $c_0 + c_1n^1 + c_2n^2 + \ldots + c_kn^k$.)

Even though we know of methods for solving the subset sum problem that are more efficient than the brute force method, we do not, at present, know of a method that is truly efficient. In particular, there is no known method capable of solving the problem in **polynomial time**. In general, an algorithm is said to be of polynomial time if, given an input of size $n$, it must perform up to $p(n)$ operations to solve the problem (rather than up to $2^{p(n)}$ operations, as in the case of exponential time).

To illustrate the difference between exponential time and polynomial time consider the polynomial function $4n^2 + 31n^{17}$ and contrast it with the exponential function $2^n$. For small values of $n$, the value of our polynomial function is greater than the value of our exponential function. (For $n = 2, 4 \cdot 2^2 + 31 \cdot 2^{17} = 4,063,248$, and $2^2 = 4$). But once $n$ gets big enough, the value of $2^n$ will always be greater than that of our polynomial function. (For $n = 1000, 4 \cdot 1000^2 + 31 \cdot 1000^{17} \approx 10^{52}$, but $2^{1000} \approx 10^{301}$.)

The class of problems that can be solved using algorithms of polynomial time is usually called $P$. This set is of considerable interest to computer scientists because any problem that can be solved using a procedure that is efficient enough to be used in practice will be in $P$.

### 9.3.2 $P = NP$

I mentioned earlier that we do not know whether there is a practical method for solving the subset sum problem. In particular, we do not know whether the problem is in *P*. In contrast, it is easy to find an efficient method for verifying whether a purported to yes solution the subset sum problem is correct. For suppose someone offers us a particular subset of the initial set and claims that its members add up to zero. We can check whether the claim is correct by adding up the numbers in the subset (a task that can certainly be performed in polynomial time).

The class of problems with this feature—problems for which purported Yes solutions can be verified in polynomial time—is often called *NP*. A fascinating property of the subset sum problem is that it is not only in *NP*; one can show that any problem in *NP* can be reduced to the subset sum problem using an algorithm that can be run in polynomial time. (*NP* problems like this are called "*NP*-complete.") This means that if there were an efficient algorithm for solving the subset sum problem, there would be an efficient algorithm for solving every problem in *NP*. In other words, it would be the case that $P = NP$.

The discovery of an efficient method for solving the problems in *NP* would change the world, since the problem of factoring a composite number of length $n$ into its prime factors is in *NP*, and many of the world's cryptographic techniques depend on there being no efficient method for solving this problem. Most mathematicians believe that it is not the case that $P = NP$, but we can't be sure until we've found a proof.

## 9.4 Optional: The Big Number Duel[2]

Imagine the following competition: two contestants, one blackboard, biggest number wins. My friend Adam Elga, a philosophy professor at Princeton University, challenged me to such a duel after reading Scott Aaronson's "Who Can Name the Bigger Number?" This optional final section is a blow-by-blow account of our mighty confrontation.

### 9.4.1 The Rules of the Game

We began by fixing some rules. The first rule was that we would take turns writing numbers on the board, and that the last person to write down a valid entry would be the winner. The second rule was that only finite numbers would be allowed. The number 17, for example, would count as a valid entry, as would the number 1010101010. But $\omega + \omega$ would not. The third rule was that semantic vocabulary—expressions such as "names" or "refers to" or "is true of"—was disallowed. This restriction was crucial. Without it, Adam would have been in a position to write "the biggest number ever

2. An earlier version of this material appeared in Rayo, "El duelo de los números grandes," *Investigación y Ciencia*, August 2008.

named by Agustín, plus one" on the board, and win on the first round. (Worse still, I would have been able to retort with "the biggest number ever named by Adam, plus one.")

Finally, we agreed not to engage in unsporting behavior. Each time one of us wrote a number on the board, the other would either recognize defeat or respond with a much bigger number. How much bigger? Big enough that it would be impossible to reach it in practice (e.g., before the audience got bored and left) using only methods that had been introduced in previous stages of the competition. That means, for example, that if Adam's last entry was 101010, it would be unsporting for me to respond with 10101010. We wanted the competition to be a war of originality, not a war of patience!

### 9.4.2 The First Few Rounds

The competition took place at MIT. As the hometown hero, I got to go first. Without thinking too much about it, I wrote down a sequence of thirty or forty ones

111111111111111111111111111111

"We're still warming up," I thought. But my first effort proved disastrous. Adam approached the board with an eraser and erased a line across all but the first two of my ones, leaving

11!!!!!!!!!!!!!!!!!!!!!!!!!!!!

"Eleven factorial, factorial, factorial ...," he declared triumphantly. To get a sense of just how big this number is, note that 11! is 39,916,800, and that 11!! is approximately $6 \times 10^{286,078,170}$ (which is much more than the number of particles in the universe). In fact, 11!!! is so big that it cannot be written using an expression of the form $10^n$ in a practical amount of time, where $n$ is a numeral in base 10. So 11!!! cannot be reached, in practice, by writing a long sequence of ones on the blackboard. And Adam's entry had many more factorials than that. There was no question that Adam had named a number much bigger than mine.

Fortunately, I was able to remember the Busy Beaver Function, $BB(n)$, and made my next entry $BB(10^{100})$: the productivity of the most productive Turing Machine with a googol states or less. How does this number compare to Adam's last entry? Well, it is possible to write a relatively short Turing Machine program that computes the factorial function and outputs the result of applying the function 30 or 40 times starting with 11—or, indeed, the result of applying the function $10^{10^{10^{10}}}$ times, starting with 11. I don't know how many states it takes to do so, but the number is significantly smaller than $10^{100}$. So $BB(10^{100})$ must be bigger than Adam's last entry.

In fact, it is much, much bigger. $BB(10^{100})$ is a truly gigantic number. Every Turing Machine program we can write in practice will have fewer than $10^{100}$ states. So no matter how big a number is, it will be smaller than $BB(10^{100})$, as long as it is possible in practice to program a Turing Machine to output it.

### 9.4.3 Beyond Busy Beaver

Over the next few rounds, the duel became a search for more and more powerful generalizations of the notion of a Turing Machine.

Imagine equipping a Turing Machine with a **halting oracle**: a primitive operation that allows it to instantaneously determine whether an ordinary Turing Machine would halt on an empty input. Call this new kind of machine a Super Turing Machine. We know that the Busy Beaver Function is not Turing-computable. But, as you'll be asked to verify in an exercise below, it can be computed using a Super Turing Machine. And, as you'll also be asked to verify below, this means that for any ordinary Turing Machine with sufficiently many states, there is a Super Turing Machine that has fewer states but is much more productive. This means that the function $BB_1(n)$—i.e., the Busy Beaver function for Super Turing Machines—can be used to express numbers that are much bigger than $BB(10^{100})$ (for instance, $BB_1(10^{100})$).

No Super Turing Machine can compute $BB_1(n)$. But we could compute this function using a Super-Duper Turing Machine: a Turing Machine equipped with a halting oracle for Super Turing Machines. $BB_2(10^{100})$—the Busy Beaver function for Super-Duper Turing Machines—can be used to express numbers that are much bigger than $BB_1(10^{100})$.

It goes without saying that by considering more and more powerful oracles, one can extend the hierarchy of Busy Beaver functions further still. After $BB(n)$ and $BB_1(n)$ come $BB_2(n), BB_3(n)$, and so forth. Then come: $BB_\omega(n)$, $BB_{\omega+1}(n)$, $BB_{\omega+2}(n),\ldots$, $BB_{\omega+\omega}(n),\ldots$, $BB_{\omega+\omega+\omega}(n),\ldots$, $BB_{\omega\times\omega}(n),\ldots$, $BB_{\omega\times\omega\times\omega}(n),\ldots$. And so forth. (It is worth noting that even if $\alpha$ is an infinite ordinal, $BB_\alpha(10^{100})$ is a finite number and therefore a valid entry to the competition.)

The most powerful Busy Beaver function that Adam and I considered was $BB_\theta(n)$, where $\theta$ is the first non-recursive ordinal—a relatively small infinite ordinal. So the next entry to the competition was $BB_\theta(10^{100})$. And although it's not generally true that $BB_\alpha(10^{100})$ is strictly larger than $BB_\beta(10^{100})$ when $\alpha > \beta$, it's certainly true that $BB_\theta(10^{100})$ is much, much bigger than $BB_1(10^{100})$, which had been our previous entry.

### 9.4.4 The Winning Entry

The last entry to the competition was a bigger number still:

> The smallest number with the property of being larger than any number that can be named in the language of set theory using $10^{100}$ symbols or less.

This particular way of describing the number was disallowed in the competition because it includes the expression "named," which counts as semantic vocabulary and is therefore ruled out. But the description that was actually used did not rely on forbidden vocabulary. It, instead, relied on a second-order language: a language that is capable of expressing not just singular quantification ("there is a number such that it is so and so") but also plural quantification ("there are some numbers such that they are so and so"). Second-order languages are so powerful that they allow us to characterize a non-semantic substitute for the notion of being named in the standard language of set theory.

And what if we had a language that was even more expressive than a second-order language? A third-order language—a language capable of expressing "super plural" quantification—would be so powerful that it would allow us to characterize a non-semantic substitute for the notion of being named in the *second-order* set theory using $10^{100}$ symbols or fewer. And that would allow one to name a number even bigger than the winning entry of our competition. Our quest to find larger and larger numbers had now morphed into a quest to find more and more powerful languages!

### Exercises

1. Give an informal description of how a Super Turing Machine might compute $BB(n)$.
2. Show there is a number $k$ such that for any ordinary Turing Machine with more than $k$ states, there is a Super Turing Machine that has fewer states but is much more productive.

## 9.5 Conclusion

We have discussed the notion of a Turing Machine and identified two functions that fail to be Turing-computable: the Halting Function and the Busy Beaver Function.

Since a function is computable by an ordinary computer if and only if it is Turing-computable, it follows that neither the Halting Function nor the Busy Beaver Function can be computed by an ordinary computer, no matter how powerful. On the assumption that the Church-Turing Thesis is true, it also follows that neither the Halting Function nor the Busy Beaver Function can be computed algorithmically.

These results have far-reaching consequences. We will discuss one of them in the next chapter, when we turn to Gödel's Theorem.

## 9.6 Further Resources

- For a more formal treatment of some of the issues described in this chapter, see Boolos, Burgess, and Jeffrey's *Computability and Logic*.
- As mentioned in the main text, the Big Number Duel was inspired by Scott Aaronson's article "Who Can Name the Bigger Number?". For further details on the Big Number Duel, including details about the winning number, see http://web.mit.edu/arayo/www/bignums.html.

## Appendix: Answers to Exercises

### Section 9.1.2

```
1. 0 _ 1 r 1
   1 _ 1 * halt

2. 0 0 1 r 0
   0 1 0 r 0
   0 _ _ * halt

3. 0 1 _ r 1
   1 1 1 r 1
   1 _ _ r 2
   2 1 1 r 2
   2 _ 1 r 3
   3 _ 1 l 4
   4 1 1 l 4
   4 _ _ l 5
   5 1 1 l 5
   5 _ _ r 6
   6 1 _ r 1
   6 _ _ r halt

4. 0 1 1 r 0
   0 _ _ l 1
   1 1 _ l 2
   1 _ _ * halt
   2 1 1 l 2
   2 _ _ l 3
   3 _ 1 r 4
   3 0 1 r 4
   3 1 0 l 3
   4 0 0 r 4
   4 1 1 r 4
   4 _ _ r 0

5. 0 0 0 r 0   ; startup
   0 1 1 r 0
   0 _ _ l 1
   1 1 0 r 2   ; subtract
   1 0 1 l 1
   1 _ _ r 5
```

2 0 0 $r$ 2 ; move right to add
2 1 1 $r$ 2
2 _ _ $r$ 3
3 _ 1 $l$ 4 ; add one
3 1 1 $r$ 3
4 1 1 $l$ 4 ; turn back
4 _ _ $l$ 1
5 1 _ $r$ 5 ; wrap things up
5 _ _ $r$ halt

**Section 9.1.5**

1. The given command line is first transformed into the sequence 0 1 0 1 1, which gets coded as the following number:

   $$2^{0+1} \cdot 3^{1+1} \cdot 5^{0+1} \cdot 7^{1+1} \cdot 11^{1+1} = 2 \cdot 9 \cdot 5 \cdot 49 \cdot 121 = 533,610$$

2. It is easy to verify that

   $$97,020 = 2^{1+1} \cdot 3^{1+1} \cdot 5^{0+1} \cdot 7^{1+1} \cdot 11^{0+1}$$

   So 97,020 codes the sequence of numerals

   1 1 0 1 0

   which corresponds to the following sequence of symbols:

   1 1 _ * 0

3. The transformation consists of two parts. The first part is to find a one-symbol analogue of an $n$-symbol *tape*. The second part is to find a one-symbol analogue of an $n$-symbol *program*.

   **Part 1** Let $k$ be the smallest integer greater than or equal to $log_2 n$. Then we can code each of the original $n$ symbols using a distinct sequence of $k$ blanks and ones. Suppose, for instance, that we start with a three-symbol machine that uses 1s, $X$s, and blanks. Then $n = 3$ and $k = 2$. So we can code each of our three symbols using a distinct sequence of 2 blanks and ones. For instance:

   _ → __
   1 → _1
   X → 1_

   To complete Part 1, we substitute a coded version of our tape for the original.

   **Part 2** Replace the program of the original $n$-symbol machine with a program that delivers an analogue effect on a (coded) one-symbol tape. This easiest way to do this is to separate the program into chunks, where each chunk is made up from command lines sharing ⟨current state⟩. Consider, for example, the following chunk:

```
0  _  _  r  2
0  1  X  l  6
0  X  1  r  3
```

We modify this chunk as follows (where two-digit states are assumed previously unused):

```
0   _  _  r  10   ; at start of sequence that codes _or 1; find out which one
0   1  _  r  20   ; at start of sequence that codes X; start writing code for 1
10  _  _  r   2   ; sequence codes _; proceed accordingly
10  1  _  l  30   ; sequence codes 1; start writing code for X
30  _  1  l  40   ; finish writing code for X, and start moving two steps left
40  _  _  l   6   ; finish moving two steps left (assuming you find a blank)
40  1  1  l   6   ; finish moving two steps left (assuming you find a 1)
20  _  1  r   3   ; finish writing code for 1, move right
```

4. Let me sketch a procedure for building $M^{\infty}$ out of $M^{\infty^2}$. The intuitive idea is that $M^{\infty}$ shifts the contents of its tape one cell to the right whenever it is at risk of reaching the end of its tape to the left.

   Here are some additional details. We will assume, for concreteness, that the tape is infinite toward the right. We also assume, with no loss of generality, that $M^{\infty^2}$ is a one-symbol machine. $M^{\infty}$, in contrast, will be a many-symbol machine. But, as we saw in the preceding exercise, it could easily be transformed into a one-symbol machine. When computing $f(n)$, $M^{\infty}$ will start out with a tape that is blank except for a sequence of $n$ ones, with the reader positioned on the left-most one. $M^{\infty}$ then proceeds as follows:

   **Step 1** $M^{\infty}$ replaces the first 1 in the sequence with a special symbol B (for "beginning"), adds a 1 at the end of the sequence (to replace the deleted 1 at the beginning of the sequence), and adds a special symbol E (for "end") to the right of the new one.

   **Step 2** For each state $s$ of $M^{\infty^2}$, $M^{\infty}$'s program will include additional command lines specifying the behavior of the machine when in state $s$ reading B or E:

   - $M^{\infty}$'s command line for $s$ when reading B will initiate a subroutine that adds a blank between the B and the symbol to its right. This is done by translating the sequence of ones and blanks that is to the right of B (along with the E at the end) one cell to the right. After doing so, $M^{\infty}$ positions its reader on the newly created blank to the right of B and goes back to state $s$.

- $M^\infty$'s command line for $s$ when reading E replaces the E with a blank and adds an E to the right of the new blank. After doing so, $M^\infty$ positions its reader on the newly created blank to the left of E and goes back to state $s$.

**Step 3** Whenever $M^{\infty^2}$ executes a command line that would cause it to halt, $M^\infty$ instead initiates a subroutine that deletes its markers, returns to its former position, and halts.

### Section 9.2.1.1

1. 0 _ _ $*$ halt
   0 1 1 $*$ 0

2. Assume $H(m,n)$ is computed by a Turing Machine $M^{H^2}$. Then it should be possible to use the $M^{H^2}$ program as a subroutine in other Turing Machine programs. But if such a subroutine were available, we would be able to use it to characterize a Turing Machine $M^H$, which computes $H(n)$ using the following procedure:

   When $M^H$ is given input $k$, it creates a second string of $k$ ones (separated from the first one by a blank), and returns to its starting position. It then uses $M^{H^2}$'s program as a subroutine to check whether Turing Machine $k$ halts on input $k$.

   This assumes that it is possible to build a Turing Machine that, given a sequence of $n$ ones as input, yields as output a sequence of $n$ ones followed by a blank, followed by a sequence of $n$ ones. Here is one way of doing so:

   0 1 $x$ $r$ 1
   1 1 1 $r$ 1
   1 _ $y$ $r$ 2
   2 _ 1 $l$ 3
   3 1 1 $l$ 3
   3 $y$ $y$ $l$ 3
   3 $x$ 1 $r$ 4
   4 $y$ _ $l$ 6
   4 1 $x$ $r$ 5
   5 1 1 $r$ 5
   5 $y$ $y$ $r$ 5
   5 _ 1 $l$ 3
   6 1 1 $l$ 6
   6 _ _ $r$ halt

3. Here is a rough description of how $M^U$ might work, given input $\langle m,n\rangle$:

**Step 1** $M^U$ uses $m$, together with the coding-scheme we use to number Turing Machines, to print out a description of the $m$th Turing Machine's program on its tape.

**Step 2** In a different region of its tape, $M^U$ prints out a sequence of ones representing the simulated machine's "current state."

**Step 3** In a different region of its tape, $M^U$ prints out a sequence of $n$ ones, preceded by a special symbol. This sequence represents the fact that the simulated machine starts out with input $n$, and the special symbol represents the position of the reader of the simulated machine on its tape. (The special symbol starts out at the beginning of the sequence of $n$ ones to represent the fact that our simulated machine starts out with its reader at the beginning of its input.)

**Step 4** The machine then gets to work on the simulation. At each stage of the simulation, it determines whether the "reader" of the simulated machine is on a one or a blank, and it uses the "state memory" to find the command from the simulated machine's description that corresponds to that symbol and that state number. The machine then carries out the action determined by that command, and repeats.

**Step 5** If the simulated machine halts, $M^U$ erases everything from its tape except for its representation of the simulated machine's tape.

4. No. We proved that the Halting Function is not computable. This means, in particular, that we cannot use a Universal Turing Machine to compute it.

   But why not? The problem is that a Universal Turing Machine works by simulating the Turing Machine whose behavior it is computing. If the simulated machine doesn't halt, the simulation won't halt either. So the Turing Machine we are trying to build will never get around to printing a zero.

## Section 9.2.2.1

1. When $k$ is 0, the problem is straightforward. When $k \geq 1$, the following Turing Machine gives us what we want:

   | | | | | |
   |---|---|---|---|---|
   | 0 | _ | 1 | $l$ | 1 |
   | 1 | _ | 1 | $l$ | 2 |
   | ... | | | | |
   | $(k-1)$ | _ | 1 | $*$ | halt |

2. We begin with some observations:
   - By using the technique of the previous exercise, one can show that it is possible to build a one-symbol Turing Machine that outputs $k$ in binary notation, on an empty input, and uses only $\lfloor \log_2 k \rfloor + 1$ states.

- As shown in exercise 5 of section 9.1.2, it is possible to build a one-symbol Turing Machine that outputs $k$, given $k$'s binary notation as input, and uses only 6 states.
- It is easy (if laborious) to build a one-symbol Turing Machine that outputs $k^2$, given input $k$. Assume that this can be done using only $d$ states, for some constant $d$.

These three observations entail that it is possible to build a one-symbol Turing Machine that outputs $k$, on an empty input, using only $|\log_k| + 7 + d$ states.

There are much more efficient ways of generating big outputs, but efficient machines tend to rely on clever tricks that are not easy to describe succinctly. Here is a famous example of a two-state machine that halts after producing a sequence of four ones:

| | | | | |
|---|---|---|---|---|
| 0 | _ | 1 | $r$ | 1 |
| 0 | 1 | 1 | $l$ | 1 |
| 1 | _ | 1 | $l$ | 0 |
| 1 | 1 | 1 | $r$ | halt |

3. (a) 0

   (b) 1

4. Since $f(n)$ is Turing-computable, it is computed by some Turing Machine $M^f$. By using $M^f$'s program as a subroutine, one can program a Turing Machine that computes $f(2k)+1$ on the basis of $k+c$ states, for $c$ a constant that does not depend on $k$. So as long as $k > c$, $BB(2k) \geq f(2k)+1$. It follows that for every even number $n$ above a certain bound, $BB(n) > f(n)$. A similar argument could be used to show that for every odd number $n$ above a certain bound, $BB(n) > f(n)$.

### Section 9.4.4

1. Here is an informal description of how such a super Turing Machine $M$ could work:

   **Step 1** $M$ enumerates all possible Turing Machine programs of $n$ symbols or less. (There are finitely many such programs.)

   **Step 2** For each such program, $M$ uses its oracle to check whether the program halts when given a blank input tape.

   **Step 3** For each program that does halt, $M$ acts like a universal Turing Machine to simulate its program and print out its output.

   **Step 4** $M$ then compares all such outputs and chooses the biggest as its own output.

2. Suppose that you have an ordinary Turing Machine $M$, with $\overline{M}$ states. We can build a super-Turing Machine $M_1$ that is much, much more productive than $M$, as follows:

**Step 1** $M_1$ writes $\overline{M}$ (or more) ones on the tape.

**Step 2** $M_1$ applies the Busy Beaver Function to the contents of the tape a couple of times.

The first of these steps can be carried out using approximately $log(\overline{M})$ steps (since $M_1$ can use $log(\overline{M})$ states to write out $log(\overline{M})$ ones, and then use a constant number of steps to apply the function $10^x$ to that number). The second of these steps requires a constant number of states, since $M_1$ can use a Super Turing Machine computing the Busy Beaver Function as a subroutine. This means that $M_1$ requires only $log(\overline{M}) + c$ states, where $c$ is a constant. So as long as $\overline{M}$ is larger than the smallest $k$ such that $k \geq log(k) + c$, $M_1$ will have fewer states than $M$.

# 10 Gödel's Theorem

## 10.1 Gödel's Incompleteness Theorem

Gödel's Incompleteness Theorem, which was proved in 1931 by the great Austrian-American mathematician Kurt Gödel, is a result of exceptional beauty and importance. There are many different formulations of the theorem, but the formulation that will be our primary focus states that no Turing Machine could output every arithmetical truth and no falsehood. A little more precisely:

**Gödel's Incompleteness Theorem** Let $\mathcal{L}$ be a (rich enough) arithmetical language. Then no Turing Machine $M$ is such that, when run on an empty input, the following would result:

1. $M$ runs forever, outputting sentences of $\mathcal{L}$.
2. Every true sentence of $\mathcal{L}$ is eventually output by $M$,
3. No false sentence of $\mathcal{L}$ is ever output by $M$.

(I'll say more about what "rich enough" means below.)

The best known formulation of the theorem is slightly different:

**Gödel's Incompleteness Theorem** No axiomatization of elementary arithmetic (in a rich-enough language) can be both consistent and complete.

Don't worry if you can't make sense of this second formulation yet. We'll talk more about it later.

Gödel's Theorem is important for both mathematical and philosophical reasons. One reason the theorem is *mathematically* important is that it teaches us something significant about the complexity of arithmetical truth. It tells us that there is a precise sense in which no finite computer program could possibly encapsulate the whole of arithmetical truth and therefore a precise sense in which arithmetical truth is too complex to be finitely specifiable. (As we'll see in section 10.3.1, this means, in particular, that no finitely specifiable set of axioms could allow us to prove every arithmetical truth and no falsehood.) And, of course, if elementary arithmetic is too complex to be finitely specifiable, mathematical theories stronger than arithmetic—such as real analysis or set theory—must also be too complex to be finitely specifiable.

One reason the theorem is *philosophically* important is that it can be used to argue that our mathematical theories can never be established beyond any possibility of doubt. It is not at all obvious that this is a consequence of the theorem, but I'll tell you more about it in section 10.4.

### Exercises

1. Show that the first of the two formulations of Gödel's Theorem mentioned above is equivalent to the following:

**Gödel's Incompleteness Theorem** Let $\mathcal{L}$ be a (rich enough) arithmetical language. Then no Turing Machine $M'$ is such that whenever it is given a sentence of $\mathcal{L}$ as input, it outputs 1 if the sentence is true and 0 if the sentence is false.

In other words, *no Turing Machine can decide the truth value of the sentences of* $\mathcal{L}$. (You may assume that there is a Turing Machine that outputs every sentence of $\mathcal{L}$ and nothing else, and that every sentence in $L$ has a negation that is also in $\mathcal{L}$.)

## 10.2 An Arithmetical Language

The first step in our proof of Gödel's Theorem is to get clear about the language we'll be working with.

An **arithmetical language** is a language in which one can talk about the natural numbers and their two basic operations, addition and multiplication. Here we will be working with an especially simple arithmetical language, $L$, which is (nearly) as simple as it can be while still being rich enough for Gödel's Theorem to hold. The sentences of $L$ are built from the following symbols:

| Arithmetical symbol | Meaning |
|---|---|
| 0 | Names the number zero |
| 1 | Names the number one |
| + | Expresses addition |
| × | Expresses multiplication |
| ∧ | Expresses exponentiation |

| Logical symbol | Meaning |
|---|---|
| $=$ | Expresses identity |
| $\neg$ | Expresses negation |
| & | Expresses conjunction |
| $\forall$ | Expresses universal quantification |
| $x_n$ (for $n \in \mathbb{N}$) | [Variable] |

| Auxiliary symbol | Meaning |
|---|---|
| ( | [Left parenthesis] |
| ) | [Right parenthesis] |

(One way that $L$ is not as simple as it could be is that the exponentiation symbol $\wedge$ is not necessary to prove the result. With some effort, the exponentiation symbol can be defined using $+$ and $\times$, as Gödel showed. I include it here because it will make proving the theorem much easier.)

### 10.2.1 Arithmetical Symbols

The arithmetical symbols work exactly as you'd expect:

- The symbols 0 and 1 name the numbers zero and one, respectively.
- The symbols $+$, $\times$ and $\wedge$ express addition, multiplication, and exponentiation, respectively.

These basic symbols can be combined to form complex symbols. For instance, one can construct the complex symbols $0+1$ and $(1+1)\times(1+1)$, which name the numbers one and four, respectively. In fact, $L$ has the expressive resources to name every natural number. Here is a partial list:

| Expression | Abbreviation | Refers to |
|---|---|---|
| 0 | - | The number zero |
| 1 | - | The number one |
| $(1+1)$ | 2 | The number two |
| $((1+1)+1)$ | 3 | The number three |
| $(((1+1)+1)+1)$ | 4 | The number four |
| $\vdots$ | $\vdots$ | $\vdots$ |

Note the abbreviations in the middle column of the table above. These symbols are not officially part of our language. Instead, they should be thought of as shorthand for the complex symbols to their left. But they'll greatly simplify our lives. For instance, they'll allow us to write 12 instead of the much more cumbersome:

$$(((((((((((1+1)+1)+1)+1)+1)+1)+1)+1)+1)+1)+1)$$

### 10.2.2 Logical Symbols

**The Identity Symbol** The arithmetical symbols we've discussed so far allow us to name numbers. But we do not yet have the resources to express any *claims* about numbers. In other words, we are not yet able to express any thoughts capable of being true or false, such as the thought that two times one equals one times two. Once the identity symbol $=$ is in place, however, this thought can be expressed as $2 \times 1 = 1 \times 2$. We can also express the false thought that two equals three: $2 = 3$.

**Negation and Conjunction** The logical operators $\neg$ and & allow us to build complex claims out of simpler ones:

- The symbol $\neg$ is read "it is not the case that" and can be used to formulate sentences like the following:

  $\neg(1 = 2)$
  (*Read:* It is not the case that one equals two.)

- The symbol & is read "and" and can be used to formulate sentences like the following:

  $(1 + 2 = 3)\ \&\ (2 + 1 = 3)$
  (*Read:* One plus two equals three, and two plus one equals three.)

**Quantifiers and Variables** None of the symbols we have discussed so far enables us to express generality. For example, we are in a position to express commutativity claims about particular numbers (e.g., $2 \times 1 = 1 \times 2$), but we are unable to state the fact that multiplication is commutative *in general*:

> For *any* numbers $n$ and $m$, the product of $n$ and $m$ equals the product of $m$ and $n$.

General claims of this kind can be expressed in $L$, using the quantifier symbol $\forall$ and the variables $x_0, x_1, x_2, \ldots$. The quantifier symbol $\forall$ expresses universal generalization over natural numbers and is read "every natural number is such that…." Each variable $x_i$ works like a name that has yet to be assigned a particular referent and is read "it." When quantifiers and variables are combined, they allow us to say all sorts of interesting things. For instance the following:

$\forall x_0(x_0 = x_0)$
(*Read:* Every natural number is such that it is identical to it.)

It is important to have variables with different indices to avoid ambiguity. Consider, for example, the following sentence:

$\forall x_1 \forall x_2(x_1 \times x_2 = x_2 \times x_1)$

When the indices are ignored, this sentence might be read as follows:

> Every natural number is such that every natural number is such that it times it equals it times it

which is highly ambiguous. With indices in place, however, we can avoid ambiguity by saying this:

> Every natural number (call it $x_1$) is such that every natural number (call it $x_2$) is such that it (i.e., $x_1$) times it (i.e., $x_2$) equals it (i.e., $x_2$) times it (i.e., $x_1$)

Or more succinctly:

> Every natural number $x_1$ and every natural number $x_2$ are such that $x_1$ times $x_2$ equals $x_2$ times $x_1$

### 10.2.3 Abbreviations

Our language $L$ is constructed on the basis of a very restricted set of symbols, and it is often cumbersome to express interesting arithmetical claims using only those symbols. Fortunately, we can introduce abbreviations. In fact, we've done so already. Recall that we introduced 2 as an abbreviation for $(1+1)$, 3 as an abbreviation for $((1+1)+1)$, and so forth. In doing so, we didn't add any new symbols to our language. All we did was introduce a notational *shortcut* to make it easier for us to keep track of certain strings of symbols on our official list.

In the remainder of this section, I'll mention a few additional abbreviations that will be useful in proving Gödel's Theorem. We start with three additions to our logical vocabulary:

**The Existential Quantifier Symbol** $L$ can be used to express existential statements of the form "there exists a natural number such that...." This is because of a happy equivalence between "there exists a natural number such that so-and-so" and "it is not the case that every natural number is such that it is not the case that so-and-so." (For example, "there exists a prime number" is equivalent to "not every number is not prime.") Accordingly, we can introduce $\exists x_i$ (read: "there exists a natural number $x_i$ such that") as an abbreviation for $\neg\forall x_i\neg$.

**The Disjunction Symbol** $L$ can be used to express disjunctive statements of the form "either $A$ is the case or $B$ is the case (or both)." This is because of a happy equivalence between "either $A$ is the case or $B$ is the case (or both)" and "it is not the case that (not-$A$ and not-$B$)." (For example, "every number is even or odd (or both)" is equivalent to "every number is not such that it is both not even and not odd.") Accordingly, we can introduce $A \vee B$ (read: "$A$ or $B$ (or both)") as an abbreviation for $\neg(\neg A \,\&\, \neg B)$.

**The Conditional Symbol** $L$ can be used to express conditional statements of the form "if $A$, then $B$." This is because of a happy equivalence (within mathematical contexts) between "if $A$, then $B$" and "either not-$A$ or $B$." (For instance, "if $a$ is even, then $a$ is divisible by two" is equivalent to "either $a$ is not even or $a$ is divisible by two.") Accordingly, we can introduce $A \supset B$ (read: "if $A$, then $B$") as an abbreviation for $\neg A \vee B$.

We can also use abbreviations to enrich our arithmetical vocabulary. Suppose, for example, that we wish to introduce the less-than symbol, $<$. Again, we can do so by taking advantage of a happy equivalence. In general, if $a$ and $b$ are natural numbers, $a$ is smaller than $b$ if and only if $b = a + c$, for some natural number $c$ distinct from 0. So $x_i < x_j$ can be defined as follows:

$$x_i < x_j \leftrightarrow_{df} \exists x_k((x_j = x_i + x_k) \mathbin{\&} \neg(x_k = 0))$$

Two nerdy observations:

1. The symbol $\leftrightarrow_{df}$ is not part of $L$. I use it to indicate that the expression to its left is to be treated as a syntactic abbreviation for the expression to its right.
2. You'll notice that I've used letters rather than numbers as variable indices. These letters should be replaced by numbers when the abbreviation is used as part of a sentence. Any numbers will do, as long as they are distinct from one another and as long as they do not already occur as indices in the context where the abbreviation is embedded.

Let me mention an additional example. Suppose we want to enrich $L$ with the expression "Prime($x_i$)," which is true of $x_i$ if and only if $x_i$ is a prime number. In general, $a$ is prime if and only if the following conditions hold: (*i*) $a$ is greater than 1, and (*ii*) $a$ is only the product of $b$ and $c$, if $a = b$ or $a = c$. So "Prime($x_i$)" can be defined as follows:

$$\text{Prime}(x_i) \leftrightarrow_{df} (1 < x_i) \mathbin{\&} \forall x_j \forall x_k((x_i = x_j \times x_k) \supset (x_i = x_j \vee x_i = x_k))$$

I'll ask you to introduce further abbreviations of this kind in the exercises below.

### Exercises

In what follows, you may avail yourself of any abbreviations previously defined.

1. Introduce the following syntactic abbreviations:
   **a)** For a given formula $\phi(x_i)$, introduce an abbreviation $\exists! x_i(\phi(x_i))$ by finding a sentence of $L$ that is true if and only if there is exactly one number $n$ such that $\phi(n)$.
   
   **b)** Introduce an abbreviation $x_i \leq x_j$ by finding a formula of $L$ that is true of $x_i$ and $x_j$ if and only if $x_i$ is smaller than or equal to $x_j$.
   
   **c)** Introduce an abbreviation $x_i | x_j$ by finding a formula of $L$ that is true of $x_i$ and $x_j$ if and only if $x_i$ divides $x_j$ with no remainder.
   
   **d)** Introduce an abbreviation Even($x_i$) by finding a formula of $L$ that is true of $x_i$ if and only if $x_i$ is an even natural number.
   
   **e)** Introduce an abbreviation Odd($x_i$) by finding a formula of $L$ that is true of $x_i$ if and only if $x_i$ is an uneven natural number.

2. Express each of the following claims in $L$:
   a) There is at least one prime number.
   b) Every number is even or odd.
   c) If the result of multiplying a number by itself is one, then that number must be one.
   d) Every even number greater than two is the sum of two primes (Goldbach's Conjecture).

### 10.2.4 Simple, Yet Mighty

Our language $L$ turns out to be extraordinarily powerful. It can be used to formulate not only mundane arithmetical claims like $2+3=5$ but also claims that have taken us hundreds of years to prove (like Fermat's Last Theorem) and hypotheses that are yet to be proved (like Goldbach's Conjecture).

The fact that $L$ is such an expressive language guarantees that if we were able to construct a Turing Machine that output every truth of $L$ (and no falsehood), we would have succeeded in concentrating a huge wealth of mathematical knowledge in a finite list of lines of code. Implicit in our program would be not just all the arithmetic we learned in school, but also remarkable mathematical results. For example, since it is possible to express Goldbach's Conjecture in $L$, such a Turing Machine would give us a way of knowing for certain whether Goldbach's Conjecture is true. Alas, Gödel's Theorem shows that the dream of constructing such a machine is not to be.

Gödel's Theorem is a very robust result, which doesn't depend on the details of one's arithmetical language. All that is required to prove the theorem is that one's language be "rich enough," which is shorthand for saying that the language must allow for the following lemma to be proved:

**Lemma** The language contains a formula (abbreviated "Halt($k$)"), which is true if and only if the $k$th Turing Machine halts on input $k$.

In an appendix at the end of this chapter, I show that our simple language $L$ satisfies this condition.

## 10.3 A Proof of Gödel's Theorem

We are now in a position to prove our main result:

**Gödel's Incompleteness Theorem** Let $\mathcal{L}$ be a rich-enough language (as it might be, $L$). Then no Turing Machine $M$ is such that when run on an empty input, the following would happen:

1. $M$ runs forever, outputting sentences of $\mathcal{L}$.
2. Every true sentence of $\mathcal{L}$ is eventually output by $M$.

3. No false sentence of $\mathcal{L}$ is ever output by $M$.

The proof proceeds by *reductio*. We will assume that $M$ is a Turing Machine that outputs all and only the true sentences of $\mathcal{L}$, and show that $M$'s program can be used as a subroutine to construct a Turing Machine $M^H$, which computes the Halting Function, $H(x)$. But we know that the Halting Function is not Turing-computable (as we saw in chapter 9). So we may conclude that $M$ is impossible.

We proceed in two steps. First, we verify that if $M$ existed, it could be used to construct $M^H$. We then verify that $M^H$ would compute the Halting Function, if it existed.

**Step 1** Here is how to construct $M^H$ on the assumption that $M$ exists. (Assume that $M^H$'s input is a sequence of $k$ ones.)

- $M^H$ starts by running $M$'s program as a subroutine. Each time $M$ outputs a sentence, $M^H$ proceeds as follows:
  — If the sentence is Halt($k$), $M^H$ deletes everything on the tape, prints a 1, and halts.
  — If the sentence is ¬Halt($k$), $M^H$ deletes everything on the tape, and halts.
  — Otherwise, $M^H$ allows $M$ to keep going.

**Step 2** We will now verify that $M^H$ would compute the Halting Function, if it existed. We will verify, in other words, that $M^H$ outputs a 1 if the $k$th Turing Machine halts on input $k$, and a 0 otherwise.

- Suppose, first, that the $k$th Turing Machine halts on input $k$. Then our lemma guarantees that Halt($k$) is true. Since $M$ will eventually output every true sentence of $\mathcal{L}$, this means that $M$ will eventually output Halt($k$). And since $M$ will never output any falsehood, it will never output ¬Halt($k$). So the construction of $M^H$ guarantees that $M^H$ will output a 1.
- Now suppose that the $k$th Turing Machine does not halt on input $k$. Then our lemma guarantees that ¬Halt($k$) is true. Since $M$ will eventually output every true sentence of $\mathcal{L}$, this means that $M$ will eventually output ¬Halt($k$). And since $M$ will never output any falsehood, it will never print out Halt($k$). So $M^H$ will output a zero.

In summary, we have seen that if $M$ existed, $M^H$ would exist too. And we have seen that $M^H$ would compute the Halting Function, which we know to be impossible. So $M$ cannot exist. We've proved Gödel's Thoerem!

### 10.3.1 A Different Formulation of the Theorem

In the preceding section, we proved that it is impossible to build a Turing Machine that outputs all and only the true statements of a rich-enough arithmetical language. That is one formulation of Gödel's Theorem. Here is another:

**Gödel's Incompleteness Theorem** No axiomatization of elementary arithmetic (in a rich-enough language) can be both consistent and complete.

In this subsection we'll see what this means, and we'll see that the theorem can be proved based on the version of the result we have been discussing so far (assuming a particular understanding of "complete").

### 10.3.2 Axiomatization

Let me start with the notion of *axiomatization*. We've talked a fair amount about having a language to express arithmetical claims. But we haven't said much about how to *prove* the arithmetical claims expressible in that language. An **axiomatization** of arithmetic is a system for proving claims within an arithmetical language. It consists of two different components: a set of axioms and a set of rules of inference.

- An **axiom** is a sentence of the language that one chooses to treat as "basic," and on the basis of which other sentences of the language are supposed to be proven. One might, for example, treat "every number has a successor" (in symbols: $\forall x_1 \exists x_2 (x_2 = x_1 + 1)$) as an axiom.
- A **rule of inference** is a rule for inferring some sentences of the language from others. An example is *modus ponens*: the rule that tells us that one can always infer the sentence $\psi$ from the sentences $\phi$ and "if $\phi$, then $\psi$."

In choosing an axiomatization, you are free to pick any list of axioms and rules of inference you like, *as long as the list is Turing-computable*. In other words, it must be possible to program a Turing Machine to determine what counts as an axiom and when one sentence follows from another by a rule of inference.

### 10.3.3 Provability, Completeness, and Consistency

Once one has an axiomatization for a given language, one can introduce a notion of provability for that language. A sentence $S$ is **provable** on the basis of a given axiomatization if there is a finite sequence of sentences of the language with the following two properties:

- Every member of the sequence is either an axiom or something that results from previous members of the sequence when a rule of inference is applied.
- The last member of the sequence is $S$.

We can now say what it means for an axiomatization to be complete and consistent. For an axiomatization to be **complete** is for every true sentence of the language to be provable on the basis of that axiomatization. For it to be **consistent** is for it never to be the case that both a sentence of the language and its negation are provable on the basis of that axiomatization.

Notice that an arithmetical falsehood can never be proven on the basis of a complete and consistent axiomatization. To see this, suppose, for *reductio*, that a complete and consistent axiomatization does entail some false sentence $S$. Since $S$ is false, its negation must be true. And since the axiomatization is complete, the negation of *s* must be provable on the basis of that axiomatization. So both a sentence and its negation must be provable on the basis of that axiomatization. So the axiomatization fails to be consistent.

(Warning: Gödel's Theorem is sometimes formulated using a slightly different notion of completeness, according to which an axiomatization is complete if it proves every sentence or its negation. The result holds on either characterization of completeness, but the version I use here has the advantage of being easily provable on the basis of our original formulation of the theorem.)

### 10.3.4 The Proof

Let us now use the claim that no Turing Machine can output all and only arithmetical truths (of a rich-enough language) to prove that no axiomatization of arithmetic (in that language) can be both consistent and complete.

We proceed by *reductio*. Let's assume that we have a consistent and complete axiomatization of arithmetic in a suitable language and use this assumption to build a Turing Machine that outputs all and only the true sentences of that language. Since we know that there can be no such Turing Machine, our assumption must be false.

The first thing to note is that it is possible to code each finite sequence of sentences of our arithmetical language as a natural number. To see this, note that we can code each symbol of the language as a natural number. Since each sentence is a finite sequence of symbols, and since we know how to code finite sequences of natural numbers as natural numbers (section 9.1.5), this means that we can code each sentence as a natural number. And once sentences have been coded as natural numbers, we can repeat the process to code each finite sequence of sentences as a natural number.

The next observation is that we can program a Turing Machine that outputs all and only the sentences of our language that are provable on the basis of our axiomatization. The machine proceeds by going through the natural numbers in order. At each stage, the machine determines whether the relevant number codes a sequence of sentences. If it doesn't, the machine goes on to the next number. If it does, the machine proceeds to determine whether the sequence constitutes a proof. (We know that this is possible because axiomatizations are Turing-computable by definition.) If the sequence does constitute a proof, the machine outputs the last member of the sequence (which is the conclusion of the proof). The machine then goes on to consider the next natural number.

Notice, finally, that if our axiomatization is consistent and complete, then a Turing Machine that outputs all and only the provable sentences of our arithmetical language must output all and only the true sentences of our arithmetical language. But we have shown this to be impossible for languages like $L$, which are rich enough to express the Halting Function. We conclude that an axiomatization for such a language cannot be both consistent and complete.

## 10.4 The Philosophical Significance of Gödel's Theorem[1]

From a mathematical point of view, Gödel's Theorem is an incredibly interesting result. But it also has far-reaching philosophical consequences. I'll tell you about one of them in this section.

### 10.4.1 Excluding All Possible Doubt

Sometimes our best theories of the natural world turn out to be mistaken. The geocentric theory of the universe turned out to be mistaken. So did Newtonian physics. We now have different theories of the natural world. But how can we be sure that they won't turn out to be mistaken too? It's hard to exclude all possibility of doubt.

Many years ago, when I was young and reckless, I used to think that mathematical theories were different: I used to think that they could be established beyond any possibility of doubt. Alas, I now believe that my youthful self was mistaken. Gödel's Theorem has convinced me that the exclusion of all possible doubt is just as unavailable in mathematical theory as it is in theory of the natural world.

There is, of course, an important disanalogy between physical theories and their mathematical counterparts. When Copernicus proposed the heliocentric theory of the universe, he defended his hypothesis by arguing that it was simpler than rival theories. But the fact that a theory is simple does not guarantee that it is true. Copernicus's theory is false as he proposed it, since he postulated circular planetary orbits, and we now know that the orbits are elliptical. In contrast, when Euclid proposed that there are infinitely many prime numbers, he justified his hypothesis with a *proof* from basic principles. Euclid's proof is valid, so it *guarantees* that if his basic principles are correct, then his conclusion is correct as well. It is therefore natural to assume—as I assumed in my youth—that Euclid, unlike Copernicus, established his result beyond any possible doubt.

Unfortunately, there's a catch. A mathematical proof is always a *conditional* result: it shows that its conclusion is true *provided* that the basic principles on which the proof is based are true. This means that in order to show conclusively that a mathematical sentence is true, it is not enough to give a proof from basic principles. We must also show that our basic principles—our axioms—are true, and that our basic rules of inference are valid.

1. An earlier version of this material appeared in Rayo, "Gödel y la verdad axiomática," *Investigación y Ciencia*, February 2014.

How could one show that an axiom is true or that a rule of inference is valid? It is tempting to answer that axioms and rules are principles so basic that they are absolutely obvious: their correctness is immediately apparent to us. There is, for example, an arithmetical axiom that says that different natural numbers must have different successors. What could possibly be more obvious?

Sadly, the fact that an axiom seems obvious is not a guarantee of its truth. Some mathematical axioms that have seemed obviously true to great mathematicians have turned out to be false. Let me give you an example.

### 10.4.2 Russell's Paradox

Mathematicians sometimes talk about the *set* of prime numbers or the *set* of functions from real numbers to real numbers. Suppose we wanted to formalize this talk of sets. What could we use as axioms? When the great German mathematician Gottlob Frege started thinking about sets, he proposed a version of the following axiom.

**Frege's Axiom** Select any objects you like (the prime numbers, for example, or the functions from real numbers to real numbers). There is a set whose members are all and only the objects you selected.

Frege's axiom seemed obviously true at the time. In fact, it is inconsistent. This was discovered by the British philosopher Bertrand Russell. Russell's argument is devastatingly simple:

> Consider the objects that are not members of themselves. (The empty set has no members, for example, so it is an object that is not a member of itself.) An immediate consequence of Frege's Axiom is that there is a set that has all and only the non-self-membered sets as members. Let us honor Russell by calling this set $R$.
>
> Now consider the following question: Is $R$ a member of $R$? The definition of $R$ tells us that $R$'s members are exactly the objects that have a certain property: the property of not being members of oneself. So we know that $R$ is a member of $R$ if and only if it has that property. In other words, $R$ is a member of itself if and only if it is not a member of itself.
>
> That's a contradiction. So $R$ can't really exist. So Frege's Axiom must be false.

The upshot of Russell's argument has come to be known as **Russell's Paradox**, which we briefly touched on in section 2.5.

### 10.4.3 A Way Out?

I told you about my youthful belief that mathematical proofs exclude any possibility of doubt. We have now seen, however, that the situation is more complicated than that. Mathematical proofs presuppose axioms and rules of inference. And, as we have seen, an axiom that appears obviously true can turn out to be false.

At the beginning of the twentieth century, the German mathematician David Hilbert suggested a program to overcome these difficulties. To a rough first approximation, the program was based on these two ideas:

**Mathematical Hypothesis** There is an algorithmic method capable of establishing, once and for all, whether a set of axioms is consistent.

**Philosophical Hypothesis** All it takes for a set of axioms to count as a true description of some mathematical structure is for it to be consistent.

If both hypotheses had turned out to be true, it would have been possible, at least in principle, to establish whether a set of axioms is a true description of some mathematical structure: all we would have to do is apply our algorithm to test for consistency.

### 10.4.4 Back to Gödel

Sadly, Hilbert's mathematical hypothesis is false. To see why this is so, I'd like to start by considering two important consequences of Gödel's Theorem:

1. As shown in Appendix A.10.1, the behavior of Turing Machines can be described within the language of arithmetic. This means that if a problem can be solved by a Turing Machine, it can also be solved by figuring out whether a certain arithmetical sentence is true. In fact, Gödel showed something stronger: if a problem can be solved by a Turing Machine, it can be proved to be true or false on the basis of the standard axiomatization of arithmetic. Assuming the Church-Turing Thesis (chapter 9), this means that *any problem that can be solved by applying an algorithmic method is settled by the standard axioms of arithmetic.*
2. Gödel showed that his Incompleteness Theorem has an important corollary:

   **Gödel's Second Theorem** If an axiomatic system is at least as strong as the standard axiomatization of arithmetic, it is unable to prove its own consistency (unless it is inconsistent, in which case it can prove anything, including its own consistency).

Let us now show that Hilbert's mathematical hypothesis is false, by bringing these two consequences of Gödel's Theorem together. Suppose, for *reductio*, that Hilbert's hypothesis is true and that some algorithmic method can establish the consistency of the standard arithmetical axioms, assuming they are indeed consistent. By the first consequence of Gödel's Theorem noted above, this means the axioms are able to prove their own consistency. But, by Gödel's Second Theorem, the axioms can only prove their own consistency if they are inconsistent. So if the axioms are consistent, their consistency cannot be established by an algorithmic method. So Hilbert's mathematical hypothesis is false.

A parallel argument shows that there can be no algorithmic method for establishing the consistency of a *strengthening* of the standard axioms of arithmetic. Since the standard axiomatizations of real analysis and set theory are both strengthenings of the standard arithmetical axioms, this means, in particular, that there can be no algorithmic method for establishing their consistency.

### 10.4.5 Mathematics without Foundations

We have discussed a startling consequence of Gödel's Theorem: when it comes to axiomatic systems strong enough to extend arithmetic, we cannot hope to establish their consistency using an algorithmic method.

This is not to say, however, that mathematicians should be worried that their preferred mathematical system might turn out to be inconsistent. When it comes to mathematical systems that are tried and tested, there is no real reason to worry. The upshot of our discussion is that this is not because we have found a method for ruling out inconsistency beyond any possibility of doubt; in some cases, at least, it is because mathematicians have a good sense for how the system works and how it is connected to other systems. But as illustrated by Frege's experience, there is always the possibility of error. (It is also worth keeping in mind that there are mathematical systems that are not tried and tested and whose consistency is genuinely in doubt. One example is a set-theoretic system referred to as *New Foundations*, which was developed by the American philosopher and logician W. V. Quine.)

The moral of our discussion is that my youthful views were mistaken. As in the case of our best theories of the natural world, we have excellent reasons for thinking that our mathematical theories are true. But, as in the case of our best theories of the natural world, our reasons are not good enough to exclude every possibility of doubt.

## 10.5 Conclusion

In this chapter we proved Gödel's Theorem. We proved, in other words, that no Turing Machine could output every arithmetical truth (of a rich-enough language) and no falsehood.

One reason the theorem is mathematically important is that it teaches us something about the complexity of arithmetical truth. It tells us that there is a precise sense in which arithmetical truth is too complex to be finitely specifiable. This means, in particular, that no finitely specifiable set of axioms could allow us to prove every arithmetical truth and no falsehood.

We also discussed an important philosophical consequence of Gödel's Theorem: when it comes to axiomatic systems strong enough to extend arithmetic, we cannot hope to establish their consistency using an algorithmic method. So even though we have excellent reasons for thinking that our mathematical theories are true, our reasons are not good enough to exclude every possibility of doubt.

## 10.6 Further Resources

- The version of Hilbert's Program that I discuss above is grossly oversimplified. If you'd like to know the real story, I recommend Richard Zach's "Hilbert's Program" in the *Stanford Encyclopedia of Philosophy*, available online.
- For a nice proof of Gödel's Second Theorem (with connections to the Surprise Exam Paradox) see Shira Kritchman and Ran Raz's "The Surprise Exam Paradox and the Second Incompleteness Theorem."
- An excellent treatment of Gödel's Theorem can be found in Boolos, Burgess, and Jeffrey's *Computability and Logic*.

## Appendix A: Proof of the Lemma

$L$ is very simple, but with some ingenuity it can be used to express all sorts of claims about Turing Machines. For instance, there is a sentence of $L$ that is true if and only if a certain Turing Machine computes a particular function. How is that possible? After all, $L$ doesn't have any symbols that refer to Turing Machines. It's possible because one can use numbers to *code* information about Turing Machines. I'll tell you all about the relevant coding tricks in this section, and we'll prove the following lemma:

**Lemma** $L$ contains a formula (abbreviated "Halt($k$)") that is true if and only if the $k$th Turing Machine halts on input $k$.

Proving this lemma is not hard, but it's a bit cumbersome.

### A.10.1 The Key Idea

The key to expressing claims about Turing Machines in $L$ is being able to express claims about *sequences* in $L$. Specifically, we need to define a formula "Seq$(c, n, a, i)$," which is true if and only if $c$ encodes a sequence of length $n$, of which $a$ is the $i$th member. Once we're equipped with such a formula, it'll be very easy to use $L$ to say things about Turing Machines because we'll be able to represent Turing Machines as number sequences of a particular sort.

As we go forward, we'll help ourselves to the abbreviations introduced in section 10.2.3, the most important of which are listed in figure A.10.1. As a further expository aid, I'll use arbitrary lowercase letters as variables, as I did in the case of Seq$(c, n, a, i)$. Finally, I will write $n^m$ instead of $n \wedge m$, to improve readability.

### A.10.2 Pairs

Before plunging into the details, I want you to get a feel for how Seq$(c, n, a, i)$ works. We'll start with the relatively simple task of using $L$ to express claims about 2-sequences (i.e., ordered pairs). In other words, we'll see how to use the natural number $x_0$ to encode the ordered pair of natural numbers $\langle x_1, x_2 \rangle$. And we'll see how that coding system can be captured in $L$. In other words: we'll see how to write down a sentence of $L$ that is true if and only if $x_0$ encodes the ordered pair $\langle x_1, x_2 \rangle$.

The first step is to specify a coding scheme, which assigns a different natural number to each ordered pair of natural numbers. There are many different such coding schemes, but the one I like best is based on the same general idea as the one we used in chapter 9, when we numbered the Turing Machines. We will code $\langle a, b \rangle$ as $2^a \times 3^b$. The reason this coding scheme gives us what we want is that, as we saw in chapter 9, the Fundamental

| Abbreviation | Read | Abbreviates |
|---|---|---|
| $A \vee B$ | $A$ or $B$ | $\neg(\neg A \,\&\, \neg B)$ |
| $A \supset B$ | If $A$, then $B$ | $\neg A \vee B$ |
| $\exists x_i$ | Some number is such that | $\neg \forall x_i \neg$ |
| $\exists! x_i$ | There is exactly one number such that | $\exists x_i(\phi(x_i) \,\&\, \forall x_j(\phi(x_j) \supset x_j = x_i))$ |
| $x_i < x_j$ | $x_i$ is smaller than $x_j$ | $\exists x_k((x_j = x_i + x_k) \,\&\, \neg(x_k = 0))$ |
| $x_i \leq x_j$ | $x_i$ is less than or equal to $x_j$ | $(x_i < x_j \vee x_i = x_j)$ |
| $x_i \vert x_j$ | $x_i$ divides $x_j$ | $\exists x_k(x_k \times x_i = x_j)$ |
| $\text{Prime}(x_i)$ | $x_i$ is prime | $(1 < x_i) \,\&\, \forall x_j \forall x_k((x_i = x_j \times x_k) \supset (x_i = x_j \vee x_i = x_k))$ |

**Figure A.10.1**
Abbreviations from section 10.2.3.

Theorem of Arithmetic guarantees that it assigns a different natural number to each ordered pair of natural numbers. (The main difference between the current scheme and the scheme we used in chapter 9 is that, in the present context, we do not add 1 to the exponents. This causes no ambiguity because we are coding sequences of *fixed length*, unlike what we did in chapter 9.)

Let us now see how our coding scheme might be captured within $L$. What we want is formula "$\text{Pair}(x_i, x_j, x_k)$," which is true if and only if $x_i$ is a code for $\langle x_j, x_k \rangle$, according to our coding scheme. This is easily done:

$$\text{Pair}(x_i, x_j, x_k) \leftrightarrow_{df} x_i = (2^{x_j} \times 3^{x_k})$$

Now that we have $\text{Pair}(x_0, x_1, x_2)$ in place, we can use it as the basis for further notational abbreviations. For example, we can define a formula "$\text{First}(x_i, x_j)$" that is true if and only if $x_i$ codes a pair of which $x_j$ is the first member:

$$\text{First}(x_i, x_j) \leftrightarrow_{df} \exists x_k(\text{Pair}(x_i, x_j, x_k))$$

### A.10.3 Sequences: Introduction

We have now seen how to use $L$ to talk about sequences of length 2. The same technique can be used for sequences of any particular length. So, for instance, one can encode the triple $\langle a, b, c \rangle$ as $2^a \times 3^b \times 5^c$ and use the formula "$x = 2^a \times 3^b \times 5^c$" to express the claim that $x$ encodes the triple $a, b, c$.

What we need to prove the theorem, however, is a single formula that allows us to talk about sequences of any finite length: we need a formula "$\text{Seq}(c, n, a, i)$" that is true if and only if $c$ encodes a sequence of length $n$, of which $a$ is the $i$th member. The definition of $\text{Seq}(c, n, a, i)$ is trickier than one might think, and I'd like to start by explaining why.

Why can't we define $\text{Seq}(c, n, a, i)$ by using the technique we've been using so far? Let's try, and see what goes wrong. Suppose we code the $n$-membered sequence $\langle a_1, a_2, \ldots, a_n \rangle$ as the number:

$$c = p_1^{a_1} \times p_2^{a_2} \times \ldots \times p_n^{a_n}$$

(where $p_1, p_2, \ldots, p_n$ are the first $n$ prime numbers).

We now need to find a formula "$\text{Seq}(c, n, a, i)$," which (relative to this coding method) expresses the claim that the $i$th member of our (encoded) $n$-membered sequence $c$ is $a$ and does so regardless of the value of $n$. The task would be straightforward if we had a straightforward way of defining a formula "$\text{Prime}(p, n)$" of $L$, which is true if and only if $p$ is the $n$th prime number; for then we would be able introduce the following notational abbreviation:

$$\text{Seq}(c, n, a, i) \leftrightarrow_{df} \forall x_1 (\text{Prime}(x_1, i) \supset ((x_1^a \mid c) \ \& \ \neg(x_1^{a+1} \mid c)))$$

Unfortunately, defining $\text{Prime}(p, n)$ turns out to be no easier than defining $\text{Seq}(c, n, a, i)$. So we'll need to follow a different route.

### A.10.4 Sequences: The Coding Scheme

The problem of defining "$\text{Seq}(c, n, a, i)$" consists of two subproblems. The first is to characterize a suitable coding scheme. The second is to identify the specific formula "$\text{Seq}(c, n, a, i)$" of $L$ that will allow us to talk about that coding scheme in $L$.

Let us start with the coding scheme. The Fundamental Theorem of Arithmetic allows us to introduce a useful definition. Suppose that $c$ is a natural number and that its (unique) decomposition into primes is as follows:

$$c = p_1^{e_1} \cdot p_2^{e_2} \cdot p_3^{e_3} \cdot \ldots$$

Then we can say that the *nontrivial exponents* of $c$ are the $e_i$ greater than 0. So, for example, the set of nontrivial exponents of 168 is $\{1, 3\}$ because $168 = 2^3 \cdot 3^1 \cdot 7^1$. (Recall that the set $\{3, 1, 1\}$ is the same as the set $\{1, 3\}$.)

The key to our coding scheme is to think of a number as a code for the set of its nontrivial exponents. So, for example, one can think of 168 as a code for the set $\{1, 3\}$, and one can think of the number $5{,}600 = 2^5 \cdot 5^2 \cdot 7^1$ as a code for the set $\{1, 2, 5\}$. (Notice, incidentally, that different numbers can code the same set. We have seen, for example, that 168 codes $\{1, 3\}$, but so does 24, since $24 = 2^3 \cdot 3^1$.)

We have found a way of using a single number to code a *set* of numbers. But what we really want is a way of using a single number to code a *sequence* of numbers. The difference between sets and sequences is that sets are unordered and sequences are ordered. (In general, one can think of a sequence as a total ordering of a set, in the sense given in chapter 2.) Since there are two different ways of totally ordering the set $\{1, 3\}$, $\{1, 3\}$

corresponds to two different sequences: $\langle 1,3\rangle$ and $\langle 3,1\rangle$. So when we think of 168 as a code for the set $\{1,3\}$, we have not yet linked it to a particular sequence.

Fortunately, there is an additional trick that can be used to get a single number to code an ordered set of numbers. For suppose that some of *c*'s nontrivial exponents coded ordered pairs and that each such pair had a different natural number as its first component. Then the first components of the pairs could be used to define an ordering of pairs and, therefore, an ordering of the pairs' second components.

Here is an example. Suppose that $c$ is the number $2^{2^2 \cdot 3^{17}} \cdot 5^{2^1 \cdot 3^7} \cdot 7^{2^3 \cdot 3^{117}}$. Then the set of *c*'s nontrivial exponents is $\{2^2 \cdot 3^{17}, 2^1 \cdot 3^7, 2^3 \cdot 3^{117}\}$. When we think of these numbers as coding ordered pairs, we get the set $\{\langle 2,17\rangle, \langle 1,7\rangle, \langle 3,117\rangle\}$. And of course, the first components of the pairs in this set naturally define an ordering of the pairs $\langle 1,7\rangle, \langle 2,17\rangle$, and $\langle 3,117\rangle$. This ordering of pairs induces an ordering of the pairs' second components: 7, 17, and 117. So $c$ can be thought of as coding the three-membered sequence $\langle 7,17,117\rangle$.

A more formal version of this basic idea is based on the following definition: ***c* encodes an *n*-sequence** if and only if for each $i$ $(1 \leq i \leq n)$, *c*'s nontrivial exponents include (the code for) exactly one pair of the form $\langle i,b\rangle$.

### A.10.5 Sequences: Expressing the Coding Scheme in *L*

Our coding scheme is now in place. The next step is to define $\text{Seq}(c,n,a,i)$, by identifying a formula that is true if and only if $c$ encodes an $n$-sequence of which the $i$th member is $a$. (All of this relative to our coding scheme, of course.) We will proceed in two steps. The first step is to introduce the formula "$\text{Seq}(c,n)$," which is true if and only if $c$ encodes an $n$-sequence. Then we'll use $\text{Seq}(c,n)$ to define $\text{Seq}(c,n,a,i)$.

In light of our definition of "encoding an $n$-sequence," we want $\text{Seq}(c,n)$ to be true if and only if for each $i$ $(1 \leq i \leq n)$, *c*'s nontrivial exponents include the code for exactly one pair of the form $\langle i,b\rangle$. Here is one way of doing so:

$$\begin{aligned}\text{Seq}(c,n) \quad \leftrightarrow_{df} \quad & \forall x_i((1 \leq x_i \ \&\ x_i \leq n) \supset \\ & \exists! x_j(\exists x_k(x_j = 2^{x_i} \times 3^{x_k}) \ \&\ \exists x_k(\text{Prime}(x_k) \ \&\ x_k^{x_j} \mid c \ \&\ \neg(x_k^{x_j+1} \mid c)))\end{aligned}$$

Now that $\text{Seq}(c,n)$ is in place, we can use it to characterize $\text{Seq}(c,n,a,i)$, which should be true if and only if $c$ encodes an $n$-sequence of which the $i$th member is $a$. Relative to our coding scheme, this will be the case if and only if each of the following three conditions are met:

- $\text{Seq}(c,n)$
- $(1 \leq i \ \&\ i \leq n)$
- *c*'s nontrivial exponents include a code for $\langle i,a\rangle$

Here is one way of defining Seq$(c, n, a, i)$, so that it captures these three conditions:

$$\begin{aligned}\text{Seq}(c,n,a,i) \quad \leftrightarrow_{df} \quad & \text{Seq}(c,n)\ \& \\ & (1 \leq i\ \&\ i \leq n)\ \& \\ & \exists x_j\big(\text{Prime}(x_j)\ \&\ (x_j^{(2^i \times 3^a)} \mid c)\ \&\ \neg(x_j^{(2^i \times 3^a)+1} \mid c)\big)\end{aligned}$$

### Exercises

1. Relative to our coding scheme, is it possible for two different numbers to encode the same $n$-sequence?
2. Relative to our coding scheme, is it possible for a single number to encode an $n$-sequence and an $m$-sequence for $n \neq m$?
3. Use Seq$(c, n, a, i)$ to define the predicate Prime$(p, n)$, which is true if and only if $p$ is the $n$th prime number (where $1 \leq n$).

## A.10.6 Describing Turing Machines in *L*

In the previous subsection we introduced a coding scheme and defined two important pieces of notation that pertain to that scheme:

- Seq$(c, n)$, which is true if and only if $c$ codes a sequence of length $n$.
- Seq$(c, n, a, i)$, which is true if and only if $c$ codes a sequence of length $n$ of which $a$ is the $i$th component.

The hardest part of proving our lemma is defining these two pieces of notation. Now that they are in place, we'll find it easy to express claims about Turing Machines in *L*. (To keep things simple, I will assume that we are working with one-symbol Turing Machines, in which anything printed on the tape is either a one or a blank.)

## A.10.7 Command Lines

Recall that each command line of a Turing Machine is a sequence of five symbols, corresponding to the following parameters:

⟨current state⟩ ⟨current symbol⟩ ⟨new symbol⟩ ⟨direction⟩ ⟨new state⟩

In chapter 9 we saw that each Turing Machine command line can be represented as a sequence of five *numbers* with the following features:

- The first and last members of the sequence are arbitrary natural numbers (since they represent possible Turing-Machine states).
- The second and third members of the sequence are natural numbers in $\{0, 1\}$ (since they represent symbols, and the only admissible symbols are ones and blanks).

- The fourth member of the sequence is a natural number in {0, 1, 2} (which represents "right," "stay put," and "left," respectively).

It is easy to define the formula "Command($x_i$)," which is true if and only if $x_i$ represents a 5-sequence that corresponds to a valid command line:

$$\begin{aligned}\text{Command}(x_i) \quad \leftrightarrow_{df} \quad & \forall x_k(\text{Seq}(x_i, 5, x_k, 2) \supset x_k \leq 1)\ \& \\ & \forall x_k(\text{Seq}(x_i, 5, x_k, 3) \supset x_k \leq 1)\ \& \\ & \forall x_k(\text{Seq}(x_i, 5, x_k, 4) \supset x_k \leq 2)\end{aligned}$$

### A.10.8 Programs

A Turing Machine program is just a sequence of command lines. Since each command line can be coded as a single number, an entire Turing Machine program can be coded as a sequence of numbers.

It is easy to define the formula "Program($s, n$)," which is true if and only if $s$ represents a sequence of $n$ valid Turing Machine command lines:

$$\begin{aligned}\text{Program}(s, n) \quad \leftrightarrow_{df} \quad & \text{Seq}(s, n)\ \& \\ & \forall x_j \forall x_k((1 \leq x_j \ \&\ x_j \leq n \ \&\ \text{Seq}(s, n, x_k, x_j)) \supset \text{Command}(x_k))\end{aligned}$$

### A.10.9 The Tape and Reader

How might one describe the contents of a Turing Machine's tape in $L$? Let me start with a couple of qualifications:

- We will focus our attention on tapes that contain at most finitely many non-blanks. This is warranted because we are only interested in the behavior of Turing Machines that start out with a finite input, and such machines must have at most finitely many non-blanks at each stage of their running processes.
- When we introduced Turing Machines in chapter 9, we stipulated that the tape was to be infinite in both directions. Here, however, we will simplify matters by assuming that tapes are only infinite to the right. But this is a harmless simplification in the present context. For, as I asked you to verify in chapter 9, any function that can be computed by a Turing Machine with a tape that is infinite in both directions can also be computed on a Turing Machine with a tape that is infinite only in one direction.

With these qualifications in place, the contents of a tape can be coded as an $n$-sequence of ones and zeroes, with ones representing ones and zeroes representing blanks. The order of the $n$-sequence corresponds to the order of cells on the tape, going from left to right, so that the first member of the sequence represents the contents of the left-most cell on the tape and the $n$th member of the sequence represents the contents of the right-most non-blank cell on the tape.

It is straightforward to define the formula "Tape$(s,n)$," which is true if and only if $s$ represents a sequence of $n$ zeroes and ones that ends with a one, as follows:

$$\begin{aligned}\text{Tape}(s,n) \quad \leftrightarrow_{df} \quad 1 \leq n \supset \\ &(\text{Seq}(s,n,1,n)\ \& \\ &\forall x_i((1 \leq x_i \,\&\, x_i < n) \supset (\text{Seq}(s,n,0,x_i) \vee \text{Seq}(s,n,1,x_i))))\end{aligned}$$

Since we are focusing on Turing Machines that are only infinite to the right, we can represent the position of the machine's reader as a positive natural number and say that the reader is at position $n$ when it is positioned at the $n$th cell from the left.

When the reader is at position $r$ of tape $t$, we'll use the formula "ReadSymbol$(y,r,t,n)$" to keep track of the symbol on which the reader is positioned. More specifically, we want "ReadSymbol$(y,r,t,n)$" to be true if and only if the symbol coded by $y$ occurs at the $r$th position of the tape coded by $t$ (whose right-most non-blank is on the $n$th cell to the right). Here is one way of doing so:

$$\begin{aligned}\text{ReadSymbol}(y,r,t,n) \quad \leftrightarrow_{df} \quad &(r \leq n \supset \text{Seq}(t,n,y,r))\ \& \\ &(n < r \supset y = 0)\end{aligned}$$

### Exercises

1. In the previous section we introduced a coding scheme that allows us to encode a Turing Machine's program as a unique natural number. In chapter 9 we introduced a different coding system to do the same thing. (The reason the coding systems are different is that they are optimized for different things. In chapter 9 we were aiming for a coding scheme that was as simple as possible. In the present context we're aiming for a coding scheme that can be perspicuously represented within $L$.)

   Use the formula "Prime$(p,n)$," which you were asked to define in section A.10.5, to define the formula "Code$(p,n,k)$," which is true if and only if $p$ codes a Turing Machine of length $n$ (relative to the coding scheme of the present section) which is coded by $k$ (relative to the coding scheme of chapter 9).

## A.10.10 Describing the Workings of a Turing Machine in *L*

We are now ready to bring together the separate pieces we have been working on and to start characterizing the running process of a Turing Machine. The first step is to define the formula "Status$(p,n,s,t,m,r)$," which expresses the following:

- Program$(p,n)$ (i.e., $p$ encodes a valid Turing Machine program consisting of $n$ command lines).
- $s$ is an arbitrary number, which will be used to represent our Turing Machine's current state.

- Tape$(t, m)$ (i.e., $t$ encodes the contents of a valid Turing Machine tape whose right-most non-blank is at the $m$th cell).
- $r$ is a positive number, which is used to represent a position of the reader on the tape.

In other words, "Status$(p, n, s, t, m, r)$" uses $\langle p, n, s, t, m, r\rangle$ to represent the status that a valid Turing Machine might have at a particular moment in time. Here is one way of defining it:

$$\text{Status}(p, n, s, t, m, r) \quad \leftrightarrow_{df} \quad \text{Program}(p, n) \;\&\; \text{Tape}(t, m)) \;\&\; 1 \leq r$$

### A.10.11 Initializing

Now that we are able to represent the status of a Turing Machine, we are in a position to define the formula "Initial$(k, i, p, n, s, t, m, r)$," which is true if and only if $\langle p, n, s, t, m, r\rangle$ represents the status of the $k$th Turing Machine at the moment in which it is initialized on input $i$. (When I speak of the $k$th Turing Machine here, I have in mind the coding system of chapter 9. We'll keep track of this coding scheme by using the formula "Code$(p, n, k)$," which I asked you to define in an exercise above.)

Formally speaking, we want "Initial$(k, i, p, n, s, t, m, r)$" to be true if and only if the following:

- Status$(p, n, s, t, m, r)$ (i.e., $\langle p, n, s, t, m, r\rangle$ represents the status of a valid Turing Machine).
- Code$(p, n, k)$ (i.e., the $k$th Turing Machine has program $p$ of length $n$).
- $s = 0$ (i.e., the machine is in state 0).
- $t$ is a sequence of $m$ ones (i.e., the tape begins with a sequence of $m$ ones and is otherwise blank).
- $r = 1$ (i.e., the reader is positioned at the left-most cell of the tape).

Here is one way of doing so:

$$\begin{aligned}\text{Initial}(k, i, p, n, s, t, m, r) \quad \leftrightarrow_{df} \quad & \text{Status}(p, n, s, t, m, r) \;\&\; \text{Code}(p, n, k) \;\&\; s = 0 \;\&\; \\ & \forall j((i \leq j \;\&\; j \leq m) \supset \text{Seq}(t, m, 1, j)) \;\&\; r = 1\end{aligned}$$

With "Initial$(k, i, p, n, s, t, m, r)$" in place, it is easy to define the formula "Initial$(k, i, a)$," which is true if and only if $a$ codes a sequence $\langle p, n, s, t, m, r\rangle$ such that Initial$(k, i, p, n, s, t, m, r)$, as follows:

$$\begin{aligned}\text{Initial}(k, i, a) \quad \leftrightarrow_{df} \quad \exists p \exists n \exists t \exists m(& \text{Seq}(a, 5, p, 1) \;\&\; \text{Seq}(a, 5, n, 2) \;\&\; \text{Seq}(a, 5, 0, 3) \;\&\; \\ & \text{Seq}(a, 5, t, 4) \;\&\; \text{Seq}(a, 5, m, 5) \;\&\; \text{Seq}(a, 5, 1, 6) \;\&\; \\ & \text{Initial}(k, i, p, n, 0, t, m, 1))\end{aligned}$$

### A.10.12 Halting

The next step is to characterize a formula expressing the thought that a Turing Machine has reached its halting state. More specifically, we will define the formula "Halted$(p, n, s, t, m, r)$," which is true if and only if program $p$ (with $n$ command lines) contains no command line whose ⟨current state⟩ parameter is filled by $s$ and whose ⟨current symbol⟩ parameter is filled by the symbol occurring at the $r$th position of the tape coded by $t$ (whose right-most non-blank is on the $m$th cell to the right). Here is one way of doing so:

$$\begin{aligned}&\text{Halted}(p, n, s, t, m, r) \leftrightarrow_{df} \text{ReadSymbol}(y, r, t, m)\ \& \\ &\quad \forall i \forall l \forall s' \forall y'((1 \leq i\ \&\ i \leq n\ \&\ \text{Seq}(p, n, l, i)\ \&\ \text{Seq}(l, 5, s', 1)\ \&\ \text{Seq}(l, 5, y', 2)) \supset \\ &\qquad (\neg(s = s') \vee \neg(y = y')))\end{aligned}$$

With "Halted$(p, n, s, t, m, r)$" in place, it is easy to define the formula "Halted$(a)$," which is true if and only if $a$ codes a sequence $\langle p, n, s, t, m, r\rangle$ such that Halted$(p, n, s, t, m, r)$, as follows:

$$\begin{aligned}\text{Halted}(a) \leftrightarrow_{df}\ &\exists p \exists n \exists s \exists t \exists m \exists r(\text{Seq}(a, 5, p, 1)\ \&\ \text{Seq}(a, 5, n, 2)\ \&\ \text{Seq}(a, 5, s, 3)\ \& \\ &\text{Seq}(a, 5, t, 4)\ \&\ \text{Seq}(a, 5, m, 5)\ \&\ \text{Seq}(a, 5, r, 6)\ \& \\ &\text{Halted}(p, n, s, t, m, r))\end{aligned}$$

### A.10.13 Steps

We are now ready to start representing the transitions in the status of a Turing Machine as it goes through each step of its running process. The first thing we need to do is find a way of expressing claims about what a Turing Machine is programmed to do when it finds itself reading a particular symbol in a particular state. More specifically, we need a formula "FirstLine$(p, n, s, y, y', d, s')$," which is true if and only if the first command line in program $p$ (of length $n$) that has $s$ as its ⟨current state⟩ and $y$ as its ⟨current symbol⟩ is such that the following is true:

- It has $y'$ as ⟨new symbol⟩.
- It has $d$ as ⟨direction⟩.
- It has $s'$ as ⟨new state⟩.

Here is one way of defining "FirstLine$(p, n, s, y, y', d, s')$":

$$\begin{aligned}&\text{FirstLine}(p, n, s, y, y', d, s') \leftrightarrow_{df} \\ &\quad \exists i \exists l (1 \leq i\ \&\ i \leq n\ \&\ \text{Seq}(p, n, l, i)\ \&\ \text{Seq}(l, 5, s, 1)\ \&\ \text{Seq}(l, 5, y, 2)\ \& \\ &\qquad \text{Seq}(l, 5, y', 3)\ \&\text{Seq}(l, 5, d, 4)\ \&\text{Seq}(l, 5, s', 5)\ \& \\ &\qquad \forall j \forall k((1 \leq j\ \&\ j \leq n\ \&\ \text{Seq}(p, n, k, j)\ \&\ \text{Seq}(k, 5, s, 1)\ \&\ \text{Seq}(k, 5, y, 2)) \supset i \leq j))\end{aligned}$$

We are now in a position to define the formula "Step$(p, n, s, t, m, r, s', t', m', r')$" which is true if and only if a Turing Machine with the status represented by $\langle p, n, s, t, m, r\rangle$

reaches the status represented by $\langle p, n, s', t', m', r' \rangle$ after completing one step of its running process. This idea can be captured formally by combining the following two claims:

- First, the claim that the machine's tape remains unchanged, with the possible exception of the $r$th cell (which is the cell at which the reader is positioned when the machine is in status $\langle p, n, s, t, m, r \rangle$).
- Second, the claim that the state of the machine, the symbol at the $r$th position of the tape, and the position of the reader all transition in accordance with the program (of length $n$) represented by $p$. This can be expressed in $L$ using "FirstLine$(p, n, s, y, y', d, s')$," where $y$ is the "old symbol" at position $r$ (i.e., ReadSymbol$(y, r, t, m)$), $y'$ is the "new symbol" at position $r$ (i.e., ReadSymbol$(y', r, t', m')$), and $d$ corresponds to the direction that the reader would need to move to get from its position on the tape according to the "old status" to its position of the tape according to the "new status." Formally, $d$ can be defined as $(r - r') + 1$, since the coding system of chapter 9 is based on the following conventions:

  $$\begin{array}{ccc} \mathrm{r} & \rightarrow & 0 \\ * & \rightarrow & 1 \\ \mathrm{l} & \rightarrow & 2 \end{array}$$

Accordingly, "Step$(p, n, s, t, m, r, s', t', m', r')$" can be defined in $L$ as follows:

$$\begin{array}{l} \text{Step}(p, n, s, t, m, r, s', t', m', r') \leftrightarrow_{df} \\ \quad \forall i((1 \leq i \,\&\, \neg(i = r)) \supset \exists z(\text{ReadSymbol}(z, i, t, m) \,\&\, \text{ReadSymbol}(z, i, t', m'))) \,\&\, \\ \quad \exists y \exists y' (\text{FirstLine}(p, n, s, y, y', (r - r') - 1, s') \,\&\, \text{ReadSymbol}(y, r, t, m) \,\&\, \\ \qquad\qquad\qquad\qquad\qquad\qquad\qquad\qquad \text{ReadSymbol}(y', r, t', m')) \end{array}$$

With "Step$(p, n, s, t, m, r, s', t', m', r')$" in place, it is easy to define a formula "Step$(a, b)$," which is true if and only if $a$ codes a sequence $\langle p, n, s, t, m, r \rangle$, $b$ codes a sequence $\langle p, n, s', t', m', r' \rangle$, and Step$(p, n, s, t, m, r, s', t', m', r')$, as follows:

$$\begin{array}{lll} \text{Step}(a, b) & \leftrightarrow_{df} & \exists p \exists n \exists s \exists t \exists m \exists r \exists s' \exists t' \exists m' \exists r' (\text{Seq}(a, 5, p, 1) \,\&\, \text{Seq}(a, 5, n, 2) \,\&\, \\ & & \qquad \text{Seq}(a, 5, s, 3) \,\&\text{Seq}(a, 5, t, 4) \,\&\, \text{Seq}(a, 5, m, 5) \,\&\, \\ & & \qquad \text{Seq}(a, 5, r, 6) \,\&\text{Seq}(b, 5, p, 1) \,\&\, \text{Seq}(b, 5, n, 2) \,\&\, \\ & & \qquad \text{Seq}(b, 3, s', 6) \,\&\text{Seq}(b, 5, t', 4) \,\&\, \text{Seq}(b, 5, m', 5) \,\&\, \\ & & \qquad \text{Seq}(b, 5, r', 6) \,\&\text{Step}(p, n, s, t, m, r, s', t', m', r')) \end{array}$$

### A.10.14 Run Sequences

We are now in a position to define the formula "RunSequence$(s, j, k, i)$," which is true just in case $s$ encodes a sequence of $j$ steps of a Turing Machine's operation. In the first member of the sequence, the $k$th Turing Machine is initialized on input $i$; in the last member of the sequence, the machine halts; and one can get from any member of the sequence to its successor by performing a step of the $k$th Turing Machine's operation. Here is one way of defining "RunSequence($s, j, k, i$)":

$$\text{RunSequence}(s, m, k, i) \leftrightarrow_{df}$$
$$\exists a \exists z(\text{Seq}(s, m, a, 1) \;\&\; \text{Seq}(s, m, z, m) \;\&\; \text{Initial}(k, i, a) \;\&\; \text{Halted}(z)) \;\&$$
$$\forall i \forall a \forall b((1 \leq i \,\&\, i < m \,\&\, \text{Seq}(s, m, a, i) \,\&\, \text{Seq}(s, m, b, i+1)) \supset \text{Step}(a, b))$$

### A.10.15 The Halting Function

At long last, we are in a position to use a formula of $L$ to describe the Halting Function, $H(k)$, which we introduced in chapter 9. Recall that $H(k) = 1$ if the $k$th Turing Machine halts on input $k$, and $H(k) = 0$ otherwise.

We will express the thought that $H(k) = 1$ by defining formula "Halt($k$)," which is true if and only if the $k$th Turing Machine halts on input $k$:

$$\text{Halt}(k) \quad \leftrightarrow_{df} \quad \exists s \exists j(\text{RunSequence}(s, j, k, k))$$

We have worked long and hard, and introduced many syntactic abbreviations along the way. But it is worth keeping mind that, at the end of the day, "Halt($k$)" is shorthand for a formula of $L$, which can be built using only variables and the symbols (,), 0, 1, $+$, $\times$, $\wedge$, $\neg$, &, and $\forall$.

## Appendix B: Answers to Exercises

### Section 10.1

1. Here is a characterization of $M'$, assuming $M$ exists. To run $M'$ with $\phi$ as input, run $M$ as a subroutine and wait until it outputs either $\phi$ or its negation. We know that it will eventually output one or the other because we know that $M$ eventually outputs every true sentence of $L$ (and because we know that one or the other is true). We also know that whichever of $\phi$ or its negation is output by $M$ must be true, since $M$ never outputs any falsehood. So $M'$ can use this information to decide whether to output 0 or 1.

   Here is a characterization of $M$, assuming $M'$ exists. We know that there is a Turing Machine $M^S$, which outputs every sentence in $L$ (and nothing else). To run $M$ on an empty input, run $M^S$ as a subroutine. Each time $M^S$ outputs a sentence, use $M$ to check whether the sentence is true or false. If it's true, output that sentence; if not, keep going.

### Section 10.2.3

1. (a) Credit for the following solution is due to Bertrand Russell:

   $$\exists! x_i(\phi(x_i)) \leftrightarrow_{df} \exists x_i(\phi(x_i) \;\&\; \forall x_j(\phi(x_j) \supset x_j = x_i))$$

   (b) $x_i \leq x_j \leftrightarrow_{df} (x_i < x_j \vee x_i = x_j)$.

   (c) $x_i | x_j \leftrightarrow_{df} \exists x_k(x_k \times x_i = x_j)$.

   (d) $\text{Even}(x_i) \leftrightarrow_{df} \exists x_j(x_i = 2 \times x_j)$.

   (e) $\text{Odd}(x_i) \leftrightarrow_{df} \exists x_j(\text{Even}(x_j) \;\&\; x_i = (x_j + 1))$.

2. (a) $\exists x_1(\text{Prime}(x_1))$.

   (b) $\forall x_1(\text{Even}(x_1) \vee \text{Odd}(x_1))$.

   (c) $\forall x_1((x_1 \times x_1 = 1) \supset (x_1 = 1)))$.

   (d) $\forall x_1((\text{Even}(x_1) \;\&\; 2 < x_1) \supset \exists x_2 \exists x_3(\text{Prime}(x_2) \;\&\; \text{Prime}(x_3) \;\&\; x_1 = x_2 + x_3)$

### Section A.10.5

1. Yes. Notice, first, that all our coding scheme cares about is the code's nontrivial exponents. It does not care about which of the code's prime factors have those nontrivial exponents. So, for instance, $2^{2^1 \cdot 3^7} \cdot 3^{2^2 \cdot 3^{17}}$ and $2^{2^2 \cdot 3^{17}} \cdot 3^{2^1 \cdot 3^7}$ encode the same 2-sequence.

   Another reason is that our coding scheme is tolerant of a certain amount of "junk." Suppose, for instance, that $n = 3$ and that $c = 2^{2^2 \cdot 3^{17}} \cdot 3^{13} \cdot 5^{2^1 \cdot 3^7}$.

$7^{2^3 \cdot 3^{117}}$. Then, as we saw above, the set of $c$'s nontrivial exponents is $\{\langle 2,17\rangle$, $13, \langle 1,7\rangle, \langle 3,117\rangle\}$. Here, 13 is junk: it doesn't encode a pair of the form $\langle i,b\rangle$ for $1 \leq i \leq n$, and therefore gets ignored by our coding scheme, which yields $\langle 7,17,117\rangle$. This means that $c$ encodes the same 3-sequence as the junk-free $2^{2^2 \cdot 3^{17}} \cdot 5^{2^1 \cdot 3^7} \cdot 7^{2^3 \cdot 3^{117}}$.

2. Yes. For instance, the number $2^{2^1 \cdot 3^7} \cdot 3^{2^2 \cdot 3^{17}} \cdot 7^{2^3 \cdot 5^{117}}$ encodes both the 2-sequence $\langle 1,17\rangle$ and the 3-sequence $\langle 1,17,117\rangle$.

3. $$\begin{aligned}\text{Prime}(p,n) \quad \leftrightarrow_{df} \quad & \exists s( \\ & \text{Seq}(s,n,2,1)\ \&\ \text{Seq}(s,n,p,n))\ \& \\ & \forall i \forall x((1 \leq i \,\&\, i \leq n \,\&\, \text{Seq}(s,n,x,i)) \supset \text{Prime}(x))\ \& \\ & \forall i \forall x \forall y((1 \leq i \,\&\, i < n \,\&\, \text{Seq}(s,n,x,i)\ \&\ \text{Seq}(s,n,y,i+1) \supset \\ & \qquad (x < y\ \&\ \neg \exists z(\text{Prime}(z)\ \&\, x < z \,\&\, z < y)))))\end{aligned}$$

## Section A.10.9

1. $$\begin{aligned}&\text{Code}(p,n,k) \leftrightarrow_{df} \\ &\quad \exists s(\text{Seq}(s, n \times 5)\ \& \\ &\qquad \forall i \forall x((1 \leq i \,\&\, i \leq n \times 5 \,\&\, \text{Seq}(s, n \times 5, x, i)) \supset \\ &\qquad\qquad \forall a \forall r(\text{Div}(i-1,5,a,r) \supset \exists z(\text{Seq}(p,n,z,a+1)\ \& \\ &\qquad\qquad \text{Seq}(z,5,x,r+1)))\ \& \exists q(\text{Prime}(q,i)\ \&\ q^{x+1} | k\ \&\ \neg(q^{x+2} | k))\ \& \\ &\qquad\qquad\qquad \forall j \forall q((i < j \,\&\, \text{Prime}(q,i)) \supset \neg(q | k))))\end{aligned}$$

where $\text{Div}(n,d,a,r)$ expresses the thought that the result of dividing $n$ by $d$ is $a$ with reminder $r$, and is defined as follows:

$$\text{Div}(n,d,a,r) \quad \leftrightarrow_{df} \quad r < d\ \&\ (d \times a) + r = n$$

# Glossary

### Arithmetical Symbols

**+:** Addition (p. 256)

**×:** Multiplication (p. 256)

**∧:** Exponentiation (p. 256)

**$n|m$:** $n$ divides $m$ with no remainder (p. 260)

### Banach-Tarski Construction

**Cayley Graph:** The set of Cayley Paths (p. 208).

**Cayley Path:** A Cayley Path is a finite sequence of steps starting from a given point, referred to as the Cayley Graph's "center." Each step is taken in one of four directions: up, down, right, or left, with the important restriction that one is forbidden from following opposite directions in adjacent steps (p. 208).

**endpoint:** The endpoint of a Cayley Path is the result of starting from the Cayley Graph's "center" and taking each step in the path. An endpoint of the Cayley Graph is the endpoint of some Cayley Path (pp. 208, 214).

### Basic Set-Theoretic Symbols

**$x \in A$:** $x$ is a member of set $A$ (p. 8)

**$\{a_1, \ldots, a_n\}$:** The set whose members are $a_1, \ldots, a_n$ (p. 8).

**$\{x : \phi(x)\}$:** The set whose members are all and only the individuals satisfying condition $\phi$ (p. 8).

**$A \subset B$:** $A$ is a subset of $B$; in other words, everything in $A$ is also in $B$ (p. 8).

**$A \subsetneq B$:** $A$ is a proper subset of $B$; in other words, $A$ is a subset of $B$ and the two are distinct (p. 35).

**$A \cup B$:** The union of $A$ and $B$; in other words, $A \cup B = \{x : x \in A \text{ or } x \in B\}$ (pp. 18, 32).

**$A \cap B$:** The intersection of $A$ and $B$; in other words, $A \cap B = \{x : x \in A \text{ and } x \in B\}$.

**$A - B$:** The set of elements in $A$ but not in $B$; in other words, $A - B = \{x : x \in A \text{ and not-}(x \in B)\}$.

**$\wp(A)$:** $A$'s powerset (p. 11).

**$\wp^n(A)$:** $\underbrace{\wp(\wp(\ldots\wp(A)\ldots))}_{n \text{ times}}$ (p. 31).

**$A \times B$:** The Cartesian product of $A$ and $B$ (pp. 19, 47).

**$|A|$:** The cardinality of set $A$ (p. 8).

**$\bigcup A$:** The set of individuals $x$ such that $x$ is a member of some member of $A$ (p. 32).

**$\overline{A}$:** When $A$ is a *set*, $\overline{A}$ is the complement of $A$ relative to a suitable universal set $U$; in other words, $\overline{A} = U - A$ (p. 182). When $A$ is a *proposition*, $\overline{A}$ is the negation of $A$ (p. 152). The abuse of notation makes sense because it is sometimes useful to think of propositions as sets of "possible worlds"; in this context, the negation of a proposition is its complement.

## Computability

**algorithm:** A finite list of instructions for solving a given problem such that: *(i)* following the instructions is guaranteed to yield a solution to the problem in a finite amount of time, and *(ii)* the instructions are specified in such a way that carrying them out requires no ingenuity, special resources, or physical conditions. For a problem to be algorithmically computable is for there to be an algorithm for that problem (p. 397).

**computable:** For a function to be computable is for there to be a (possible) computer that computes it. For a function to be Turing-computable is for there to be a (possible) Turing Machine that computes it. It can be shown that a function is Turing-computable if and only if it is computable by an ordinary computer (i.e., a register machine), assuming unlimited memory and running time (p. 233).

**compute:** For a computer to compute a function $f$, from natural numbers to natural numbers, is for the computer to be such that it always returns $f(n)$ as output when given $n$ as input (p. 232).

**one-symbol:** A one-symbol Turing Machine is a machine in which the only symbol allowed on the tape (not counting blanks) is 1. One can show that every function computed by a many-symboled Turing Machine can also be computed by some one-symbol Machine (p. 235).

**oracle:** In the context of Turing Machines, an oracle is a primitive operation that allows a Turing Machine to instantaneously determine the answer to a given question. For example, a halting oracle is a primitive operation that allows a Turing Machine to instantaneously determine whether a given Turing Machine would halt on an empty input (p. 246).

**Turing Machine:** A computer of a particularly simple sort. Its hardware consists of a memory tape and a tape-reader. The tape is divided into cells and is assumed to be infinite in both directions. The reader is able to read and write symbols on the tape and move between cells. The machine's software consists of a finite list of command lines. Each command line is a sequence of five symbols: ⟨current state⟩ ⟨current symbol⟩ ⟨new symbol⟩ ⟨direction⟩ ⟨new state⟩ (p. 229).

### Conditionals

**indicative:** An indicative conditional is a conditional of the form *if A, then B*; its probability is the conditional probability of its consequent on its antecedent (p. 132).

**subjunctive:** A subjunctive conditional is a conditional of the form *if it were that A, it would be that B*. There is a connection between subjunctive conditionals and causal dependence, but it is fairly subtle (p. 132).

### Decision Theory

**causal dependence:** Two events are causally independent of one another if neither of them is a cause of the other; otherwise, the effect is causally dependent on the cause (p. 130).

**dominance:** Option *A* (strictly) dominates option *B* if you'll be better off choosing *A* than choosing *B* regardless of how matters you have no control over turn out (p. 126).

**Expected Value:** According to Evidential Decision Theory, the expected value of an option *A* is the weighted average of the value of the outcomes that *A* might lead to, with weights determined by the conditional probability of the relevant state of affairs, given that you choose *A* (p. 124). According to Causal Decision Theory, the expected value of an option *A* is the weighted average of the value of the outcomes that *A* might lead to, with weights determined by the probability of the subjunctive conditional: *were you to do A, the relevant outcome would come about* (p. 134).

**Expected Value Maximization:** The principle that one should choose an option whose expected value is at least as high as that of any rival option (p. 121).

**probabilistic dependence:** Two events are probabilistically independent of one another if the assumption that one of them occurs does not affect the probability that the other one will occur; otherwise, each of them is probabilistically dependent on the other (p. 130).

### Determinism

**determinism:** For a system of laws to be determinisitic is for it to entail, on the basis of a full specification of the state of the world at any given time, a full specification of the state of the world at any later time (pp. 101, 110).

### Functions

**bijection:** A bijection from set *A* to set *B* is a function from *A* to *B* such that: (*i*) each member of *A* is assigned to a different member of *B*, and (*ii*) no member of *B* is left without an assignment from *A* (p. 5).

**composite function:** The composite function $g \circ f$ is the function $g(f(x)))$ (p. 23).

**domain:** If $f$ is a function from $A$ to $B$, then $A$ is the domain of $f$ (p. 18).

**function:** A function from $A$ to $B$ is an assignment of one member of $B$ to each member of $A$ (p. 5).

**injection:** An injection from $A$ to $B$ is a bijection from $A$ to a subset of $B$ (p. 9).

**inverse function:** When $f$ is a bijection from $A$ to $B$, $f$'s inverse function $f^{-1}$ is the function $g$, such that for each $b \in B$, $g(b)$ is the $a \in A$ such that $f(a) = b$ (p. 23).

**range:** If $f$ is a function from $A$ to $B$, then the range of $f$ is the subset $\{f(x) : x \in A\}$ of $B$ (p. 18).

**surjection:** A surjection from $A$ to $B$ is a function $f$ from $A$ to be $B$ such that there is no member of $B$ to which $f$ fails to assign some member of $A$ (p. 10).

## Logic and Arithmetic

**arithmetical language:** A language in which one can talk about the natural numbers and their two basic operations, addition and multiplication (p. 256).

**axiom:** A sentence on the basis of which other sentences are to be proved (p. 263).

**axiomatization:** A system of axioms and rules of inference for proving claims within a suitable language. Axiomatizations are assumed to be Turing-computable; in other words, it must be possible to program a Turing Machine to determine what counts as an axiom and when one sentence follows from another by a rule of inference. (p. 263).

**complete:** In the sense of completeness presupposed in chapter 10, an axiomatization is said to be complete if and only if every true sentence of the language is provable on the basis of that axiomatization (p. 263). There is a different sense of completeness according to which an axiomatization is said to be complete if and only if it proves every sentence or its negation (p. 263).

**consistent:** An axiomatization is consistent if and only if it is never the case that both a sentence of the relevant language and its negation are provable on the basis of that axiomatization (p. 263).

**provable:** A sentence $S$ is provable on the basis of a given axiomatization if and only if there is a finite sequence of sentences of the language such that: (i) every member of the sequence is either an axiom, or something that results from previous members of the sequence by applying a rule of inference, and (ii) the last member of the sequence is $S$ (p. 263).

**rule of inference:** A rule for inferring some sentences from others (p. 263).

## Logical Symbols

**=:** Identity (p. 257)

**¬:** Negation (p. 257)

**&:** Conjunction (p. 257)

**$\vee$:** Disjunction (p. 259)

**$\supset$:** Conditional (p. 259)

**$\forall$:** Universal quantification (p. 257)

**$\exists$:** Existential quantification (p. 259)

**$\exists!$:** $\exists ! x_i(\phi(x_i))$ is true if and only if there is exactly one object $x$ such that $\phi(x)$ (p. 260).

### Measure

**Countable Additivity:** A function $\mu$ is countably additive if and only if $\mu\left(\bigcup\{A_1, A_2, A_3, \ldots\}\right) = \mu(A_1) + \mu(A_2) + \mu(A_3) + \ldots$ whenever $A_1, A_2, \ldots$ is a countable family of disjoint sets for each of which $\mu$ is defined (p. 183).

**Lebesgue measurable:** A set $A \subset \mathbb{R}$ is Lebesgue measurable if and only if $A = A^B \cup A^0$, for $A^B$, a Borel Set, and $A^0$, a subset of some Borel Set of Lebesgue Measure 0 (p. 315).

**Lebesgue Measure:** The unique measure satisfying Length on Line Segments (p. 185), and the unique function on the Borel Sets that satisfies Length on Line Segments, Countable Additivity, and Non-Negativity (p. 184).

**Length on Segments:** A function $\mu$ satisfies Length on Segments if and only if $\mu([a, b]) = b - a$ for every $a, b \in \mathbb{R}$ (p. 183).

**measure:** A function on the Borel Sets is a measure if and only if it satisfies Countable Additivity, satisfies Non-Negativity, and assigns the value of 0 to the empty set (p. 185).

**Non-Negativity:** A function $\mu$ satisfies Non-Negativity if and only if $\mu(A)$ is either a non-negative real number or the infinite value $\infty$, for any set $A$ in the domain of $\mu$ (p. 184).

**Uniformity:** A function $\mu$ satisfies Uniformity if and only if $\mu(A^c) = \mu(A)$, whenever $\mu(A)$ is well-defined and $A^c$ is the result of adding $c \in \mathbb{R}$ to each member of $A$ (p. 185).

### Miscellaneous Symbols

**$\leftrightarrow_{df}$:** Indicates that the expression to its left is to be treated as a syntactic abbreviation for the expression to its right (p. 260).

### Numerical Systems

**binary:** A number's binary expansion is a name for that number in binary notation. One names a number in decimal notation by writing out a finite sequence of binary digits (which are the symbols 0 and 1), followed by a point, followed by an infinite sequence of decimal digits. The decimal expansion $a_n a_{n-1} \ldots a_0.b_1 b_2 b_3 \ldots$ represents the number $a_n 2^n + a_{n-1} 2^{n-1} + \ldots + a_0 2^0 + b_1 2^{-1} + b_2 2^{-2} + b_3 2^{-3} + \ldots$. The number 3/2, for example, has two decimal expansions: 1.1000... and 1.0111... (p. 19).

**decimal:** A number's decimal expansion is a name for that number in decimal notation. One names a number in decimal notation by writing out a finite sequence of decimal digits (which are the symbols 0, 1,...,9) followed by a point, followed by an infinite sequence of decimal digits. The decimal expansion $a_n a_{n-1} \ldots a_0.b_1 b_2 b_3 \ldots$ represents the number $a_n 10^n + a_{n-1} 10^{n-1} + \ldots + a_0 10^0 + b_1 10^{-1} + b_2 10^{-2} + b_3 10^{-3} + \ldots$. The number 17, for example, has two decimal expansions: 17.0000... and 16.9999... (p. 13).

## Orderings

**$\omega$-sequence:** A sequence of items whose ordering is isomorphic to the natural ordering of the natural numbers: $0 < 1 < 2 < 3 < 4 < \ldots$ (p. 61).

**ordering:** A set $A$ is ordered by the two-place relation $R$ if $R$ satisfies asymmetry and transitivity for every member of $A$ (p. 33).

**partial ordering:** A set $A$ is (non-strictly) partially ordered by the two-place relation $R$ if $R$ satisfies reflexivity, anti-symmetry, and transitivity for every member of $A$ (p. 9).

**reverse $\omega$-sequence:** A sequence of items whose ordering is isomorphic to the reverse of the natural ordering of the natural numbers: $\ldots 4 < 3 < 2 < 1 < 0$ (p. 61).

**total ordering:** A set $A$ is (strictly) totally ordered by the two-place relation $R$ if $A$ is ordered by $R$ and $R$ satisfies totality for every member of $A$ (p. 35).

**well-ordering:** A set $A$ is well-ordered by the two-place relation $R$ if $A$ is totally ordered by $R$ and $R$ satisfies the following additional condition: every non-empty subset $S$ of $A$ has some member $x$ with the property that $xRy$ for every $y$ in $S$ other than $x$ (p. 36).

**well-order type:** A class that consists of an ordering $<$, and every ordering $<$ is isomorphic to (p. 37).

## Ordinals

**beth hierarchy:** The hierarchy of sets $\beth_\alpha$ for $\alpha$, an ordinal, where $\beth_\alpha$ is the smallest ordinal of cardinality $|\mathfrak{B}_\alpha|$ (p. 50).

**cardinal:** An initial ordinal used to represent its own cardinality (p. 51).

**cardinal addition:** An operation that takes two cardinalities as input and yields a cardinality as output. It is defined as follows: $|A| \oplus |B| = |A \cup B|$, assuming $A$ and $B$ have no members in common. If they do, find sets $A'$ and $B'$ with no members in common such that $|A| = |A'|$ and $|B| = |B'|$, and let $|A| \oplus |B| = |A' \cup B'|$ (p. 47).

**cardinal multiplication:** An operation that takes two cardinalities as input and yields a cardinality as output. It is defined as follows: $|A| \otimes |B| = |A \times B|$ (p. 47).

**initial ordinal:** An ordinal that precedes all other ordinals of the same cardinality (p. 50).

**limit ordinal:** An ordinal that is not a successor ordinal (p. 41).

**ordinal:** Ordinals are sets that are used to represent types of well-orderings. Informally, the ordinals are built in stages, in accordance with the Open-Endedness Principle and the Construction Principle (p. 40). Formally, an ordinal is a (pure) set that is set-transitive and well-ordered by $<_o$ (p. 43).

**ordinal addition:** An operation that takes two ordinals as input and yields an ordinal as output. Here is the intuitive idea: a well-ordering of type $(\alpha + \beta)$ is the result of starting with a well-ordering of type $\alpha$ and appending a well-ordering of type $\beta$ at the end (p. 44). For an official definition, see p. 46.

**ordinal multiplication:** An operation that takes two ordinals as input and yields an ordinal as output. Here is the intuitive idea: a well-ordering of type $(\alpha \times \beta)$ is the result of starting with a well-ordering of type $\beta$ and replacing each position in the ordering with a well-ordering of type $\alpha$ (p. 44). For an official definition, see p. 46.

**successor ordinal:** An ordinal $\alpha$ such that $\alpha = \beta'$ for some ordinal $\beta$ (p. 41).

### Other

**logical impossibility:** For a hypothesis to be logically impossible is for it to contain an absurdity (p. 64).

**paradox:** An argument that appears to be valid and goes from seemingly true premises to a seemingly false conclusion (p. 61).

### Probabilistic Symbols

**$p(A)$:** The probability of $A$ (pp. 134, 151).

**$p(A|B)$:** The conditional probability of $B$, given $A$ (pp. 124, 152).

### Probability

**Additivity:** A probability function $p$ satisfies Additivity if and only if $p(A \text{ or } B) = p(A) + p(B)$ whenever $A$ and $B$ are incompatible propositions (p. 151). Further, $p$ satisfies Countable Additivity if and only if for $A_1, A_2, A_3, \ldots$, a countable list of propositions, $p(A_1 \text{ or } A_2 \text{ or } A_3 \text{ or } \ldots) = p(A_1) + p(A_2) + p(A_3) + \ldots$, assuming $A_i$ and $A_j$ are incompatible whenever $i \neq j$ (p. 64).

**credence:** A degree of belief (p. 151).

**credence function:** A function that assigns to each proposition in a suitable range a real number between 0 and 1, representing the degree to which *S* believes that proposition (p. 151).

**frequentism:** The view that what it is for the objective probability that an event of type *E* yields outcome *O* to be *n*% is for *n*% of events of type *E* to yield outcome *O* (p. 158).

**localism:** A variety of rationalism according to which the notion of perfect rationality is only well-defined in certain special circumstances; for example, circumstances in which there is an unproblematic way of deploying a Principle of Indifference (p. 161).

**Necessity:** A probability function $p$ satisfies Necessity if and only if $p(A)=1$ whenever $A$ is a necessary truth (p. 151).

**objective:** The objective probability of *p* is meant to be a fact about the probability that *p*, which is independent of the beliefs of any particular subject (p. 156). There are many different views about what objective probability consists of (p. 158).

**primitivism:** The view that the notion of objective probability resists full elucidation and is best regarded as a theoretical primitive (p. 163).

**probability function:** An assignment of real numbers between 0 and 1 to propositions that satisfy Necessity and Additivity (p. 151).

**rationalism:** The view that there is nothing more to objective probability than the Objective-Subjective Connection of sections 6.1.1 and 6.3.2 (p. 160).

**subjective:** *S*'s subjective probability in *p* is the degree to which *S* believes that *p* (p. 151).

### Relations

**anti-symmetry:** A relation $R$ on $A$ is anti-symmetric if and only if, for each $a, b \in A$, if $aRb$ and $bRa$, then $a=b$ (pp. 34, 9).

**asymmetry:** A relation $R$ on $A$ is asymmetric if and only if, for each $a, b \in A$, if $aRb$, then not-($bRa$) (p. 33).

**equivalence:** A two-place relation $R$ on $A$ is an equivalence if and only if it is reflexive, symmetric, and transitive (p. 9).

**irreflexivity:** A relation $R$ on $A$ is irreflexive if and only if, for each $a \in A$, not-($aRa$) (p. 34).

**isomorphism:** Let $R_A$ be a relation on $A$ and $R_B$ be a relation on $B$. Then $R_A$ is isomorphic to $R_B$ if and only if there is a bijection $f$ from $A$ to $B$ such that, for every $x$ and $y$ in $A$, $xR_Ay$ if and only if $f(x)R_Bf(y)$ (p. 37).

**reflexivity:** A relation $R$ on $A$ is reflexive if and only if, for each $a \in A$, $aRa$ (pp. 9, 9, 34).

**relation:** A relation $R$ on set $A$ is a subset of $A \times A$. If $R$ is a relation on $A$ and $a, b \in A$, we say that $aRb$ holds (or that $Rab$ holds) if and only if $\langle a, b \rangle \in R$.

**symmetry:** A relation $R$ on $A$ is symmetric if and only if, for each $a, b \in A$, if $aRb$, then $bRa$ (p. 9).

**totality:** A relation $R$ on $A$ is total if and only if, for each $a, b \in A$, one of $aRb$ or $bRa$ or $a = b$ (p. 35).

**transitivity:** A relation $R$ on $A$ is transitive if and only if, for each $a, b \in A$, it satisfies $aRc$ whenever it satisfies $aRb$ and $bRc$ (pp. 9, 9, 34).

## Sets of Numbers

**Borel Set:** Any member of the smallest set $\mathcal{B}$ such that (*i*) every line segment is in $\mathcal{B}$, (*ii*) if a set is in $\mathcal{B}$, then so is its complement, and (*iii*) if a countable family of sets is in $\mathcal{B}$, then so is its union (p. 182).

**infinitesimal:** An infinitesimal number is a number that is greater than zero but smaller than every positive real number (p. 165).

**integer:** An integer is one of $\ldots -2, -1, 0, 1, 2, \ldots$ (p. 6).

**line segment:** Where $a$ and $b$ are real numbers, the line segment $[a, b]$ is the set of real numbers $x$ such that $a \leq x \leq b$ (p. 182).

**natural:** A natural number is one of the non-negative finite integers $0, 1, 2, \ldots$ (pp. 3, 6).

**rational:** A rational number is a number $a/b$, where $a$ and $b$ are integers and $b \neq 0$ (p. 6).

**real:** A real number is a number that can be written down in decimal notation by starting with a numeral (e.g., 17), adding a decimal point (e.g., 17.), and then adding an infinite sequence of digits (e.g., 17.8423...) (p. 13).

**square:** A square number is a number of the form $n^2$, for $n$ a positive integer (pp. ix, 4).

## Set Theory

**cardinality:** The cardinality (or size) of a set is a measure of how many objects it has (p. 8).

**Cartesian product:** The Cartesian product of $A$ and $B$ is the set of pairs $\langle a, b \rangle$ such that $a \in A$ and $b \in B$ (pp. 19, 47).

**complement:** The complement of set $A$ and set $B$ is the set of everything in both $A$ and $B$: $\{x : x \in A \text{ and } x \in B\}$ (p. 182).

**countability:** A set $A$ is countable if $|A| \leq |\mathbb{N}|$ (p. 18).

**disjoint:** For the family of sets $A_1, A_2, \ldots$ to be disjoint is for $A_i$ and $A_j$ to have no members in common whenever $i \neq j$ (p. 183).

**infinity:** A set $A$ is infinite if $|\mathbb{N}| \leq |A|$ (p. 18).

**intersection:** The intersection of set $A$ and set $B$ is the set of everything in both $A$ and $B$: $\{x : x \in A \text{ and } x \in B\}$. The intersection of the countable family of sets $A_1, A_2, A_3, \ldots$ is the set of everything in each of the $A_1, A_2, A_3, \ldots$ (p. 183).

**partition:** A partition of a set $A$ is a set $P$ of non-empty subsets of $A$ such that: (i) $\bigcup P = A$ and (ii) $a \cap b = \{\}$ for any distinct $a, b \in P$. One way to generate a partition $P$ of $A$ is to define an equivalence relation $R$ on $A$ and let $P = \{S : S = \{x : xRa\} \text{ for } a \in A\}$ (pp. 78, 190, 209, 218).

**powerset:** The powerset of a set $A$ is the set of all and only $A$'s subsets (p. 11).

**proper subset:** Set $A$ is a proper subset of set $B$ if and only if every member of $A$ is also a member of $B$, but not vice-versa (p. 8).

**subset:** Set $A$ is a subset of set $B$ if and only if every member of $A$ is also a member of $B$ (p. 8).

**union:** The union of set $A$ and set $B$ is the set of everything in either in $A$ or $B$: $\{x : x \in A \text{ or } x \in B\}$ (pp. 18, 32). The union of the countable family of sets $A_1, A_2, A_3, \ldots$ is the set of everything in at least one of the $A_1, A_2, A_3, \ldots$ (p. 182).

## Symbols for Functions

$f^{-1}$: When $f$ is a bijection from $A$ to $B$, $f^{-1}$ is $f$'s inverse; i.e., the function $g$ such that for each $b \in B$, $g(b)$ is the $a \in A$ such that $f(a) = b$ (p. 23).

$g \circ f$: The function $g(f(x))$ (p. 23).

$H(n, m)$: The two-place Halting Function. In general, $H(n, m) = 1$ if the $n$th Turing Machine (on a given coding scheme) halts on input $m$; otherwise, $H(n, m) = 0$ (p. 238).

$H(n)$: The one-place Halting Function. In general, $H(n) = H(n, n)$ (p. 238).

$BB(n)$: The Busy Beaver Function. In general, $BB(n)$ is the productivity of the most productive (one-symbol) Turing Machine with $n$ states or fewer (p. 240).

## Symbols for Ordinals and Cardinals

**0:** The smallest ordinal, defined as the set $\{\}$ (p. 41).

$\alpha'$: The successor of ordinal $\alpha$, defined as $\alpha \cup \{\alpha\}$ (p. 41).

$\omega$: The smallest infinite ordinal, defined as the set $\{0, 0', 0'', \ldots\}$ (p. 41).

$\alpha + \beta$: The result of applying the operation of ordinal addition to ordinals $\alpha$ and $\beta$ (p. 44).

$\alpha \times \beta$: The result of applying the operation of ordinal multiplication to ordinals $\alpha$ and $\beta$ (p. 44).

$\alpha^\beta$: The result of applying the operation of ordinal exponentiation to ordinals $\alpha$ and $\beta$ (p. 46).

$|A| \oplus |B|$: The result of applying the operation of cardinal addition to the cardinalities of sets $A$ and $B$ (p. 47).

$|A| \otimes |B|$: The result of applying the operation of cardinal multiplication to the cardinalities of sets $A$ and $B$ (p. 47).

$\mathfrak{B}_\alpha$:

$$\mathfrak{B}_\alpha = \begin{cases} \mathbb{N}, \text{ if } \alpha = 0 \\ \wp(\mathfrak{B}_\beta), \text{ if } \alpha = \beta' \\ \bigcup\{\mathfrak{B}_\gamma : \gamma <_o \alpha\} \text{ if } \alpha \text{ is a limit ordinal greater than 0 (p. 49).} \end{cases}$$

$\aleph_\alpha$: The first infinite ordinal of cardinality greater than every $\aleph_\beta$, for $\beta <_o \alpha$ (p. 51).

$\beth_\alpha$: The first ordinal of cardinality $|\mathfrak{B}_\alpha|$ (p. 50).

## Symbols for Sets of Numbers

$\mathbb{N}$: The set of natural numbers (p. 6).

$\mathbb{Z}$: The set of integers (p. 6).

$\mathbb{Z}^+$: The set of positive integers (p. 167).

$\mathbb{Q}$: The set of rational numbers (p. 6).

$\mathbb{Q}^{\geq 0}$: The set of non-negative rational numbers (p. 6).

$\mathbb{Q}^{[0,1)}$: The set of rational numbers in $[0, 1)$ (p. 190).

$\mathbb{R}$: The set of real numbers (pp. 12, 13).

$\mathfrak{F}$: The set of finite sequences of natural numbers (p. 12).

$(a, b)$: The set of real numbers $x$ such that $a < x < b$ (p. 17).

$[a, b]$: The set of real numbers $x$ such that $a \leq x \leq b$ (pp. 17, 181).

$[a, b)$: The set of real numbers $x$ such that $a \leq x < b$ (pp. 15, 182).

$(-\infty, \infty)$: The set of real numbers (p. 184).

$[0, \infty)$: The set of non-negative real numbers (pp. 184, 205).

## Symbols in Banach-Tarski Construction

$X^e$: When $X$ is a set of Cayley Paths, $X^e$ is the set of endpoints of paths in $X$ (p. 208).

$\overleftarrow{X}$: When $X$ is a set of Cayley Paths, $\overleftarrow{X}$ is the set that results from eliminating the first step from each of the Cayley Paths in $X$ (p. 211).

# Bibliography


Aaronson, Scott. "Who Can Name the Bigger Number?" Accessed January 28, 2017. http://www.scottaaronson.com/writings/bignumbers.html.

Adams, Ernest W. "Subjunctive and Indicative Conditionals." *Foundations of Language* 6, no. 1 (1970): 89–94.

Ahmed, Arif. "Dicing with Death." *Analysis* 74, no. 14 (2014): 587–592.

Arntzenius, Frank. "Time Travel: Double Your Fun." *Philosophy Compass* 1, no. 6 (2006): 599–616.

Arntzenius, Frank, Adam Elga, and John Hawthorne. "Bayesianism, Infinite Decisions, and Binding." *Mind* 113 (2004): 251–283.

Arntzenius, Frank, and Tim Maudlin. "Time Travel and Modern Physics." Winter 2013 ed. In *The Stanford Encyclopedia of Philosophy*. Online ed. Edited by Edward N. Zalta. http://plato.stanford.edu/archives/win2013/entries/time-travel-phys/.

Ash, Robert B. *Probability and Measure Theory*. 2nd ed. With contributions from Catherine Doleans-Dade. San Diego, CA: Harcourt Academic Press, 2000.

Bacon, Andrew. "A Paradox for Supertask Decision Makers." *Philosophical Studies* 153, no. 2 (2001): 307–311.

Bagaria, Joan. "Set Theory." Fall 2016 ed. In *The Stanford Encyclopedia of Philosophy*, edited by Edward N. Zalta. Stanford, CA: Metaphysics Research Lab, Stanford University, 2016.

Benacerraf, Paul. "Tasks, Super-Tasks, and the Modern Eleatics." *Journal of Philosophy* 59, no. 24 (1962): 765–784.

Benacerraf, Paul, and Hilary Putnam, eds. *Philosophy of Mathematics*, 2nd ed. Cambridge: Cambridge University Press, 1983.

Bennett, Jonathan. *A Philosophical Guide to Conditionals*. Oxford: Oxford University Press, 2003.

Bernadete, J., ed. *Infinity: An Essay in Metaphysics*. Oxford: Clarendon Press, 1964.

Blitzstein, Joseph K. *Introduction to Probability: CRC Texts in Statistical Science*. Boca Raton, FL: CRC Press, 2015.

Boolos, George. "Iteration Again." *Philosophical Topics* 17, no. 2 (1989): 5–21.

Boolos, George. "The Iterative Conception of Set." *Journal of Philosophy* 68 (1971): 215–31. Reprinted in Boolos (1998).

Boolos, George. *Logic, Logic and Logic*. Cambridge, MA: Harvard University Press, 1998.

Boolos, George, John Burgess, and Richard Jeffrey. *Computability and Logic*. Cambridge: Cambridge University Press, 2007.

Bostrom, Nick. "The Meta-Newcomb Problem." *Analysis* 61, no. 4 (2001): 309–310.

Broome, John. "The Two-Envelope Paradox." *Analysis* 55, no. 1 (1995): 6–11.

Burgess, John. *Fixing Frege*. Princeton, NJ: Princeton University Press, 2005.

Cartwright, Richard. *Philosophical Essays*. Cambridge, MA: MIT Press, 1987.

Chalmers, David J. "The St. Petersburg Two-Envelope Paradox." *Analysis* 62, no. 274 (2002): 155–157.

Drake, F. R., and D. Singh, eds. *Intermediate Set Theory*. Chichester, UK: Wiley, 1996.

Easwaran, Kenny. "Bayesianism I: Introduction and Arguments in Favor." *Philosophy Compass* 6, no. 5 (2011a): 312–320.

Easwaran, Kenny. "Bayesianism II: Applications and Criticisms." *Philosophy Compass* 6, no. 5 (2011b): 321–332.

Easwaran, Kenny. "Conditional Probability." In Hájek and Hitchcock, *The Oxford Handbook of Probability and Philosophy* (forthcoming).

Easwaran, Kenny. "Why Countable Additivity?" *Thought* 1 (2013): 53–61.

Egan, Andy. "Some Counterexamples to Causal Decision Theory." *Philosophical Review* 116, no. 1 (2007): 93–114.

Field, Hartry. *Saving Truth from Paradox*. Oxford: Oxford University press, 2008.

Frankfurt, Harry G. "Alternate Possibilities and Moral Responsibility." *Journal of Philosophy* 66, no. 3 (1969): 829–839.

Frankfurt, Harry G. "Freedom of the Will and the Concept of a Person." *Journal of Philosophy* 68, no. 1 (1971): 5–20.

Galilei, Galileo, ed. *Dialogues Concerning Two New Sciences*. English translation by Henry Crew and Alfonso de Salvio. New York: Cosimo Classics, 2010.

Gibbard, Allan, and William Harper. "Counterfactuals and Two Kinds of Expected Utility." In *Foundations and Applications of Decision Theory*, edited by A. Hooker, J. J. Leach, and E. F. McClennen, 125–162. Dordrecht: Springer, 1978.

Hájek, Alan, and Chris Hitchcock, eds. *The Oxford Handbook of Probability and Philosophy*. Oxford: Oxford University Press, forthcoming.

Hawthorne, John. "Before-Effect and Zeno Causality." *Noûs* 34, no. 4 (2000): 622–633.

Hilbert, David. "On the Infinite." Originally delivered before a congress of the Westphalian Mathematical Society, 1925. Translation printed in Benacerraf and Putnam, *Philosophy of Mathematics*, 2nd ed. Cambridge: Cambridge University Press, 1983.

Hill, B. M., and D. Lane. "Conglomerability and Countable Additivity." *Sankhyā: The Indian Journal of Statistics, Series A* 47 (1985): 366–379.

Horgan, Terence. "Counterfactuals and Newcomb's Problem." *Journal of Philosophy* 78, no. 6 (1981): 331–356.

Hrbáek, Karel, and Thomas Jech. *Introduction to Set Theory*. 3rd ed. Oxford: Clarendon Press, 1999.

Jeffrey, Richard. *The Logic of Decision*. Chicago: University of Chicago Press, 1983.

Jeffrey, Richard, ed. *Studies in Inductive Logic and Probability*. Vol. 2 of *Proceedings of the Symposia on Pure Mathematics*. Berkeley: University of California Press, 1980.

Joyce, James. *The Foundations of Causal Decision Theory*. Cambridge: Cambridge University Press, 1999.

Kripke, Saul. "Outline of a Theory of Truth." *Journal of Philosophy* 72 (1975): 690–716.

Kritchman, Shira, and Ran Raz. "The Surprise Exam Paradox and the Second Incompleteness Theorem." *Notices of the AMS* 57 (1975): 1,454–1,458.

Kunen, Kenneth. *Set Theory*. Revised ed. *Studies in Logic: Mathematical Logic and Foundations*. London: College Publications, 2011.

Lewis, David. "Causal Decision Theory." *Australasian Journal of Philosophy* 59, no. 1 (1981): 5–30.

Lewis, David. "Causation." *Journal of Philosophy* 70, no. 17 (1973): 556–567.

Lewis, David. "Humean Supervenience Debugged." *Mind* 103 (1994): 473–490.

Lewis, David. *Philosophical Papers, Volume II*. Oxford: Oxford University Press, 1986.

Lewis, David. "Prisoner's Dilemma Is a Newcomb Problem." *Philosophy and Public Affairs* 8, no. 3 (1979): 235–240.

Lewis, David. "A Subjectivist's Guide to Objective Chance." In *Studies in Inductive Logic and Probability*, edited by Richard Jeffrey, 263–293. Vol. 2 of *Proceedings of the Symposia on Pure Mathematics*. Berkeley: University of California Press, 1980. Reprinted with postscripts in *Philosophical Papers, Volume II*, edited by Richard Lewis, 83–132. Oxford: Oxford University Press, 1986.

Linnebo, Øystein. "The Potential Hierarchy of Sets." *Review of Symbolic Logic* 6, no. 2 (2013): 205–228.

Maudlin, Tim. "Three Roads to Objective Probability1." In *Probabilities in Physics*, edited by Claus Beisbart and Stephan Hartmann, 293–319. Oxford: Oxford University Press, 2011.

McGee, Vann. *Truth, Vagueness and Paradox: An Essay on the Logic of Truth*. Indianapolis: Hacket, 1990.

McKenna, Michael, and D. Justin Coates. "Compatibilism." Summer 2015. ed. In *The Stanford Encyclopedia of Philosophy*. Online ed. Edited by Edward N. Zalta. http://plato.stanford.edu/archives/sum2015/entries/compatibilism/.

McManus, Denis. *Wittgenstein and Scepticism*. London: Routledge, 2004.

Paseau, A. C. "An Exact Measure of Paradox." *Analysis* 73, no. 1 (2013): 17–26.

Pereboom, Derk. "Determinism *Al Dente*." *Noûs* 29, no. 1 (1995): 21–45.

Priest, Graham. "What Is So Bad about Contradictions?" *Journal of Philosophy* 95, no. 8 (1998): 410–426.

Rayo, Agustín. "A Plea for Semantic Localism." *Noûs* 47, no. 4 (2013): 647–679.

Robinson, Avery. "The Banach-Tarski Paradox." Accessed January 9, 2015. http://www.math.uchicago.edu/ low may/REU2014/REUPapers/Robinson.pdf.

Rucker, Rudy. *Infinity and the Mind: The Science and Philosophy of the Infinite*. Princeton, NJ: Princeton University Press, 1982.

Schervish, M. J., T. Seidenfeld, and J. B.Kadane. "The Extent of Non-Conglomerability of Finitely Additive Probabilities." *Zeitschrift für Wahrscheinlichkeitstheorie und verwandte Gebiete* 66 (1984): 205–226.

Spencer, Jack, and Ian Wells. "Why Take Both Boxes?" *Philosophy and Phenomenological Research*, forthcoming.

Tarski, Alfred. "The Concept of Truth in Formalized Languages." In *Logic, Semantics and Metamathematics*, 2nd ed., 152–278. Indianapolis: Hacket, 1983.

Tarski, Alfred. *Logic, Semantics and Meta-mathematics*, 2nd ed. Indianapolis: Hacket, 1983.

Van Fraassen, B. C. *Laws and Symmetry*. Oxford: Oxford University Press, 1989.

Weirich, Paul. "Causal Decision Theory." Winter 2012 ed. In *The Stanford Encyclopedia of Philosophy*. Online ed. Edited by Edward N. Zalta. http://plato.stanford.edu/archives/win2012/entries/decision-causal/.

Weston, Tom. "The Banach-Tarski Paradox." Accessed July 23, 2018. http://citeseerx.ist.psu.edu/viewdoc/summary?doi=10.1.1.187.548.

Williamson, Timothy. "How Probable Is an Infinite Sequence of Heads?" *Analysis* 67, no. 3 (2007): 173–180.

Wittgenstein, Ludwig. *On Certainty*. Edited by G. E. M. Anscombe and G. H. von Wright. Translated by G. E. M. Anscombe and D. Paul. Oxford: Blackwell, 1969.

Wright, Crispin. "Wittgensteinian Certainties." In McManus, *Wittgenstein and Scepticism*, 22–55. London: Routledge, 2004.

Yablo, Stephen. "Paradox without Self–Reference." *Analysis* 53, no. 4 (1993): 251–252.

Yablo, Stephen. "A Reply to New Zeno." *Analysis* 60, no. 2 (2000): 148–151.

Zach, Richard. "Hilbert's Program." Spring 2016 ed. In *The Stanford Encyclopedia of Philosophy*, edited by Edward N. Zalta. Stanford, CA: Metaphysics Research Lab, Stanford University, 2016.

# Index